T0202973

Statistical Modeling for Biological Systems

Anthony Almudevar • David Oakes • Jack Hall
Editors

Statistical Modeling for Biological Systems

In Memory of Andrei Yakovlev

 Springer

Editors
Anthony Almudevar
Department of Biostatistics and
Computational Biology
University of Rochester Medical Center
Rochester, NY, USA

David Oakes
Department of Biostatistics and
Computational Biology
University of Rochester Medical Center
Rochester, NY, USA

Jack Hall *(deceased)*
Rochester, NY, USA

ISBN 978-3-030-34677-5   ISBN 978-3-030-34675-1   (eBook)
https://doi.org/10.1007/978-3-030-34675-1

This Springer imprint is published by the registered company Springer Nature Switzerland AG.
The registered company address is: Gewerbestrasse 11, 6330 Cham, Switzerland

Preface

November 30, 2019, was the 75th anniversary of the birth of Andrei Yakovlev, an extraordinary scholar and inspirational leader. To mark the anniversary, we present this volume of papers dedicated to his memory. The contributors include collaborators, colleagues, and other members of the statistical community influenced by him. Many of the papers grew out of a conference in his honor, held in Rochester in June 2009.

Here is a summary sketch of his life and achievements, drawn extensively from his obituary in the IMS Bulletin.[1] A full list of publications is included in this volume as an Appendix. Andrei Yakovlev was born in 1944 in Leningrad (now St. Petersburg), Russia; his father was a mathematician and senior naval officer. Although engrossed in mathematics—often carrying a mathematics book about as a child—he sought a different career. He earned an MD (Leningrad, 1967), followed by a PhD in cell biology (Pavlov Institute of Physiology, 1973) and a DSc in

---

[1] Hall, W. J. and Hanin, L. IMS Bulletin, Vol. 37.5:16–17 (March 2008).

mathematics (Moscow State University, 1981). He set out on a career of research, applying mathematics and statistics to basic questions in biology and medical science. His unusual combination of training positioned him to transcend boundaries and cultures of the worlds of mathematics, statistics, biology, and medicine, and he did so consistently throughout his career.

Andrei Yakovlev was a natural leader. He chaired departments during 26 of his 41-year post-MD career. He founded and chaired the Department of Biomathematics in the Central Research Institute of Roentgenology and Radiology (Leningrad, 1978–1988), followed by the Department of Applied Mathematics of Leningrad State Polytechnical University (1988–1992). Finding academic science under tight bureaucratic restrictions frustrating, he left Russia in 1992, visiting universities and research institutes in Germany, France, Australia, California, and Ohio before settling at the Huntsman Cancer Institute in Salt Lake City as chair of Biostatistics (1996–2002). His final position was as chair at the University of Rochester (2003–2008), School of Medicine and Dentistry, with a mandate to expand the activities of Biostatistics into Computational Biology. Under his leadership, the faculty was tripled and grant support expanded manyfold. Along the way, he was honored extensively in Europe and the United States.

Andrei was foremost a researcher. He authored four research monographs and over 200 publications, done jointly with literally dozens of collaborators—including ten with 10–40 joint papers each, and among them biomathematicians, probabilists, statisticians, biologists, geneticists, and medical scientists from varied fields. Every publication was directed towards solution of a biological or medical science problem, but included innovative mathematical developments. Even a largely probabilistic or statistical paper was identified as having such a purpose. He made definitive and lasting contributions to a stunning number of fields: branching stochastic processes—especially multi-type age-dependent processes and associated statistical inference, stochastic models in radiation biology, cell population dynamics, carcinogenic risk assessment, stochastic models of carcinogenesis, cancer growth/progression/detection, optimal schedules of cancer surveillance and screening, optimization of radiation cancer treatment, cure models in survival analysis, statistical inference from microarray gene expression data, genetic regulatory networks, and so on. His collaborators without exception attribute the innovative direction of their joint work to Andrei, their unquestioned leader.

In 2008, at age 63, he was at the height of his scholarly activity, with seemingly endless energy, when he died at night in his sleep of a heart attack.

Dr. Yakovlev's scientific legacy does not consist only of ideas, methods, and results. Most importantly, he left behind scientific principles and research paradigms that will guide his co-workers and students for decades to come. He forged them through incisive analysis, trial and error, and excruciating, sometimes heated and fierce, debates with his colleagues, disciples, and co-workers. He never articulated these principles explicitly, perhaps leaving this endeavor for a future occasion, but followed them religiously in his own work. We believe that with the passage of time the broad scientific community will come to realize that in Dr. Yakovlev we have lost one of the greatest scientific visionaries of our epoch. It is heartwrenching that he did not live long enough to witness this recognition. He is survived by his wife Nina and son Yuri.

The preface to this volume ends with an essay by Dr. Hartmut Land of the Department of Biomedical Genetics of the University of Rochester, giving his personal views of Andrei's philosophy.

The contributions to this volume are divided into two parts. Part A consists of original research articles, which can be roughly grouped into four thematic areas, (1) branching processes, especially as models for cell kinetics [19, 8, 2], (2) multiple testing issues as they arise in the analysis of biologic data [16, 4, 1], (3) applications of mathematical models and of new inferential techniques in epidemiology [14, 20], and (4) contributions to statistical methodology, with an emphasis on the modeling and analysis of survival time data [3, 6, 17, 11, 5, 15].

Part B consists of methodological research reported as a short communication [21, 13], ending with some personal reflections on research fields associated with Andrei [18, 9] and on his approach to science [7]. The Appendix contains a list of Andrei's publications, complete as far as we know.

We have many people to thank: Nina Yakovlev, for her encouragement and support throughout; the University of Rochester School of Medicine and Dentistry, where our department is located, for providing vital resources for holding the conference—we are particularly grateful to the Dean at that time, Dr. David Guzick, for his interest in the project; staff in the Department of Biostatistics and Computational Biology, particularly the "two Cheryls", Cheryl Bliss-Clark and Cheryl Cicero, for their work in organizing and ensuring the smooth running of the conference; the speakers and contributors of course; and the anonymous referees, who were responsible for many improvements in the quality of the submissions. Significant assistance in setting up the final latex files for compilation was provided by Katie Evans, Yu Han, Fei Ma, and Jason Morrissette, graduate students in our department. We also wish to thank our colleagues at Springer for supporting this publication and for their editorial help.

Most of all we thank Andrei for his inspiration.

Rochester, NY, USA   Anthony Almudevar
Rochester, NY, USA   David Oakes
Rochester, NY, USA   Jack Hall[2]

---

[2]Jack Hall died on October 14, 2012. Until a few weeks before his death he had been deeply involved in planning and preparing this volume including soliciting, reviewing, and editing manuscripts. We greatly appreciate his contribution.—A.A., D.O.

Cancer Cell Vulnerabilities: Finding Needles in a Haystack with Andrei Yakovlev: A Personal Observation

When Andrei Yakovlev arrived at the University of Rochester Medical Center, work in my laboratory towards developing rational approaches to finding molecular targets for cancer intervention was showing some promise, but proceeded at snail's pace. We were interested in finding cancer cell vulnerabilities downstream of oncogenic mutations that might be suitable for drug targeting, as drug intervention at the level of oncogenic mutations remains an exception rather than the rule. Identification of such vulnerabilities, however, appeared notoriously difficult and unpredictable because neoplastic progression is associated with profound and complex changes in the genetic and metabolic networks that control the functioning of the cell. Pinpointing changes in cell regulation with causal relevance to malignant transformation thus turned out to be a particularly vexing problem. We therefore began to focus on searching for rules of change at the molecular level that drive the transformation process. This required generation and careful analysis of large, genome-scale data sets, one of Andrei's main areas of research interest. Andrei was a bio-statistician with a natural affinity towards biological problems and thus keen to begin collaborations.

In previous work we and others had established that carcinogenesis is a multi-step process driven by multiple cooperating oncogenic mutations with many features of the cancer cell phenotype only emerging as a result of the interplay between these mutations. Moreover, we discovered that regulatory processes driving the transition from the normal to the cancer phenotype were synergistic in nature. These findings gave rise to the hypothesis that synergistic gene regulation may be a useful indicator for the detection of drivers in malignant cell transformation downstream of oncogenic mutations. Our experiments carried out in collaboration with Andrei Yakovlev indeed confirmed this idea. In fact, further work now suggests that synergistic gene regulation by cooperating oncogenic mutations provides a blueprint for a genome-scale landscape of largely unexpected cancer cell vulnerabilities.

When we began collaborating with Andrei, genomic gene expression profiling had become available. On the other hand, high-quality methods for data analysis and statistical evaluation and, perhaps more importantly, an appreciation of the importance of the quality of such analysis methods still had to be developed among biologists.

Andrei not only created rigorous methods for microarray data analysis but also passionately taught and fought to introduce us molecular biologists to the statistical rigor that must be applied to complex data sets. Andrei was faced with an uphill battle. In the field of molecular biology, data that require statistical analysis, for the most part, were considered suspect because of the high levels of associated noise. On the other hand, we were keen to have access to genome-scale data sets that told a story. To put it mildly, our arrogance and ignorance regarding statistical methods was "significant." Moreover, Andrei's never-ending and at times noisy requests to produce more of the very costly replicates, with little idea of a number that would be sufficient to produce a meaningful data set, did not make things easier. Given time and persistence, however, he got through, making us painfully aware of the reality/threat that complex data sets derived from microarrays were subject to false discovery rates at levels where signal could not be distinguished from noise, unless.... One can only imagine how difficult and frustrating this must have been for Andrei. In fact, Andrei's effectiveness as a mathematician and scientist, I feel, was equally due to his brilliant mind and his persistence in challenging us biologists to invest into our experiments appropriately, as (hopefully) to avoid drowning in the dreaded noise.

To find cancer cell vulnerabilities downstream of oncogenic mutations and to test our prediction that synergistic gene regulation in response to multiple oncogenic insults would be a useful indicator for their detection, we first needed to create cancer cells from normal cells in a molecularly defined manner. We used cultured mouse colon epithelial cells in which an activated Ras oncogene cooperates with mutant p53, a disabled tumor suppressor gene, to induce malignant cell transformation. The two oncogenic alterations, a constitutively active signaling protein and a dominant-negative transcription factor, cooperate to simultaneously stimulate distinct cell behaviors such as cell cycle progression, survival, motility, invasiveness, and cell metabolism in these cells. In contrast, activated Ras or mutant p53 alone has little effect. We were thus able to derive and compare genomic gene expression profiles from parental non-transformed mouse colon epithelial cells, cells with either activated Ras or mutant 53 alone and cells that had undergone transition to malignancy in response to the combination of both oncogenic proteins. Genes differentially expressed between non-transformed cells and cells expressing both oncogenic mutations were identified by the step-down Westfall-Young procedure in conjunction with the permutation N-test [10], FWER < 0.01, using 10 microarray replicates for each cell population. Among the differentially expressed genes, we identified a set of 95 genes that are synergistically regulated by activated Ras and mutated p53. These genes, termed cooperation response genes (CRGs), were selected by the following procedure. Let $a =$ mean expression value for a given gene in mp53 cells, $b =$ mean expression value for the same gene in Ras cells, and $d =$ mean expression value for this gene in mp53/Ras cells. Then, the criterion defines CRGs as $(a + b)/d < 0.9$ for genes overexpressed in mp53/Ras cells and as $(d/a) + (d/b) < 0.9$ for genes underexpressed in mp53/Ras cells, as compared to controls. To assess robustness of synergy scores, jackknife subsampling was used to generate estimated p-values for these scores. In addition, quantitative PCR was used to independently test gene expression differences observed between the various cell populations by conventional gene arrays.

A definitive answer as to whether CRGs were functionally relevant to the malignant state had to come from gene perturbation experiments. For this purpose we genetically manipulated the expression of 24 CRGs in the transformed cells in a manner that would re-establish their level of expression in normal parental cells and test their potential to grow as tumors following subcutaneous transplantation into mice. To our surprise we found that CRGs are essential for expression and maintenance of the cancer cell phenotype at an unexpectedly high frequency of over 50%. We also showed that CRGs contribute to the malignant state of a variety of human cancer cells. In contrast, only one out of fifteen differentially expressed non-CRGs identified in our experiments modulated the cancer phenotype. Synergistic regulation of downstream mediators by multiple oncogenic mutations thus emerges as an architectural feature of biological significance in malignant cell transformation [12].

Notably, CRGs can act as mediators in the control of multiple and diverse cellular processes, such as proliferation, survival, differentiation, motility, invasiveness, and cell metabolism, suggesting that cooperating oncogenic mutations simultaneously can affect multiple cancer cell traits. Moreover, our data indicate a surprisingly large number of hitherto unknown cancer cell vulnerabilities. Our findings thus reveal a general approach to discovery of cancer cell-specific drug targets, a problem that without Andrei's contributions, I am sure, we would not have been able to solve.

Department of Biomedical Genetics   Hartmut Land
University of Rochester
Rochester, NY, USA

References

1. Almudeyar, A. (2019). Applications of sequential methods in multiple hypothesis testing. In *Statistical modeling for biological systems* (pp. 97–115). New York: Springer.
2. Chen, L., Klebanov, L., Almudeyar, A., Proschel, C., & Yakovlev, A. (2019). A study of the correlation structure of microarray gene expression data based on mechanistic modeling of cell population kinetics. In *Statistical modeling for biological systems* (pp. 47–61). New York: Springer.
3. De Rycke, Y., & Asselain, B. (2019). A latent time distribution model for the analysis of tumor recurrence data: Application to the role of age in breast cancer. In *Statistical modeling for biological systems* (pp. 157–167). New York: Springer.
4. Gordon, A. Y. (2019). Multiple testing procedures: monotonicity and some of its implications. In *Statistical modeling for biological systems* (pp. 81–96). New York: Springer.
5. Gu, W., Wu, H., & Xue, H. (2019). Parameter estimation for multivariate nonlinear stochastic differential equation models: a comparison study. In *Statistical modeling for biological systems* (pp. 245–258). New York: Springer.
6. Hall, W. J., & Wellner, J. A. (2019). Estimation of mean residual life. In *Statistical modeling for biological systems* (pp. 169–189). New York: Springer.

7. Hanin, L. (2019). Principles of mathematical modeling in biomedical sciences: An unwritten gospel of Andrei Yakovlev. In *Statistical modeling for biological systems* (pp. 321–331). New York: Springer.
8. Hyrien, O., & Yanev, N. M. (2019). Age-dependent branching processes with non-homogeneous Poisson immigration as models of cell kinetics. In *Statistical modeling for biological systems* (pp. 21–46). New York: Springer.
9. Jagers, P. (2019). Branching processes: A personal historical perspective. In *Statistical modeling for biological systems* (pp. 311–319). New York: Springer.
10. Klebanov, L., Gordon, A., Xiao, Y., Land, H., & Yakovlev, A. (2005). A permutation test motivated by microarray data analysis. *Computational Statistics and Data Analysis, 50*, 3619–3628.
11. Kunz, C., & Edler, L. (2019). On the application of flexible designs when searching for the better of two anticancer treatments. In *Statistical modeling for biological systems* (pp. 211–243). New York: Springer.
12. McMurray, H. R., Sampson, E. R., Compitello, G., Kinsey, C., Newman, L., Smith, B., et al. (2008). Synergistic response to oncogenic mutations defines gene class critical to cancer phenotype. *Nature, 453*, 1112–1116.
13. Miecznikowski, J., Wang, D., Ren, X., Wang, J., & Liu, S. (2019). Analyzing gene pathways from microarrays to sequencing platforms. In *Statistical modeling for biological systems* (pp. 289–296). New York: Springer.
14. Moolgavkar, S., & Luebeck, G. (2019). Multistage carcinogenesis: A unified framework for cancer data analysis. In *Statistical modeling for biological systems* (pp. 117–136). New York: Springer.
15. Oakes, D. (2019). On frailties, Archimedean copulas and semi-invariance under truncation. In *Statistical modeling for biological systems* (pp. 259–278). New York: Springer.
16. Qiu, X., & Hu, R. (2019). Correlation between the true and false discoveries in a positively dependent multiple comparison problem. In *Statistical modeling for biological systems* (pp. 63–79). New York: Springer.
17. Tsodikov, A., Liu, L. X., & Tseng, C. (2019). Likelihood transformations and artificial mixtures. In *Statistical modeling for biological systems* (pp. 191–209). New York: Springer.
18. von Collani, E. (2019). A new approach for quantifying uncertainty in epidemiology. In *Statistical modeling for biological systems* (pp. 297–309). New York: Springer.
19. Yanev, N. M. (2019). Stochastic models of cell proliferation kinetics based on branching processes. In *Statistical modeling for biological systems* (pp. 3–20). New York: Springer.
20. Young, J., Hubbard, A. E., Eskenazi, B., & Jewell, N. (2019). A machine-learning algorithm for estimating and ranking the impact of environmental risk factors in exploratory epidemiological studies. In *Statistical modeling for biological systems* (pp. 137–156). New York: Springer.
21. Zhang, H., & Tu, X. (2019). The generalized ANOVA: A classic song sung with modern lyrics. In *Statistical modeling for biological systems* (pp. 281–287). New York: Springer.

Contents

Part I Research Articles

**Stochastic Models of Cell Proliferation Kinetics Based on Branching
Processes** 3
Nikolay M. Yanev
1 Introduction.. 3
2 Subsequent Generations of Cells Induced to Proliferation and Some
   Characteristics of Cell Cycle Temporal Organization 5
3 Distributions of Pulse-Labeled Discrete Marks in Branching
   Populations of Cells.. 6
4 Continuous Label Distributions in Proliferating Cell Populations.......... 7
5 Limiting Age and Residual Lifetime Distributions for
   Continuous-Time Branching Processes 9
6 Limit Theorems and Estimation Theory for Multitype Branching
   Populations and Relative Frequencies with a Large Number
   of Ancestors 13
7 Age-Dependent Branching Populations with Randomly Chosen
   Paths of Evolution 17
8 Concluding Remarks 18
References 19

**Age-Dependent Branching Processes with Non-homogeneous
Poisson Immigration as Models of Cell Kinetics**............................ 21
Ollivier Hyrien and Nikolay M. Yanev
1 Introduction... 21
2 A Biological Motivation 23
3 Model, Notation, and Basic Equations 24
4 Asymptotics for First- and Second-Order Moments......................... 29
5 Statistical Inference 37

6 Examples and Applications 39
   6.1 An Age-Dependent Branching Process with Homogeneous
      Poisson Immigration 39
   6.2 A Simulation Study 42
7 Dedication 44
References 45

**A Study of the Correlation Structure of Microarray Gene
Expression Data Based on Mechanistic Modeling of Cell Population
Kinetics**... 47
Linlin Chen, Lev Klebanov, Anthony Almudevar, Christoph Proschel,
and Andrei Yakovlev
1 Introduction.. 48
2 The Effect of Cell Mixtures on Observed Correlations...................... 48
3 A Comprehensive Study of Correlation Based on Mechanistic
   Modeling of Cell Population Kinetics.. 53
   3.1 Gene Expression Levels During the Cell Cycles 55
   3.2 Simulation Study 56
4 Heterogeneity of Subjects.. 58
5 Discussion 59
References 60

**Correlation Between the True and False Discoveries in a Positively
Dependent Multiple Comparison Problem** 63
Xing Qiu and Rui Hu
1 Background and Introduction... 63
2 Methods.. 66
   2.1 A Parametric Model with Two t-Tests 66
   2.2 Correlation Between the Two t-Statistics 67
   2.3 Correlation Between the True and False Positives 68
   2.4 A General Multiple Testing Design Motivated by Microarray
      Analysis 73
3 Simulation Results 74
4 Discussion 75
References 78

**Multiple Testing Procedures: Monotonicity and Some
of Its Implications** 81
Alexander Y. Gordon
1 Introduction.. 81
2 Basic Notions.. 83
   2.1 Uninformed MTPs 83
   2.2 Monotonicity 84
   2.3 Step-Down and Step-Up Procedures 84
   2.4 Threshold Step-Up-Down Procedures 86
   2.5 Generalized Family-Wise Error Rates 86

2.6   Per-Family Error Rate   87
2.7   Comparison of Procedures..   88
3   Some Implications of Monotonicity.......................................   88
3.1   Optimality of the Holm Procedure   88
3.2   Extensions of the Holm Procedure   89
3.3   Extensions of the Bonferroni Procedure   90
3.4   Quasi-Thresholds   90
3.5   Some Sharp Inequalities   91
3.6   Bounds on Generalized Family-Wise Error Rates....................   92
3.7   An "All-or-Nothing" Theorem   94
References   95

Applications of Sequential Methods in Multiple Hypothesis Testing.......   97
Anthony Almudevar
1   Introduction..   98
2   The Empirical Hypothesis Test as Stopped Binary Process.................   98
2.1   Monte Carlo Hypothesis Tests as SBPs.............................   99
2.2   Estimation of Significance Level....................................   100
2.3   Fixed Level Tests..   100
2.4   Hybrid Test   102
3   Overview of the Sequential Probability Ratio Test   102
4   Application of the SPRT to Hypothesis Tests Based on Simulated
Replications of an Accept–Reject Rule   103
4.1   Single Hypothesis Test...   104
4.2   Multiple Hypothesis Tests   106
5   Optimal Design of Stopping Times Based on SPRTs   109
5.1   Constrained Optimization   110
5.2   Solution Method   111
5.3   Numerical Example...   112
6   Examples   113
6.1   Gene Set Analysis...   113
6.2   Confidence Sets in Statistical Genetics   114
7   Conclusion..   114
References   115

**Multistage Carcinogenesis: A Unified Framework for Cancer Data
Analysis**   117
Suresh Moolgavkar and Georg Luebeck
1   Introduction..   118
2   Brief Review of Mathematical Issues   120
3   Construction of Likelihoods   122
4   Number and Size Distribution of Intermediate (Premalignant) Lesions....   127
5   Model for Colon Cancer   127
5.1   Applications of the Model   128

6   Discussion 131
References 132

A Machine-Learning Algorithm for Estimating and Ranking
the Impact of Environmental Risk Factors in Exploratory
Epidemiological Studies 137
Jessica G. Young, Alan E. Hubbard, Brenda Eskenazi,
and Nicholas P. Jewell
1   Introduction.. 138
2   Background... 139
3   Data Structure and Parameters of Interest................................. 141
4   CHAMACOS Data Description 143
5   Methods.. 145
    5.1   Estimation: DR-IPW Estimation of the Population
          Intervention Model... 145
    5.2   Single Exposure Inference: A Modified Conditional
          Permutation Test 146
    5.3   Joint Inference: Quantile Transformation Method 148
6   Results 149
7   Discussion 153
References 155

A Latent Time Distribution Model for the Analysis of Tumor
Recurrence Data: Application to the Role of Age in Breast Cancer 157
Yann De Rycke and Bernard Asselain
1   Introduction.. 158
2   Material and Methods 159
    2.1   Cox Model.. 159
    2.2   Cox Model with Time-Dependent Covariates 160
    2.3   Yakovlev Models.. 160
    2.4   Breast Cancer Data... 162
3   Results 162
    3.1   Dataset... 162
    3.2   Cox Models 163
    3.3   Yakovlev Model.. 165
4   Discussion 165
References 167

Estimation of Mean Residual Life 169
W. J. Hall and Jon A. Wellner
1   Introduction and Summary... 169
2   Convergence on $\mathbb{R}^+$: Covariance Function of the Limiting Process 171
3   Alternative Sufficient Conditions: $Var[\mathbb{Z}(x)]$ as $x \to \infty$ 176
4   Examples 178
5   Confidence Bands for e 179
6   Illustration of the Confidence Bands 182

7 Further Developments.. 184
    7.1 Confidence Bands and Inference 184
    7.2 Censored Data.. 185
    7.3 Median and Quantile Residual Life Functions 185
    7.4 Semiparametric Models for Mean and Median Residual Life 186
    7.5 Monotone and Ordered Mean Residual Life Functions 186
    7.6 Bivariate Residual Life 187
References 187

Likelihood Transformations and Artificial Mixtures 191
Alex Tsodikov, Lyrica Xiaohong Liu, and Carol Tseng
1 Artificial Mixtures... 191
2 The Quasi-Expectation Operator and the Quasi-EM (QEM)
   Algorithm 193
3 Multinomial Regression.. 197
4 Nonlinear Transformation Models (NTM)................................... 199
5 Copula Models 200
6 Example.. 203
    6.1 A Composition of Gamma and Positive Stable Shared Frailty
        Models 203
    6.2 Retinopathy Application... 205
    6.3 Numerical Experiments and Simulations 206
7 Discussion 208
References 209

**On the Application of Flexible Designs When Searching
for the Better of Two Anticancer Treatments** 211
Christina Kunz and Lutz Edler
1 Introduction 212
2 Flexible Two-Stage Designs in Survival Trials............................. 214
    2.1 Two-Stage Group Sequential Survival Trials 215
    2.2 Adaptive Design Approaches for Two-Stage Survival Trials 217
3 Simulation Studies 221
    3.1 Design of Simulation Studies... 222
    3.2 Simulation Results 223
    3.3 Evaluation of Simulation Results 229
    3.4 Application of Simulation Results 230
4 Discussion 231
References 241

**Parameter Estimation for Multivariate Nonlinear Stochastic
Differential Equation Models: A Comparison Study** 245
Wei Gu, Hulin Wu, and Hongqi Xue
1 Introduction... 245
2 Estimation Methods for Nonlinear SDEs 247
    2.1 Euler Method... 247
    2.2 Simulated Maximum Likelihood Method.............................. 248

3   Improved Local Linearization (ILL) Method 250
4   Numerical Example: HIV Dynamic Model 255
5   Conclusion 256
References 257

**On Frailties, Archimedean Copulas and Semi-Invariance Under
Truncation** 259
David Oakes
1   Introduction 259
2   Basics 259
3   Uniqueness 262
4   Some Examples 265
    4.1   Clayton's Model—Gamma Distributed Frailties 265
    4.2   Inverse Gaussian Model 266
    4.3   Logarithmic Series Distribution: Frank's Model 267
    4.4   Poisson Model 267
    4.5   Positive Stable Model 267
    4.6   Exterior Power Families 269
    4.7   Interior Power Families 269
5   The Kendall Distribution and Kendall's Tau 270
6   Truncation Invariance and Semi-Invariance 271
7   The Kendall Distribution Under Truncation 274
8   Higher Dimensions 276
References 277

Part II   Short Communications

The Generalized ANOVA: A Classic Song Sung with Modern Lyrics 281
Hui Zhang and Xin Tu
1   Introduction 281
2   Generalized ANOVA for Mean and Variance 282
3   Inference for Generalized ANOVA 283
4   Extension to a Longitudinal Data Setting with Missing Data 284
5   Simulation 285
6   Discussion 286
References 287

Analyzing Gene Pathways from Microarrays to Sequencing Platforms ... 289
Jeffrey Miecznikowski, Dan Wang, Xing Ren, Jianmin Wang,
and Song Liu
1   Introduction 289
2   Pathway Methods for Microarrays 290
    2.1   The GSA Method 291
3   Case Study 293
4   Next Generation Sequencing Tests 293
5   Discussion and Conclusions 295
References 295

A New Approach for Quantifying Uncertainty in Epidemiology 297
Elart von Collani
1 Preamble 297
2 Uncertainty and Epidemiology 297
   2.1 Quantification of Randomness 299
   2.2 Quantification of Ignorance 300
3 Ambiguity of Epidemiological Results 300
   3.1 Statistics and Ambiguity 302
   3.2 Epidemiological Measurements 303
4 A Stochastic Model of Uncertainty 304
   4.1 The Bernoulli Space 304
   4.2 Estimating Probabilities 305
References 309

Branching Processes: A Personal Historical Perspective 311
Peter Jagers
1 Introduction and Summary 311
References 318

Principles of Mathematical Modeling in Biomedical Sciences:
An Unwritten Gospel of Andrei Yakovlev 321
Leonid Hanin
1 Mathematics and Biology: A Tough Marriage 322
2 Mathematical Models as Axiomatic Systems 324
3 A Mathematical Modeler's Dilemma: Deterministic or Stochastic? 326
4 A Devil in the Corner: Model Non-identifiability 327
5 Mathematical Models and Biological Reality 329
References 330

Appendix: Publications of Andrei Yakovlev 333

Index 349

Contributors

Anthony Almudevar University of Rochester, Rochester, NY, USA

Bernard Asselain Institut Curie, Paris, France

Linlin Chen Rochester Institute of Technology, Rochester, NY, USA

Elart von Collani *(deceased)* Würzburg, Germany

Yann De Rycke Institut Curie, Paris, France

Lutz Edler German Cancer Research Center, Heidelberg, Germany

Brenda Eskenazi University of California at Berkeley, Berkeley, CA, USA

Alexander Y. Gordon *(deceased)* Charlotte, NC, USA

Wei Gu Zhongnan University of Economics and Law, Wuhan, Hubei, People's Republic of China

W. J. Hall *(deceased)* Rochester, NY, USA

Leonid Hanin Idaho State University, Pocatello, ID, USA

Alan E. Hubbard University of California at Berkeley, Berkeley, CA, USA

Rui Hu University of Rochester, Rochester, NY, USA

Ollivier Hyrien Fred Hutchinson Cancer Research Center, Seattle, WA, USA

Peter Jagers Chalmers University of Technology and University of Gothenburg, Gothenburg, Sweden

Nicholas P. Jewell University of California at Berkeley, Berkeley, CA, USA

Lev Klebanov Charles University, Prague, Czech Republic

Christina Kunz German Cancer Research Center, Heidelberg, Germany

Lyrica Xiaohong Liu Amgen, South San Francisco, CA, USA

Song Liu Roswell Park Comprehensive Cancer Center, Buffalo, NY, USA

H. Land University of Rochester Medical Center, Rochester, NY, USA

Georg Luebeck Fred Hutchinson Cancer Research Center, Seattle, WA, USA

Jeffrey Miecznikowski SUNY University at Buffalo, Department of Biostatistics, Buffalo, NY, USA
Roswell Park Comprehensive Cancer Center, Buffalo, NY, USA

Suresh Moolgavkar Fred Hutchinson Cancer Research Center, Seattle, WA, USA

David Oakes University of Rochester, Rochester, NY, USA

Christoph Proschel University of Rochester, Rochester, NY, USA

Xing Qiu University of Rochester, Rochester, NY, USA

Xing Ren SUNY University at Buffalo, Buffalo, NY, USA

Carol Tseng H2O Clinical, Hunt Valley, MD, USA

Alex Tsodikov University of Michigan, Ann Arbor, MI, USA

Xin Tu UC San Diego School of Medicine, CA, USA

Jon A. Wellner University of Washington, Seattle, WA, USA

Dan Wang Roswell Park Comprehensive Cancer Center, Buffalo, NY, USA

Jianmin Wang Roswell Park Comprehensive Cancer Center, Buffalo, NY, USA

Hulin Wu University of Texas Health Sciences Center, Houston, Texas, USA

Hongqi Xue University of Rochester, New York, NY, USA

Nikolay M. Yanev Bulgarian Academy of Sciences, Sofia, Bulgaria

Andrei Yakovlev University of Rochester, Rochester, NY, USA

Jessica G. Young Harvard Medical School & Harvard Pilgrim Health Care Institute, Boston, MA, USA

Hui Zhang North Western University Chicago, IL, USA

Part I
Research Articles

Stochastic Models of Cell Proliferation Kinetics Based on Branching Processes

Nikolay M. Yanev

Abstract The aim of this memorial survey paper is to present some joint work with Andrei Yu. Yakovlev (http://www.biology-direct.com/content/3/1/10) focused on new ideas for the theory of branching processes arising in cell proliferation modeling. The following topics are considered: some basic characteristics of cell cycle temporal organization, distributions of pulse-labeled discrete markers in branching cell populations, distributions of a continuous label in proliferating cell populations, limiting age and residual lifetime distributions for continuous-time branching processes, limit theorems and estimation theory for multitype branching populations and relative frequencies with a large number of ancestor, age-dependent branching populations with randomly chosen paths of evolution. Some of the presented results have not been published yet.

Keywords Branching processes · Cell proliferation · Discrete and continuous labels · Label distributions · Immigration · Age and residual lifetime distributions · Age-dependent processes · Large number of ancestors · Multitype branching processes · Limiting distributions · Asymptotic normality · Statistical inference

AMS 2000 Subject Classifications Primary 60J80, 60J85; Secondary 62P10, 92D25

1 Introduction

New ideas and results in the theory of branching stochastic processes obtained by modeling of cell proliferation kinetics and based on some joint papers with Andrei Yakovlev [21–29], [32–34] are presented in this memorial survey paper.

N. M. Yanev (✉)
Institute of Mathematics and Informatics, Bulgarian Academy of Sciences, Sofia, Bulgaria
e-mail: yanev@math.bas.bg

© Springer Nature Switzerland AG 2020
A. Almudevar et al. (eds.), *Statistical Modeling for Biological Systems*,
https://doi.org/10.1007/978-3-030-34675-1_1

Note that branching stochastic processes arise as models of population dynamics of objects having different nature, for example, photons, electrons, neutrons, protons, atoms, molecules, cells, microorganisms, plants, animals, humans, prices, information. Thus, many real situations in physics, chemistry, biology, demography, ecology, epidemiology, economy, and so on could be modeled by different types of branching processes.

The theory of branching processes has a long history of biological applications. It is worth pointing out that the first asymptotic result for branching processes was obtained by Kolmogorov [12] while considering some biological problems. Recall that the terminology "branching processes" was first introduced by Kolmogorov and Dmitriev [14] and Kolmogorov and Sevastyanov [15] in proposing multitype branching processes, which received much attention in biological applications. For a further development of the theory of branching processes and their applications in biology we refer the reader to several sources [1, 4–6, 10, 11, 17, 19, 23].

The paper is organized as follows. Some basic models and characteristics of cell proliferation kinetics using branching processes are considered in Sect. 2 [22, 23, 32]. In Sect. 3 [23, 33, 34] the distribution of the discrete markers (labels) using a model with infinitely many types of Bellman–Harris branching processes is presented. Generalizations in the case of continuous labels are given in Sect. 4 [25, 27]. It is assumed that the mitotic division results in a random distribution of the label among daughter cells in accordance with some bivariate probability distribution. In the event of cell death the label borne by that cell disappears. The main focus is on the label distribution as a function of the time elapsed from the moment of label administration. Explicit expressions for this distribution are derived in some particular cases which are of practical interest in the analysis of the cell cycle. The Markov branching process with the same evolution of a continuously distributed label is considered as well. Note that processes with continuous labels were first considered by Kolmogorov [13]. In Sect. 5 new models of renewing cell populations (in vivo) using age-dependent branching processes with non-homogeneous Poisson immigration are proposed. An interesting and important problem arising from cell proliferation kinetics is considered: the definition and limiting behavior of age and residual lifetime distributions for branching processes [24, 26]. From a mathematical point of view it can be considered as a generalization of classical renewal theory. Asymptotic normality for multitype branching processes with a large number of ancestors is given in Sect. 6 [28, 29]. The relative frequencies of distinct types of cells are also investigated. The results are useful in cell kinetics studies where the relative frequencies but not the absolute cell counts are accessible to measurement. In [28] some relevant statistical applications are discussed in the context of asymptotic maximum likelihood inference for multitype branching processes. In [29] the asymptotic behavior of multitype Markov branching processes with discrete or continuous time is investigated in the positive regular and nonsingular case when both the initial number of ancestors and the time tend to infinity. The results from [28] and [29] have specific applications in cell proliferation kinetics. Multitype age-dependent branching processes with randomly chosen paths of evolution are proposed in Sect. 7 as models of progenitor cell populations (in vitro). This model, in two-type case, is used to estimate the offspring distributions using real data as well as bootstrap methods [21].

Finally it is worth pointing out once again that the main purpose of this review is to present some new directions and results in the theory of branching processes obtained in the case of the stochastic modeling of cell proliferation kinetics. On the other hand, the proposed models describe more realistically the real cell proliferation processes and allow one to estimate successfully the basic parameters.

2 Subsequent Generations of Cells Induced to Proliferation and Some Characteristics of Cell Cycle Temporal Organization

It is assumed in [22] that the evolution of cells follows the classical Bellman–Harris branching process. It means that every cell has a duration of a lifetime τ with a c.d.f. $G(x) = P(\tau \leq x)$ and produces new cells at the end of a lifetime according to a p.g.f. $h(s) = \sum_{k=0}^{\infty} p_k s^k$. Let $Z(t)$ denote the number of cells at the moment $t \geq 0$, assuming the classical independence of the cell evolutions. Then the p.g.f. $F(t; s) = E\{s^{Z(t)}\}$ satisfies the equation

$$F(t; s) = \int_0^t h(F(t - u; s))dG(u) + s(1 - G(t)), \quad F(0; s) = s.$$

This basic equation can be obtained by conditioning on the evolution of the first cell and applying the law of the total probability (see [1, 6, 10, 17, 19], where one can find also some conditions for a regularity, i.e., $P(Z(t) < \infty) = 1$ for any $t > 0$).

In [22], together with the process $Z(t)$, the following characteristics are considered, which are of interest to biologists:

1. $\mu_n(t) =$ the number of cells in the n-th generation alive at time t;
2. $u_n(t) =$ the number of cells in the first $n - 1$ generations, which are still alive at time t;
3. $N_k(t, x) =$ the number of cells in the kth generation which are younger than x at moment t;
4. $\mu^i(t) = (\mu_1^i(t), \ldots, \mu_d^i(t))$, where $\mu_j^i(t)$ is the number of cells of type T_j born up to the moment t by an initial cell of type T_i in a d-type age-dependent branching process.

Equations for the corresponding p.g.f. can be obtained and these results are used in [32] to investigate some characteristics of the mitotic cycle. In [32] the distribution of the number of cells which synthesize DNA (i.e., are in S phase) under the action of certain proliferation agents is investigated. The processes of main interest are as follows:

5. $N_S^C(t)$ = the number of cells which went into the S phase up to time t;
6. $N_S(t)$ = the number of cells which are in S phase at time t.

All these results are used for further applications [23].

3 Distributions of Pulse-Labeled Discrete Marks in Branching Populations of Cells

Among many applied problems for which methods of stochastic branching processes hold much promise is the analysis of labeling experiments. These experimental techniques are intended for making quantitative inference on the mitotic cycle parameters in renewing cell populations from observed dynamics of cells after a fraction of the cell population is labeled with specially designed molecular markers. DNA precursors labeled either with radioactive isotopes (e.g., U-456 or H-$thymidine$) or with fluorescent antibodies are typically used for this purpose. Such labeling of the cells occurs during their progression through the S-phase of the mitotic cycle. When using U-456 or H-$thymidine$ and an autoradiographic technique, one can obtain data on grain counts, the latter being interpreted as discrete marks attached to each labeled cell. The distribution of such marks as a function of the time elapsed from the administration of a pulse label yields the needed information on the structure of the mitotic cycle to be extracted by methods of mathematical modeling. Assuming that the initial distribution of marks is Poisson, and treating the evolution of labeled cells as a Bellman–Harris age-dependent branching process with infinitely many cell types, Yanev and Yakovlev [33] derived an analytic form of this distribution. On the other hand, analyzing the kinetics of cells that have been pulse-labeled with CFSE (a continuous label) on a fluorescence-activated cell sorter has become a method of choice in this field of research. This technique calls for modeling the distribution of CFSE intensity and its variations with time. However, little attention has been given to this problem within the framework of stochastic branching processes.

The problem for the distribution of discrete labels was solved completely in [33] for the case of pulse labels at $t = 0$ and initial Poisson distribution $P_0(\theta)$ of the label among cells. It was proved that the states of the system can be described by a Bellman–Harris branching process with infinitely many types of particles $\tilde{Z} = (Z_0(t), Z_1(t), \ldots, Z_j(t), \ldots)$, where $Z_j(t)$ is the number of cells with label j at time t. For the distribution $\Pi_j(t) = \mathbf{E}\{Z_j(t)\}/\mathbf{E}\{\sum_{k=0}^{\infty} Z_k(t)\}$ for $j = 0, 1, 2, \ldots$ and $t \geq 0$, it was shown (in the case of synchronized populations) that $\Pi_j(t) = (\theta^j/j!) \sum_{k=0}^{\infty} (p2^{1-j})^k \exp(-\theta/2^k)(\bar{G}*G^{*k})(t)/\sum_{k=0}^{\infty}(2p)^k(\bar{G}*G^{*k})(t)$, where $G(t)$ is the c.d.f. of the mitotic cycle and p is the probability of successful division. Here $\bar{G} = 1 - G$ and $*$ means a classical convolution, where

$$G^{*k}(t) = \int_0^t G^{*(k-1)}(t-u))dG(u), \quad k = 1, 2, \ldots$$

$$G^{*0}(t) = 1, \quad t \geq 0, \text{ and}$$

$$G^{*0}(t) = 0, \quad t < 0.$$

The generalization of these results and further applications were followed up in [34].

A comprehensive review of biological applications of branching processes can be found in the book [23].

4 Continuous Label Distributions in Proliferating Cell Populations

Suppose that every cell has a lifetime τ with distribution function $G(x) = P(\tau \leq x)$ and at the end of its life it either divides into two cells with probability p ($0 < p \leq 1$) or it dies with probability $1-p$. If a cell divides, its label L is distributed randomly between daughter cells so that their labels L_1 and L_2 satisfy the condition $L_1 + L_2 \leq L$. The case $L_1 + L_2 < L$ can be explained as a loss of label. For cell proliferation the most important case is $L_1 + L_2 = L$, but the further developed methods are valid in the general case. Introduce the conditional distribution $P(L_1 \leq y_1, L_2 \leq y_2 \mid L = y) = K(y_1/y, y_2/y), 0 \leq y_1 \leq y, 0 \leq y_2 \leq y$, where the bivariate distribution function $K(x_1, x_2)$ is symmetric, that is $K(x_1, x_2) = K(x_2, x_1), 0 \leq K(x_1, x_2) \leq 1$ for $0 \leq x_1 \leq 1, 0 \leq x_2 \leq 1$. Let $K(x) = K(x, 1) = K(1, x)$ be the one-dimensional distribution that defines both marginal distributions of $K(x_1, x_2)$. In the event of cell death the label borne by the cell disappears.

We assume for simplicity that the process begins with one cell of age zero at time $t = 0$ and the initial cell bears a certain amount L_0 of label. The results can readily be generalized to include an arbitrary initial distribution of the random variable L_0 and then the resultant formulas can be compounded with respect to this distribution. The initial distribution can be estimated nonparametrically from the data on label intensity available at time $t = 0$.

The case of a randomly distributed continuous label is more complicated than its discrete counterpart and involves modeling of a branching process with states characterized by a real-valued parameter (see [25, 27]). Recall that Kolmogorov [13] was the first to consider a branching process (particle splitting) of this type with the continuous parameter being the particle size.

Let $Z(t, x)$ be the number of cells at time $t > 0$ with the label intensity of $x \leq L_0$. It is clear that $Z(t, x) = 0$ if $x > L_0$. Introduce the notation $P_n(t, x \mid L_0) = P(Z(t, x) = n)$. Then $P_n(t, x \mid L_0) = P_n(t, x/L_0 \mid 1), x \leq L_0$, and in what follows, we will use the notation $P_n(t, x) = P_n(t, x \mid 1), 0 \leq x \leq 1$. Note that $P_n(t, x) = 0$ for $n > 1/x$.

Introducing the p.g.f. $\Psi(t, x, s) = E\{s^{Z(t,x)}\}$ one can obtain the equation

$$\Psi(t, x, s) = (1 - p)G(t) + s[1 - G(t)]$$
$$+ p \int_0^t \left\{ \int_x^1 \int_x^1 \Psi(t - y, x/u_1, s)\Psi(t - y, x/u_2, s)K(du_1, du_2) \right\}$$
$$\times dG(y),$$

which has a unique solution in the class of all p.g.f.'s [25, 27].

Introduce the notation:

$$A(t, x) = E\{Z(t, x)\} = \frac{\partial}{\partial s}\Psi(t, x, s)|_{s=1},$$

$$B(t, x) = E\{Z(t, x)[Z(t, x) - 1]\} = \frac{\partial^2}{\partial s^2}\Psi(t, x, s)|_{s=1}.$$

Then one can obtain the following equations:

$$A(t, x) = 2p \int_0^t \left\{ \int_x^1 A(t - y, x/u)dK(u) \right\} dG(y) + 1 - G(t),$$

$$B(t, x) = 2p \int_0^t \left\{ \int_x^1 B(t - y, x/u)dK(u) \right\} dG(y)$$
$$+ 2p \int_0^t \left\{ \int_x^1 \int_x^1 A(t - y, x/u_1)A(t - y, x/u_2)K(du_1, du_2) \right\} dG(y).$$

Setting $x = 0$ one has

$$\Psi(t, 0, s) = (1 - p)G(t) + s[1 - G(t)] + p \int_0^t \Psi^2(t - y, 0, s)dG(y),$$

$$A(t, 0) = 2p \int_0^t A(t - y, 0)dG(y) + 1 - G(t),$$

$$B(t, 0) = 2p \int_0^t B(t - y, 0)dG(y) + 2p \int_0^t A^2(t - y, 0)dG(y),$$

describing an age-dependent binary branching process considered by Bellman and Harris.

Definition 4.1 The label distribution is defined as $D_t(x) = 1 - A(t, x)/A(t, 0)$.

Additionally, $\overline{D}_t(x) = 1 - D_t(x)$. A closed form solution can be obtained in the special case where one of the daughter cells receives a fixed fraction c ($0 < c < 1$) of the mother label while the complement $1 - c$ goes to the second daughter cell. By a symmetry argument we have the condition: $0 < c \leq 1/2$.

Theorem 4.1 *Under condition $K_c(u) = 0$ for $u < c$ and $K_c(u) = 1$ for $u \geq c$, the following label distribution holds:*

(i) *For $x < c \leq 1/2$, $\overline{D}_t(x) = \{\sum_{k=0}^{N}(2p)^k(\overline{G} * G^{*k})(t)\}/\sum_{k=0}^{\infty}(2p)^k(\overline{G} * G^{*k})(t)$, where $N = N(x, c) = \langle(\ln(x/c))/\ln c\rangle$ and $\langle z \rangle$ denotes the smallest integer greater or equal to z;*

(ii) *If $x \geq c$, then $\overline{D}_t(x) = \overline{G}(t)/\sum_{k=0}^{\infty}(2p)^k(\overline{G} * G^{*k})(t)$ for every $c \in (0, 1/2]$.*

Corollary 4.1 *The distribution given by Theorem 4.1 assumes a particularly simple form in the biologically plausible case of $c = 1/2$. In this case, $N = N(x, 1/2) = \langle-(\ln 2x)/\ln 2\rangle$ for $x < 1/2$.*

Let $G(x) = 1 - \exp(-\lambda x)$, $\lambda > 0$, which means that the considered process is a Markovian one.

Theorem 4.2 *In the Markov case $\overline{D}_t(x) = \sum_{n=0}^{\infty}\Pi_n(2p\lambda t)R^{*n}(-\log x)$, where $R(x) = 1 - K(\exp(-x))$ and $\Pi_n(x) = n^x\exp(-x)/n!$ is the Poisson distribution.*

Corollary 4.2 *Assuming additionally that $K(u) = u$ for $0 \leq u \leq 1$, then $\overline{D}_t(x) = \sum_{n=0}^{\infty}\Pi_n(2p\lambda t)\Gamma_n(-\log x)$, where $\Gamma_n(y) = \int_0^y z^{n-1}\exp(-z)dz/(n - 1)!$.*

Theorem 4.3 *Let $\alpha = \int_0^1 \log(1/x)dK(x) < \infty$, $\beta = \int_0^1 \log^2(1/x)dK(x) < \infty$ and $\Delta_t(z) = \exp\left(-2p\lambda\alpha t - z(2p\lambda t\beta)^{1/2}\right)$. Then in the Markov case for every $z \in (-\infty, \infty)$*

$$\lim_{t \to \infty} \overline{D}_t(\Delta_t(z)) = 1/(2\pi)^{1/2} \int_{-\infty}^{z} \exp(-u^2/2)du.$$

The label distribution can be generalized by replacing $A(t, x)$ and $A(t, 0)$ from other pertinent models of cell proliferation kinetics. In particular, age-dependent branching processes with immigration are gaining in importance in conjunction with recent advancements in experimental approaches to cell proliferation kinetics in analysis of renewing cell populations [24]. These advancements have made it possible to distinguish many cell types by antibody labeling so that cells of different types can be counted in the dissociated tissue by using flow cytometry. Finally, it is interesting to point out that the validity of the asymptotic results in Theorem 4.3 remains an open problem in the non-Markov case.

5 Limiting Age and Residual Lifetime Distributions for Continuous-Time Branching Processes

Consider now proliferating cell populations with the following evolution. Every cell with probability p has a random lifetime ξ with c.d.f. $G(x) = P(\xi \leq x)$, $x \geq 0$, or with probability $1 - p$ it differentiates into another cell type. At the end of its life, every cell gives rise to ν offsprings (of the same cell type) with discrete distribution

$p_k(u) = P(v = k \mid \xi = u)$, $\sum_{k=0}^{\infty} p_k(u) = 1$, $u \geq 0$. It takes a random time η with c.d.f. $L(x) = P(\eta \leq x)$, $x \geq 0$, for the event of differentiation to actually occur. If $p = 1$, the stochastic process thus defined reduces to the Sevastyanov branching process [19]. The mixture-type branching (allowing non-identical distributions of the time to division and the time to differentiation) was introduced by Jagers [10]. Denote the offspring p.g.f. by $h(s; u) = E\{s^v \mid \xi = u\} = \sum_{k=0}^{\infty} p_k(u)s^k$, $|s| \leq 1$, $u \geq 0$. The most representative example is given by $h(u, s) = 1 - \beta(u) + \beta(u)s^2$, implying that the cell divides with probability $\beta(u)$ or dies with probability $1 - \beta(u)$. In what follows, our focus will be on the general case.

Let $F(t; s) = E\{s^{Z(t)}\}$ be the p.g.f. of the number of cells $Z(t)$ at time t produced by one cell of zero age and let $Z(t, x)$ be the number of cells of age $\leq x$ that are present at time t. The latter process has p.g.f. $F(t, x; s) = E\{s^{Z(t,x)}\}$, $t, x \geq 0$. Note that $Z(t) = Z(t, x)$ if $x \geq t$. The p.g.f.'s $F(t; s)$ and $F(t, x; s)$ satisfy the following integral equations:

$$F(t; s) = p \int_0^t h(u; F(t - u; s))dG(u)$$
$$+ s\{p[1 - G(t)] + (1 - p)[1 - L(t)]\} + (1 - p)L(t),$$

$$F(t, x; s) = p \int_0^t h(u; F(t - u, x; s))dG(u)$$
$$+ [s\delta(x - t) + 1 - \delta(x - t)]\{p[1 - G(t)] + (1 - p)[1 - L(t)]\}$$
$$+ (1 - p)L(t),$$

with $F(0; s) = F(0, 0; s) = s$, where $\delta(z) = 1$ for $z \geq 0$ and $\delta(z) = 0$ for $z < 0$.

The equations are obtained by conditioning on the evolution of the first cell and applying the law of total probability. It is not difficult to obtain equations for the corresponding expectations by taking partial derivatives with respect to s at the point $s = 1$. Denote $M(t) = E[Z(t)] = \partial F(t; s)/\partial s|_{s=1}$ and $M(t, x) = E[Z(t, x)] = \partial F(t, x; s)/\partial s|_{s=1}$, respectively.

Definition 5.1 For the process $Z(t)$, the average age distribution at time $t \geq 0$ is given by $A_t(x) = M(t, x)/M(t)$, $x \geq 0$. The limiting average age distribution is defined as $A(x) = \lim_{t\to\infty} A_t(x)$.

Let $\overline{Z}_t(y)$ be the number of cells at time t with residual lifetime greater than y.

Definition 5.2 For the process $Z(t)$ the average residual lifetime distribution at time $t \geq 0$ is given by $R_t(y) = 1 - \overline{R}_t(y) = 1 - E[\overline{Z}_t(y)]/M(t)$, $y \geq 0$. Then the limiting average residual lifetime distribution is defined as $R(y) = \lim_{t\to\infty} R_t(y)$.

Note that $E[\overline{Z}_t(y)] = M(t + y) - M(t + y, y)$. Therefore, one can write

$$\overline{R}_t(y) = \frac{M(t + y) - M(t + y, y)}{M(t)} = \frac{M(t + y)}{M(t)}\left[1 - \frac{M(t + y, y)}{M(t + y)}\right],$$

which offers some advantages.

Remark Since every renewal process is a particular case of the age-dependent branching process in which every individual produces exactly one offspring, the introduced notions can be considered as a generalization of the age and residual lifetime distributions encountered in the renewal theory. However, these characteristics are more difficult to handle in the context of branching processes.

Let α be the Malthus parameter of the equation $p \int_0^\infty \exp(-\alpha u)m(u)dG(u) = 1$, where $m(u) = \partial h(s; u)/\partial s|_{s=1}$ is the offspring mean. Assume the conditions $\int_0^\infty u\exp(-\alpha u)m(u)dG(u) < \infty$, $\int_0^\infty \exp(-\alpha u)[1 - G(u)]du < \infty$ and $\int_0^\infty \exp(-\alpha u)[1 - L(u)]du < \infty$. Note that in the supercritical case $\alpha > 0$, the last two conditions are automatically met.

Theorem 5.1 *Under the conditions the following limiting age and residual lifetime distributions hold:*

(i) $A(x) = \dfrac{p \int_0^x \exp(-\alpha u)[1 - G(u)]du + (1 - p) \int_0^x \exp(-\alpha u)[1 - L(u)]du}{p \int_0^\infty \exp(-\alpha u)[1 - G(u)]du + (1 - p) \int_0^\infty \exp(-\alpha u)[1 - L(u)]du}$,

(ii) $R(y) =$

$$1 - \exp(\alpha y)\frac{p \int_y^\infty \exp(-\alpha u)[1 - G(u)]du + (1 - p) \int_y^\infty \exp(-\alpha u)[1 - L(u)]du}{p \int_0^\infty \exp(-\alpha u)[1 - G(u)]du + (1 - p) \int_0^\infty \exp(-\alpha u)[1 - L(u)]du}.$$

Note that in the critical case $\alpha = 0$

$$A(x) = R(y) = \frac{p \int_0^x [1 - G(u)]du + (1 - p) \int_0^x [1 - L(u)]du}{p \int_0^\infty [1 - G(u)]du + (1 - p) \int_0^\infty [1 - L(u)]du}.$$

Corollary 5.1 *Let us now set $p = 1$. If we assume in addition that $h(s) \equiv h(u; s)$, $m = h'(1)$, then $Z(t)$ will be the classical Bellman–Harris branching process. Hence one obtains the limiting age distribution*

$$A(x) = \frac{\int_0^x \exp(-\alpha u)(1 - G(u))du}{\int_0^\infty \exp(-\alpha u)(1 - G(u))du},$$

which can be found in Harris [6] but only for $\alpha > 0$. Note that in this case $R(y) = 1 - \exp(\alpha y)(1 - A(y))$.

Corollary 5.2 *Consider now the Markov case $G(t) = 1 - \exp(-\lambda t)$, $t \geq 0$, with $h(u, s) \equiv h(s)$. Assuming additionally that $p = 1$, one obtains the age distribution $A_t(x) = 1 - \exp(-m\lambda x)$. Note that the age distribution for the Markov branching process is stationary, but it depends on the critical parameter m. It is not difficult to see also that $R_t(y) = 1 - \exp(-\lambda y)$, i.e., the residual lifetime distribution in the Markov case is also stationary and identical to the lifetime distribution.*

It is worth explaining that Corollary 5.2 is related to the memoryless property of the exponential distribution.

Let now $Y(t)$ be an age-dependent process with the cell evolution as previously defined and with the following immigration component. The notation $\Lambda(t)$ is used for the cumulative rate $\Lambda(t) = \int_0^t r(u)du$ of a non-homogeneous Poisson process $\xi(t)$, generated by a sequence of the time points $0 = S_0 < S_1 < S_2 < S_3 < \ldots$, which are in fact times of immigrant arrivals. If one uses the notation $T_i = S_i - S_{i-1}$ then $S_k = \sum_{i=1}^k T_i$, $k = 1, 2, \ldots$. Associated with every point S_k is an independent immigration component I_k, where $\{I_k\}$ are i.i.d. r.v.'s with p.g.f. $g(s) = E[s^{I_k}] = \sum_{i=0}^\infty q_i s^i$.

This process can be represented as $Y(t) = \sum_{k=1}^{\xi(t)} Z_{(k)}(t - S_k)$ if $\xi(t) > 0$ and $Y(t) = 0$ if $\xi(t) = 0$, where $Z_{(k)}(t)$ are i.i.d. branching processes with the same evolution as $Z(t)$ but originated from a random number of ancestors I_k. Each of the processes $Z_{(k)}(t)$ has p.g.f. $F^*(t; s) = g(F(t; s))$.

Let $Y(t, x)$ be the number of cells of age $\le x$ at time t in this branching model with immigration. Introduce the p.g.f.s $\Psi(t; s) = E\{s^{Y(t)} \mid Y(0, 0) = 0\}$, $\Psi(t, x; s) = E\{s^{Y(t,x)} \mid Y(0, 0) = 0\}$. Note that if $x \ge t$ then $Y(t) = Y(t, x)$. Let $\overline{Y}_t(y)$ be the number of cells at time t whose residual lifetime is greater than y.

Theorem 5.2 *The p.g.f.'s of the processes $Y(t)$ and $Y(t, x)$ are given by*

$$\Psi(t; s) = \exp\left\{-\int_0^t r(t - u)[1 - F^*(u; s)]du\right\}, \quad \Psi(0; s) = 1,$$

$$\Psi(t, x; s) = \exp\left\{-\int_0^t r(t - u)[1 - F^*(u, x; s)]du\right\}, \quad \Psi(0, 0; s) = 1,$$

where $F^*(t; s) = g(F(t; s))$ and $F^*(t, x; s) = g(F(t, x; s))$.

Further on we will use Definitions 5.1 and 5.2, where the process Z is replaced by the process Y.

Theorem 5.3 *Assume conditions of Theorem 5.1 and let $\lim_{t\to\infty} r(t) = r > 0$. Then the limiting age distribution $A(x)$ is given by Theorem 5.1 and for the limiting residual lifetime one has: $R(y) = A(y)$ for $\alpha \le 0$ and $R(y) = 1 - \exp(\alpha y)(1 - A(y))$ for $\alpha > 0$.*

Remark It is well known that the asymptotic behavior of the branching processes without and with immigration is quite different. But we obtained an interesting new phenomenon: the limiting age distributions for the processes with and without immigration are identical. The same statement is valid for the limiting residual lifetime distributions.

Let us consider the Markov case with $p = 1$ and $r(t) \equiv r$. Sevastyanov [19] was the first to study this Markov branching process with homogeneous Poisson immigration (MBPHPI).

Theorem 5.4 *In the MBPHPI there exists a stationary age distribution* $A_t(x) = N(t, x)/N(t) = 1 - \exp(-m\lambda x)$, $x \geq 0$, *and the limiting residual lifetime distribution is given as follows:* $R(y) = 1 - \exp(-m\lambda y)$, $m < 1$ *and* $R(y) = 1 - \exp(-\lambda y)$, $m \geq 1$.

Remark The residual lifetime distributions associated with the Markov branching process with homogeneous Poisson immigration depend on the critical parameter m. It is interesting to note that in the critical and supercritical cases the limiting residual lifetime distribution is the same as the lifetime distribution (just as it comes about when considering the Markov process without immigration), while in the subcritical case it is exponential with parameter $m\lambda$.

6 Limit Theorems and Estimation Theory for Multitype Branching Populations and Relative Frequencies with a Large Number of Ancestors

Usually branching processes with discrete or continuous time are investigated separately but in this section we will treat them together. We will consider first a multitype Markov branching process $\mathbf{Z}(t; N) = (Z_1(t; N), Z_2(t; N), \ldots, Z_d(t; N))$, where $Z_k(t; N)$ denotes the number of cells of type k ($k = 1, 2, \ldots, d$) at time $t \in \mathbf{T}$ with $\mathbf{Z}(0) = (N, 0, \ldots, 0)$. The time may be discrete ($\mathbf{T} = \mathbf{N}_0 = \{0, 1, 2, \ldots\}$) or continuous ($\mathbf{T} = \mathbf{R}^+ = [0, \infty)$), but the process is assumed to be positive regular and nonsingular. Let R and r denote Perron-Frobenius eigenvalues, respectively, for a discrete or continuous time process and with a left eigenvector $\mathbf{v} = (v_1, \ldots, v_d)$.

The relative frequencies (fractions, proportions) of types can be defined on the non-extinction set as follows:

$$\Delta_i(t; N) = Z_i(t; N)/\sum_{i=1}^{d} Z_i(t; N), \quad i = 1, 2, \ldots, d.$$

The investigation of the relative frequencies is very important for the applications (especially in the cell biology) because there are a lot of situations when it is impossible to observe the number of cells but only their relative proportions. In what follows, we will also need the following deterministic proportions:

$$p_i(t) = A_{1i}(t)/M(t), \quad \text{where}$$

$$M(t) = \sum_{j=1}^{d} A_{1j}(t) \quad \text{and}$$

$$A_{1j}(t) = \mathrm{E}\{Z_j(t; 1)\}, \quad i = 1, 2, \ldots, d; \quad t \in \mathbf{T}.$$

Let us introduce:

Condition A In the discrete time case $R < 1$, $NR^t \to \infty$, or in the continuous time case $r < 0$, $N\exp(rt) \to \infty$.

Condition B In the discrete time case $R = 1$, $N/t \to \infty$, or in the continuous time case $r = 0$, $N/t \to \infty$.

Condition C Supercritical case: $R > 1$ (discrete time) or $r > 0$ (continuous time).

Theorem 6.1 *(See [29]) Let $p_k = v_k/V$, $k = 1, \ldots, d$, $V = \sum_{k=1}^{d} v_k$, and $N, t \to \infty$.*

(i) *Assume Condition A or Condition B. Then $\Delta_k(t; N) \to p_k$ in probability, $k = 1, 2, \ldots, d$;*
(ii) *Under Condition C one has $\Delta_k(t; N) \to p_k$ a.s., $k = 1, 2, \ldots, d$.*

Consider now $\mathbf{X}(t; N) = (X_1(t; N), \ldots, X_d(t; N))$, $t \in \mathbf{T}$, where

$$X_k(t; N) = [Z_k(t; N) - A_{1k}(t; N)]/[D_k(t)\sqrt{N}]$$

and $A_{1k}(t; N) = \mathrm{E}\{Z_k(t; N)\}$ and $D_k(t)$ are some normalization functions.

Theorem 6.2 *(See [29]) Propose finite second moments of the offspring distribution and $N, t \to \infty$.*

(i) *Assume Condition A and let $D_k(t) \equiv D(t)$, where $D(t) \sim \sqrt{R^t}$ in the discrete time case or $D(t) \sim \sqrt{\exp(rt)}$ in the continuous time case. Then $\mathbf{X}(t; N) \xrightarrow{d} \xi$ and $\xi = (\xi_1, \ldots, \xi_d)$ has a multivariate normal distribution with $\mathrm{E}\{\xi_i\} = 0$ and a covariance matrix $C = \|C_{jk}\|$, where $C_{jk} = \lim B_{jk}^{(1)}(t)/R^t$, $t \in \mathbf{N}_0$ or $C_{jk} = \lim B_{jk}^{(1)}(t)/\exp(rt)$, $t \in \mathbf{R}^+$, $B_{jk}^{(1)}(t) = \mathrm{E}\{Z_j(t; 1)Z_k(t; 1)\}$.*
(ii) *Assume Condition B and let $D_k(t) \sim v_k\sqrt{u_1 Bt}$ in the discrete time case or $D_k(t) \sim v_k\sqrt{u_1 bt}$ in the continuous time case. Then $\mathbf{X}(t; N) \xrightarrow{d} \eta = (\eta_1, \ldots, \eta_d)$, where $\eta_1 = \ldots = \eta_d$ a.s. and $\eta_1 \in N(0, 1)$.*
(iii) *Assume Condition C and let $D_k(t) \sim v_k R^t$ in the discrete time case or $D_k(t) \sim v_k\exp(rt)$ in the continuous time case. Then $\mathbf{X}(t; N) \xrightarrow{d} \zeta = (\zeta_1, \ldots, \zeta_d)$, where $\zeta_1 = \ldots = \zeta_d$ a.s., $\zeta_1 \in N(0, \tau^2)$ and $\tau^2 = \mathrm{E}\{W^2 \mid \mathbf{Z}(0) = \delta_1\}$.*

Theorem 6.3 *(See [29]) Assume $N, t \to \infty$ and Condition (i) of Theorem 6.2. Then:*

(a) *$\Phi_k(t; N) = u_1 V\sqrt{ND(t)}[\Delta_k(t; N) - p_k(t)] \xrightarrow{d} Y_k$, where Y_k is a normally distributed r.v. with $\mathrm{E}\{Y_k\} = 0$ and*

$$Var\{Y_k\} = C_{kk} - 2p_k\sum_{j=1}^{d}C_{kj} + p_k^2\sum_{i,j=1}^{d}C_{ij}, \quad k = 1, 2, \ldots, d.$$

(b) $(\Phi_{n_1}(t; N), \ldots, \Phi_{n_k}(t; N)) \overset{d}{\to} (Y_{n_1}, \ldots, Y_{n_k})$, where for every $k = 2, 3, \ldots, d - 1$ and every subset $(n_1, n_2, \ldots, n_k)$ with nonrecurring elements from $\{1, 2, \ldots, d\}$ the r.v. $(Y_{n_1}, \ldots, Y_{n_k})$ have a multivariate normal distribution with

$$Cov(Y_i, Y_j) = C_{ij} - p_i \sum_{k=1}^{d} C_{kj} - p_j \sum_{l=1}^{d} C_{li} + p_i p_j \sum_{k,l=1}^{d} C_{kl}.$$

Remark Surprisingly in the critical and supercritical cases there are no analogs of Theorem 6.3 as it is shown in [29].

Introduce the following notion:

$$\sigma_i^2(t) = Var\{Z_i(t, 1)\}, \quad i = 1, 2, \ldots, d,$$

$$\mathbf{R}^{(d)}(t) = ||r_{ij}(t)||, \quad r_{ij}(t) = Cor\{Z_i(t, 1), Z_j(t, 1)\}, i, j = 1, 2, \ldots, d,$$

$$\mathbf{C}^{(d)}(t) = ||c_{ij}(t)||, \quad c_{ii}(t) = \sigma_i(t)(1 - p_i(t)),$$

$$c_{ij}(t) = -\sigma_i(t)p_j(t) \quad \text{for } i \neq j; i, j = 1, 2, \ldots, d.$$

By contrast of the previous results the following theorem is valid for any kind of branching processes with discrete or continuous time (Markov or non-Markov, reducible or irreducible) assuming only the usual independence of the individual evolutions.

Let

$$V_i(t; N) = [Z_i(t; N) - A_{1i}(t; N)]/[\sigma_i(t)\sqrt{N}], \quad \text{and}$$

$$W_i(t; N) = M(t)\sqrt{N}[\Delta_i(t; N) - p_i(t)], \quad i = 1, 2, \ldots, d.$$

Theorem 6.4 *(See [28]) If t is fixed and $N \to \infty$ then the following statements are valid:*

(i) $(V_1(t; N), \ldots, V_d(t; N)) \overset{d}{\to} (X_1(t), \ldots, X_d(t))$, which have a joint normal distribution with $E\{X_i(t)\} = 0$, $Var\{X_i(t)\} = 1$ and $Cor\{X_i(t), X_j(t)\} = r_{ij}(t)$;

(ii) $W_i(t; N) \overset{d}{\to} Y_i(t)$, where $Y_i(t)$ is a normally distributed r.v. with $E\{Y_i(t)\} = 0$ and $Var\{Y_i(t)\} = \sum_{j,k=1}^{d} r_{jk}(t)c_{ji}(t)c_{ki}(t)$, $i = 1, 2, \ldots, d$;

(iii) $(W_1(t; N), \ldots, W_k(t; N)) \overset{d}{\to} (Y_1(t), \ldots, Y_k(t))$, where $(Y_1(t), \ldots, Y_k(t))$ have a multivariate normal distribution (for every $k=2, 3, \ldots, d-1$) with a covariance matrix $\mathbf{D}^{(k)}(t) = ||Cov\{Y_i(t), Y_j(t)\}|| = [\mathbf{C}_{d \times k}(t)]^T \mathbf{R}^{(d)}(t)\mathbf{C}_{d \times k}(t)$, where $[\mathbf{C}_{d \times k}(t)]^T = ||c_{ji}(t)||$, $j = 1, 2, \ldots, k$; $i = 1, 2, \ldots, d$, is the corresponding transposed matrix of $[k \times d]$ dimensions.

In particular, the following observation process is directly relevant to quantitative studies of proliferation, differentiation, and death of cells. Suppose that the process under study begins with $N = \sum_{k=1}^{n} N_k$ cells of type T_1 and the values of N_k are all large, i.e., $N_0 = \min\{N_1, N_2, \ldots, N_n\} \to \infty$. The descendants of the first N_1 ancestors are examined only once at time t_1 to determine the observations of $Z_i(t_1; N_1)$ or $\Delta_i(t_1; N_1), i = 1, 2, \ldots, d$, whereupon the observation process is discontinued (i.e., the cells under examination are destroyed). At the next moment $t_2 \geq t_1$, the process $Z_i(t_2; N_2)$ or the fractions $\Delta_i(t_2; N_2), i = 1, 2, \ldots, d$, related to the descendants of the second N_2 ancestors are observed, and so on. This procedure results in n independent observations of the form: $\zeta_k = \mathbf{Z}(t_k; N_k) = (Z_1(t_k; N_k), \ldots, Z_d(t_k; N_k))$ or $\zeta_k = \Delta(t_k; N_k) = (\Delta_1(t_k; N_k), \ldots, \Delta_d(t_k; N_k))$, $k = 1, 2, \ldots, n, t_1 \leq t_2 \leq \ldots \leq t_n$, where each vector ζ_k is asymptotically normal in accordance with Theorems 6.2–6.4.

Denoting the corresponding contribution to the log-likelihood function as $L_k(\zeta_k; t_k, N_k)$, the overall log-likelihood is given by $\Lambda_n(\zeta_1, \ldots, \zeta_n) = \sum_{k=1}^{n} L_k(\zeta_k; t_k, N_k)$. The log-likelihood depends on the individual parameters only, which are of primary interest in applications and especially in cell kinetics studies. Finally the parameters can be estimated from the data on the process or the relative frequencies by maximizing the log-likelihood. In this way, the asymptotic results give a new direction toward statistical inference and applications of branching processes in cell proliferation kinetics [28].

Feller [3] considered first branching process with a large number of ancestors. For a classical Bienaymé–Galton–Watson (BGW) process he showed a diffusion approximation in the near-critical case. Lamperti [16] derived also some interesting limiting distributions for BGW process. These results were summarized and discussed by Jagers [10] . Statistical inference for BGW processes with an increasing number of ancestors as well as limiting distributions when N and t tend to infinity were developed by Yanev [30] and Dion and Yanev [2] (see also a review chapter by Yanev [31]).

Jagers [9] was probably the first to consider relative frequencies (proportions, fractions) of cells within the framework of multitype branching processes. He studied asymptotic (as $t \to \infty$) properties of a reducible age-dependent branching process with two types of cells and proved convergence of their relative frequencies to non-random limits in mean square and almost surely on the non-extinction set. The usefulness of such frequencies in cell cycle analysis was further demonstrated by Mode [18] who considered a four-type irreducible age-dependent branching process. Mode built his cell cycle analysis on a model of multitype positively regular age-dependent branching process. In the supercritical case, he proved that $\lim \Delta_k(t) = \delta_k$ *a.s.* as $t \to \infty$, providing the population does not become extinct. In his monograph, Mode [17] also considered the utility of fractions and reported a similar result for the BGW process.

7 Age-Dependent Branching Populations with Randomly Chosen Paths of Evolution

Consider a multitype population with the following cell evolution (see [21]). Every newborn cell of type T_k ($k = 1, 2, \ldots, d$) with probability $r_{k,i}$ ($\sum_{i=1}^{d} r_{k,i} = 1$) has a lifetime c.d.f. $F_{k,i}(t)$ and at the end of its life produces progeny in accordance with p.g.f. $h_{k,i}(\mathbf{s})$, $\mathbf{s} = (s_1, s_2, \ldots, s_d)$. Let $\mathbf{Z}(t) = (Z_1(t), Z_2(t), \ldots, Z_d(t))$, where $Z_k(t)$ denotes the number of cells of type T_k at time t. Introduce p.g.f.'s $\Phi_k(t; \mathbf{s}) = \mathrm{E}\{\mathbf{s}^{\mathbf{Z}(t)} \mid \mathbf{Z}(0) = \mathbf{e}_k\}$ and $\Phi(t; \mathbf{s}) = (\Phi_1(t; \mathbf{s}), \Phi_2(t; \mathbf{s}), \ldots, \Phi_d(t; \mathbf{s}))$, where $\mathbf{e}_k = (0, \ldots, 0, 1, 0, \ldots 0)$ with 1 at the k-th place . Then the following system of integral equations holds:

$$\Phi_k(t; \mathbf{s}) = \sum_{i=1}^{d} r_{k,i} \int_0^t h_{k,i}(\Phi(t-u; \mathbf{s})) dF_{k,i}(u) + s_k \sum_{j=1}^{d} r_{k,j} \overline{F}_{k,j}(t), \quad k = 1, 2, \ldots, d,$$

which is equivalent to the system of equations

$$\Phi_k(t; \mathbf{s}) = \int_0^t h_k(u; \Phi(t - u; \mathbf{s})) dG_k(u) + s_k \overline{G}_k(t), \quad k = 1, 2, \ldots, d,$$

where

$$\overline{F}_{k,j}(t) = 1 - F_{k,i}(t),$$

$$G_k(t) = \sum_{i=1}^{d} r_{k,i} F_{k,i}(t), \quad \overline{G}_k(t) = 1 - G_k(t),$$

$$h_k(u; \mathbf{s}) = \sum_{i=1}^{d} \alpha_{k,i}(u) h_{k,i}(\mathbf{s}),$$

$$\alpha_{k,i}(u) = r_{k,i} / \sum_{j=1}^{d} r_{k,j} (dF_{k,j}(t)/dF_{k,i}(t)),$$

$$\sum_{i=1}^{d} \alpha_{k,i}(u) = 1.$$

We use the notation $dF_{k,j}(t)/dF_{k,i}(t)$ for the corresponding Radon–Nikodym derivatives. The above model represents a special case of the d-type Sevastyanov branching process.

In [21], two new models of an age-dependent branching process with two types of cells were proposed to describe the kinetics of progenitor cell populations cultured in vitro. Another approach is given in Hyrien et al. [7, 8], Yakovlev et al. [20], and Zorin et al. [35]. Our models considered in [21] with two cell types can be derived by the general case setting $d = 2$, $r_{1,1} = p$, $r_{1,2} = 1 - p$, $F_{1,1}(t) = F_1(t)$, $F_{1,2}(t) = F_3(t)$, $h_{1,2}(s_1, s_2) \equiv 1$, $r_{2,1} = 0$, $r_{2,2} = 1$, $F_{2,2}(t) = F_2(t)$, $h_{2,2}(s_1, s_2) \equiv 1$. The most biologically relevant example of the p.g.f. $h_{1,1}(s_1, s_2)$ is given by $h_{1,1}(s_1, s_2) = p_0 + p_1 s_1^2 + p_2 s_2$, $h(1, 1) = p_0 + p_1 + p_2 = 1$. This form of $h_{1,1}(s_1, s_2)$ implies that every cell of type T_1 either dies with probability p_0, or divides into two new T_1 cells with probability p_1, or differentiates into a new cell type T_2 with probability p_2, all the transformations occurring upon completion of the cell's mitotic cycle.

The main focus in [21] was on the estimation of the offspring distribution from data on individual cell evolutions. Such data are typically provided by time-lapse video-recording of cultured cells. Some parameters of the life cycle of progenitor cells, such as the mean (median) and variance of the mitotic cycle time, can be estimated nonparametrically without resorting to any mathematical model of cell population kinetics. For other parameters, such as the offspring distribution, a model-based inference is needed. Age-dependent branching processes have proven to be useful models for this purpose. Within the framework of branching processes, the loss of cells to follow-up can be modeled as a process of emigration that precludes other cell transformations from occurring. This motivates the development of branching models with emigration and associated estimation techniques. The work [21] develops the needed methodology and put it to practical use.

The basic idea behind our approach is to consider a hidden discrete process that would have been observed in the absence of emigration. The net reproduction probabilities of such a process represent the parameters of interest and the developed in [21] procedure makes possible to estimate them from experimental observations. This approach allows us to employ the theory of statistical inference for branching processes with a large number of ancestors (see Yanev [30, 31] and Dion and Yanev [2]).

The proposed new models of a two-type age-dependent branching process with emigration are motivated by a specific biological application. In this application, the first type of cells represents immediate precursors of oligodendrocytes while the second type represents terminally differentiated oligodendrocytes. The proposed methodology can be readily extended to branching processes with any finite number of cell types.

8   Concluding Remarks

The presented results were planned as a part of a joint book with Andrei Yakovlev[1] on branching processes as models in cell proliferation kinetics. Unfortunately the

---

[1] https://biologydirect.biomedcentral.com/articles/10.1186/1745-6150-3-10.

book will never appear. This paper is a tribute to the stimulating ideas of Andrei Yakovlev and the friendship that we shared in our collaboration.

Acknowledgements The author is very grateful to the referee for the useful remarks. The paper is supported by NIH/NINDS grant NS39511, NIH/NCI R01 grant CA134839, NIH grant N01-AI-050020 and grant KP-6-H22/3 from the NSF of the Ministry of Education and Science of Bulgaria.

References

1. Athreya, K. B., & Ney, P. E. (1972). *Branching processes.* Berlin: Springer.
2. Dion, J.-P., & Yanev, N. M. (1997). Limit theorems and estimation theory for branching processes with an increasing random number of ancestors. *Journal of Applied Probability, 34,* 309–327.
3. Feller, W. (1951). Diffusion processes in genetics. *Proceedings of Second Berkeley Symposium on Mathematical Statistics and Probability* (pp. 227–246). Berkeley: University of California Press.
4. Guttorp, P. (1991). *Statistical inference for branching processes.* New York: Wiley.
5. Haccou, P., Jagers, P., & Vatutin, V. A. (2005). *Branching processes: Variation, growth and extinction of populations.* Cambridge: Cambridge University Press.
6. Harris, T. E. (1963). *Branching processes.* New York: Springer.
7. Hyrien, O., Mayer-Proschel, M., Noble, M., & Yakovlev, A. Yu. (2005). Estimating the lifespan of oligodendrocytes from clonal data on their development in cell culture. *Mathematical Biosciences, 193,* 255–274.
8. Hyrien, O., Mayer-Proschel, M., Noble, M., & Yakovlev, A. Yu. (2005). A stochastic model to analyze clonal data on multi-type cell populations. *Biometrics, 61,* 199–207.
9. Jagers, P. (1969). The proportions of individuals of different kinds in two-type populations. A branching process problem arising in biology. *Journal of Applied Probability, 6,* 249–260.
10. Jagers, P. (1975). *Branching processes with biological applications.* London: Wiley.
11. Kimmel, M., & Axelrod, D. E. (2002). *Branching processes in biology.* New York: Springer.
12. Kolmogorov, A. N. (1938). Zur Lösung einer biologischen Aufgabe. *Proceedings of Tomsk State University, 2,* 1–6.
13. Kolmogorov, A. N. (1941). On the lognormal distribution of the particle sizes in splitting (in Russian). *Doklady Akademii Nauk (Proceedings of the Academy of Sciences USSR), 31,* 99–101.
14. Kolmogorov, A. N., & Dmitriev, N. A. (1947). Branching random processes (in Russian). *Doklady Akademii Nauk (Proceedings of the Academy of Sciences USSR), 56,* 7–10.
15. Kolmogorov, A. N., & Sevastyanov, B. A. (1947). Calculation of final probabilities of branching random processes (in Russian). *Doklady Akademii Nauk (Proceedings of the Academy of Sciences USSR), 56,* 783–786.
16. Lamperti, J. (1967). Limiting distributions for branching processes. In *Proceedings of Fifth Berkeley Symposium on Mathematical Statistics and Probability* (pp. 225–241). Berkeley: University of California Press.
17. Mode, C. J. (1971). *Multitype branching processes: Theory and applications.* New York: Elsevier.
18. Mode, C. J. (1971). Multitype age-dependent branching processes and cell cycle analysis. *Mathematical Biosciences, 10,* 177–190.
19. Sevastyanov, B. A. (1971). *Branching processes* (in Russian). Moscow: Nauka.
20. Yakovlev, A. Yu., Boucher, K., Mayer-Proschel, M., & Noble, M. (1998). Quantitative insight into proliferation and differentiation of oligodendrocyte type 2 astrocyte progenitor cells in vitro. *Proceedings of the National Academy of Sciences of the United States of America, 95,* 14164–14167.

21. Yakovlev, A. Yu., Stoimenova, V. K., & Yanev, N. M. (2008). Branching processes as models of progenitor cell populations and estimation of the offspring distributions. *Journal of the American Statistical Association, 103*, 1357–1366.
22. Yakovlev, A. Yu., & Yanev, N. M. (1980). The dynamics of induced cell proliferation within the models of branching stochastic processes: 1. Numbers of cells in successive generations. *Cytology, 22*, 945–953.
23. Yakovlev, A. Yu., & Yanev, N. M. (1989). *Transient processes in cell proliferation kinetics. Lecture notes in biomathematics* (Vol. 82). New York: Springer.
24. Yakovlev, A. Yu., & Yanev, N. M. (2006). Branching stochastic processes with immigration in analysis of renewing cell populations. *Mathematical Biosciences, 203*, 37–63.
25. Yakovlev, A. Yu., & Yanev, N. M. (2006). Distributions of continuous labels in branching stochastic processes. *Proceedings of Bulgarian Academy of Sciences, 60*, 1123–1130.
26. Yakovlev, A. Yu., & Yanev, N. M. (2007). Age and residual lifetime distributions for branching processes. *Statistics & Probability Letters, 77*, 503–513.
27. Yakovlev, A. Yu., & Yanev, N. M. (2007). Branching populations of cells bearing a continuous label. *Pliska Studia Mathematica Bulgarica, 18*, 387–400.
28. Yakovlev, A. Yu., & Yanev, N. M. (2009). Relative frequencies in multitype branching processes. *The Annals of Applied Probability, 19*, 1–14.
29. Yakovlev, A. Yu., & Yanev, N. M. (2010). Limiting distributions in multitype branching processes. *Stochastic Analysis and Applications, 28*, 1040–1060.
30. Yanev, N. M. (1975). On the statistics of branching processes. *Theory of Probability and its Applications, 20*, 612–622.
31. Yanev, N. M. (2008). Statistical inference for branching processes. In M. Ahsanullah & G. P. Yanev (Eds.), *Records and branching processes* (pp. 143–168). New York: NOVA Science Publishers.
32. Yanev, N. M., & Yakovlev, A. Yu. (1983). The dynamics of induced cell proliferation within the models of branching stochastic processes: 2. Some characteristics of cell cycle temporal organization. *Cytology, 25*, 818–825.
33. Yanev, N. M., & Yakovlev, A. Yu. (1985). On the distributions of marks over a proliferating cell population obeying the Bellman-Harris branching processes. *Mathematical Biosciences, 75*, 159–173.
34. Yanev, N. M., Yakovlev, A. Yu., & Tanushev, M. S. (1987). Bellman-Harris branching processes and distribution of marks in proliferating cell populations. *Proceedings of the 1stWord Congress of the Bernoulli Society, 2*, 725–728.
35. Zorin, A. V., Yakovlev, A. Yu., Mayer-Proschel, M., & Noble, M. (2000). Estimation problems associated with stochastic modeling of proliferation and differentiation of O-2A progenitor cells in vitro. *Mathematical Biosciences, 67*, 109–121.

Age-Dependent Branching Processes with Non-homogeneous Poisson Immigration as Models of Cell Kinetics

Ollivier Hyrien and Nikolay M. Yanev

Abstract This article considers age-dependent branching processes with non-homogeneous Poisson immigration as models of cell proliferation kinetics. Asymptotic approximations for the expectation and variance–covariance functions of the process are developed. Estimators relying on the proposed asymptotic approximations are constructed, and their performance investigated using simulations.

Keywords Bellman–Harris branching process · Renewal equation · Quasi-likelihood · Pseudo-likelihood · Flow cytometry · Leukemia · Stem cells · Oligodendrocytes

1 Introduction

This paper is concerned with developing quantitative tools to analyze the dynamics of renewing cell populations—such as precursor cell populations or their terminally differentiated progenies—using the theory of branching processes. This problem raises two distinct issues: the first issue is that of constructing tractable models able to capture the most salient features of the temporal development of cellular systems of such complexity. The second issue has to do with performing statistical inference and parameter estimation, given the complexity of the modeling framework under consideration and given the nature of the data that are typically collected during biological experiments.

O. Hyrien (✉)
Vaccine and Infectious Disease and Public Health Science Divisions, Fred Hutchinson Cancer Research Center, Seattle, WA, USA
e-mail: ohyrien@fredhutch.org

N. M. Yanev
Department of Probability and Statistics, Institute of Mathematics and Informatics, Bulgarian Academy of Sciences, Sofia, Bulgaria
e-mail: yanev@math.bas.bg

© Springer Nature Switzerland AG 2020
A. Almudevar et al. (eds.), *Statistical Modeling for Biological Systems*,
https://doi.org/10.1007/978-3-030-34675-1_2

We approach the modeling problem by resorting to age-dependent branching processes, which successful research accomplished over the past decades has firmly established as a powerful theoretical means to study cell proliferation kinetics. For a comprehensive overview on branching processes and their biological applications we refer the reader to well-known monographs in the field: Jagers [17], Yakovlev and Yanev [24], Kimmel and Axelrod [19], Haccou et al. [7]. Additional relevant publications specially focused on cell proliferation kinetics include Yakovlev et al. [22], Boucher et al. [2], Zorin et al. [29], Hyrien et al. [11–15], Yanev et al. [28], Yakovlev and Yanev [25], Hyrien and Zand [16], Yakovlev et al. [23], Chen et al. [4], to name a few.

A feature that is common to many in vivo experimental studies on cell proliferation is that the cells that are experimentally observable are supplemented by an influx of cells that result from the differentiation of stem or precursor cells. This influx is typically unobservable, and so is the population of stem/precursor cells. In order to account for this feature of cell kinetics, our model includes an immigration component formulated as a Poisson process. The rate of this Poisson process is assumed to be time non-homogeneous to accommodate situations, where the influx of cells immigrating into the observed population is non-stationary. The first contribution of this paper is to provide asymptotic approximations to the expectation and variance–covariance functions for this class of branching processes in all cases: critical, subcritical, and supercritical. As a by-product, we also extend existing results on the Bellman–Harris branching process.

This paper considers also the problem of estimation when observations consist of the size of multiple (independent) populations observed at discrete time points, as is typical in biological experiments. To date, and to the best of these authors' knowledge, the literature devoted to parameter estimation for age-dependent branching processes with non-homogeneous immigration remains scarce, if existing at all (see Guttorp [6] and Yanev [27]). Several estimators have been proposed for continuous-time branching processes without immigration. The main difficulty encountered in this simpler context is that neither the distribution function nor the moments of the process usually admit closed-form expressions. Hoel and Crump [10] used an asymptotic approximation to the expectation of the process to estimate parameters of the distribution of the cell cycle time. Nedleman et al. [21] obtained a maximum likelihood estimator for a particular branching process describing the growth of mast cells in culture. Simulated maximum and pseudo-maximum likelihood estimators have been investigated by Zorin et al. [29], Hyrien et al. [14, 15], Hyrien [11], and Chen et al. [4] for a broader class of branching processes. Based on saddlepoint approximations, Hyrien et al. [13] constructed a quasi-likelihood estimator. Chen and Hyrien [3] proposed conditional and unconditional pseudo- and quasi-likelihood estimators in the Markov case. By comparison, virtually no work has focused on estimation in the context of age-dependent branching processes with non-homogeneous immigration. The asymptotic approximations of the first two moments of the process will offer an opportunity for developing estimators in this context.

The paper is organized as follows. Section 2 presents the biological motivation. The branching process with non-homogeneous Poisson immigration along with equations satisfied by the associated probability generating functions (p.g.f.) and first- and second-order moments is formulated in Sect. 3. The asymptotic behavior of the expectation and variance–covariance functions of the process with and without immigration is investigated in Sect. 4. Estimation of model parameters relying on the proposed asymptotic approximations of the moments is developed in Sect. 5. Section 6 presents some examples. Throughout special attention shall be given to the Markov case and to the process with homogeneous Poisson immigration.

2   A Biological Motivation

The present paper addresses a modeling problem that appeared when investigating two distinct cellular systems (namely, the population of terminally differentiated oligodendrocytes of the central nervous system, and that of leukemia blast cells) that develop in vivo, and in which the cell population is sustained or expands not only through self-renewing division of existing cells, but also through arrival of new (immigrating) cells as a result of the differentiation of unobservable, upstream stem, or precursor cells. This problem motivated the formulation of age-dependent branching processes with non-homogeneous Poisson immigration. The basic structure of this model is depicted in Fig. 1. It is defined from the following set of modeling assumptions:

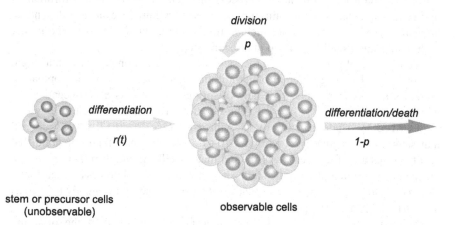

Fig. 1 Schematic representation of the dynamic of a typical cell population that develops in vivo. Unobservable stem or precursor cells differentiate at rate $r(t)$ and turn into cells that are observable. Observable cells either divide with probability p or disappear by differentiating or dying with probability $1 - p$

- The process begins at time zero with N_0 cells. When modeling tissue development during the earliest embryonic stages it is reasonable to set $N_0 = 0$.
- Immigrating cells (simply referred to as immigrants thereafter) arrive in the population in accordance with a non-homogeneous Poisson process with instantaneous rate $r(t)$. Upon arrival, these immigrants are assumed to be of age zero.
- Upon completion of its lifespan, every cell of the population either divides into two new cells, or it leaves the population (as a result of either differentiation, death, or migration). These two events occur with probability p and $q = 1 - p$, respectively.
- The time to division or differentiation/death of any cell is described by a non-negative random variable (r.v.) τ with cumulative distribution function (c.d.f.) $G(x) = P(\tau \leq x)$ that satisfies $G(0+) = 0$.
- Cells are assumed to evolve independently of each other.

The above example motivated us to study a class of age-dependent branching processes with non-homogeneous Poisson immigration and to develop associated methods of statistical inference for estimating cell kinetics parameters from flow cytometry experiments, which recent technological developments allow now experimentalists to measure simultaneously various properties of individual cells isolated from specific tissues.

3 Model, Notation, and Basic Equations

This section considers a branching process that generalizes the model formulated in Sect. 2. In order to define this process, we first introduce an age-dependent branching process that does not include the immigration component. The model with immigration shall be defined in a second step.

Without loss of generality, the process without immigration begins with a single cell of age 0. Upon completion of its mitotic cycle, every cell of the population produces a random number ξ of new offsprings, where ξ is a non-negative r.v. with p.g.f. $h(s) = E\{s^\xi\}$, $|s| \leq 1$. The duration of the mitotic cycle of any cell is described by a non-negative r.v. T with c.d.f. $G(x) = P(T \leq x)$, $x \geq 0$ that satisfies the regularity condition $G(0+) = 0$. We shall finally assume that all offsprings are of age zero at birth, and that all cells accomplish their evolution independently of that of any other cell. The above-formulated process is therefore a Bellman–Harris branching process that we shall sometimes conveniently refer to as a (G, h)-Bellman–Harris process.

Define the expectation and second-order factorial moment of the offspring distribution

$$m = E\xi = \frac{\partial h(s)}{\partial s} \Big|_{s=1} \quad \text{and} \quad m_2 = E\xi(\xi - 1) = \frac{\partial^2 h(s)}{\partial s^2} \Big|_{s=1},$$

as well as the expected mitotic cycle duration

$$\mu = \int_0^\infty x \, dG(x).$$

These parameters are assumed to be finite: $m < \infty$, $m_2 < \infty$, and $\mu < \infty$.

An important particular case of the above model that is relevant in cell kinetics studies is when the p.g.f. of the offspring distribution takes the form

$$h(s) = 1 - p + ps^2.$$

Under this particular model, every cell either divides into two new cells with probability p, or it dies with probability $1 - p$. It follows that $m = 2p = m_2$. We shall give special attention to this model later on, but, for now, we concentrate on the general process.

Introduce the process $Z(t)$, which denotes the number of cells at time t under the (G, h)-Bellman–Harris process, and define the corresponding p.g.f.

$$F(t; s) = E\{s^{Z(t)} \mid Z(0) = 1\}.$$

It is easy to show that $F(t; s)$, $t \geq 0$ and $|s| \leq 1$, satisfies the following nonlinear integral equation (Harris [8]):

$$F(t; s) = \int_0^t h\{F(t - u; s)\} dG(u) + s\{1 - G(t)\}, \tag{3.1}$$

with the initial condition $F(0; s) = s$. Introduce the expectation of the process $Z(t)$

$$A(t) = \frac{\partial}{\partial s} F(t; s) \mid_{s=1} = E\{Z(t) \mid Z(0) = 1\},$$

and its second-order factorial moment

$$B(t) = \frac{\partial^2}{\partial s^2} F(t; s) \mid_{s=1} = E\{Z(t)(Z(t) - 1) \mid Z(0) = 1\}.$$

Write $V(t) = \mathrm{Var}\{Z(t) \mid Z(0) = 1\}$ for the variance of the process, which relates to $A(t)$ and $B(t)$ through

$$V(t) = B(t) + A(t)\{1 - A(t)\}.$$

Using (3.1) one can prove that $A(t)$ and $B(t)$ satisfy the renewal-type integral equations

$$A(t) = m \int_0^t A(t - u) \, dG(u) + 1 - G(t), \tag{3.2}$$

and

$$B(t) = m \int_0^t B(t - u)dG(u) + m_2 \int_0^t A^2(t - u)dG(u), \qquad (3.3)$$

with the initial conditions $A(0) = 1$ and $B(0) = 1$.

Let us now define the process with immigration. To do this, introduce the ordered r.v.s

$$0 = S_0 < S_1 < S_2 < S_3 < \ldots$$

which denote the sequence of time points at which new immigrants arrive in the population. We shall assume that this sequence forms a non-homogeneous Poisson process with rate $r(t)$ that is possibly time-dependent. This Poisson process, denoted as $\{\Pi(t), t \geq 0\}$, is formally defined as $\Pi(t) = \sum_{k=1}^{\infty} 1_{\{S_k \leq t\}}$. Assume that $r(t) \geq 0$ for every $t \geq 0$, and let

$$R(t) = \int_0^t r(u)du$$

denote the cumulative rate of the process. For every $k = 1, 2 \ldots$, let $U_k = S_k - S_{k-1}$ be the inter-arrival times so that $S_k = \sum_{i=1}^{k} U_i$. The number of immigrants that arrive in the population at time S_k is described by a non-negative r.v. I_k. These I_k new immigrants are all of age zero at time S_k, and we shall further assume that $\{I_k\}_{k=1,2\ldots}$ forms a sequence of independent and identically distributed (i.i.d.) r.v.s with p.g.f. $g(s) = E\{s^{I_k}\} = \sum_{i=0}^{\infty} g_i s^i, |s| \leq 1$. Let $\gamma = E\{I_k\} = dg(s)/ds \mid_{s=1}$ be the expected number of immigrants that arrive at any time S_k, and introduce the second-order factorial moment $\gamma_2 = d^2 g(s)/ds^2 \mid_{s=1} = E\{I_k(I_k - 1)\}$. The evolution of every cell of the population, including those that immigrated, is assumed to satisfy the assumptions of the (G, h)-Bellman–Harris process.

Notice that in the particular case where the $\{U_k\}_{k=1,2\ldots}$ are i.i.d. exponentially distributed r.v.s with c.d.f. $G_0(x) = P(U_k \leq x) = 1 - \exp(-rx), x \geq 0$, the process $\Pi(t)$ reduces to an ordinary (homogeneous) Poisson process with cumulative rate $R(t) = rt$.

Let $Y(t)$ denote the number of cells at time t under the process with immigration. Assume that $Y(0) = 0$. Clearly, the process can be expressed as

$$Y(t) = \begin{cases} \sum_{k=1}^{\Pi(t)} Z^{I_k}(t - S_k) & \text{if } \Pi(t) > 0 \\ 0 & \text{if } \Pi(t) = 0, \end{cases} \qquad (3.4)$$

where $Z^{I_k}(t)$ denotes the total number of cells at time t generated by the pool of immigrants that arrived in the population at time S_k. The $Z^{I_k}(t)$s are i.i.d. copies of the process $Z(t)$, with the distinction, however, that $Z^{I_k}(t)$ was started from I_k ancestors (instead of a single one).

Introduce the p.g.f. $\Psi(t; s) = E\{s^{Y(t)} \mid Y(0) = 0\}$. Using the representation (3.4), Yakovlev and Yanev [26]; Theorem 1 proved that $\Psi(t; s)$ is given by

$$\Psi(t; s) = \exp\left\{-\int_0^t r(t - u)[1 - g\{F(u; s)\}]du\right\}, \qquad (3.5)$$

where $\Psi(0, s) = 1$, and where the p.g.f. $F(u; s)$ satisfies the nonlinear integral equation (3.1). It is quite clear from its definition that $\{Y(t), t \geq 0\}$ is a time non-homogeneous, non-Markov process.

One of the main objectives of this paper is to obtain asymptotic approximations to the expectation and variance–covariance functions of age-dependent branching processes with non-homogeneous Poisson immigration. Thus define the expectation and the second-order factorial moments of $Y(t)$

$$M(t) = E\{Y(t) \mid Y(0) = 0\} = \frac{\partial}{\partial s}\Psi(t; s)\mid_{s=1},$$

and

$$M_2(t) = E[Y(t)\{Y(t) - 1\} \mid Y(0) = 0] = \frac{\partial^2}{\partial s^2}\Psi(t; s)\mid_{s=1},$$

and let $W(t) = \text{Var}\{Y(t) \mid Y(t) = 0\}$ denote the variance of the process with immigration started from 0 cell at time 0. We have that

$$W(t) = M_2(t) + M(t)\{1 - M(t)\}.$$

Firstly, it follows from (3.5) that the expectation of $Y(t)$ satisfies the integral equation

$$M(t) = \gamma \int_0^t r(t - u)A(u)du, \qquad (3.6)$$

with $M(0) = 0$, whereas $M_2(t)$ is given by

$$M_2(t) = \gamma \int_0^t r(t - u)B(u)du + \left[\gamma \int_0^t r(t - u)A(u)du\right]^2$$

$$+ \gamma_2 \int_0^t r(t - u)A^2(u)du, \qquad (3.7)$$

with $M_2(0) = 0$, and where $A(t)$ and $B(t)$ satisfy Eqs. (3.2) and (3.3).

In order to derive approximations for the covariance function of the process, define the bivariate p.g.f. of the vector $\{Z(t), Z(t + \tau)\}$

$$F(s_1, s_2; t, \tau) = E\left\{s_1^{Z(t)} s_2^{Z(t+\tau)} \mid Z(0) = 1\right\}, \quad t, \tau \geq 0.$$

By conditioning on the evolution of the initiator cell and applying the law of total probability we obtain the nonlinear integral equation

$$F(s_1, s_2; t, \tau) = \int_0^t h\{F(s_1, s_2; t - u, \tau)\}dG(u)$$

$$+ s_1 \int_t^{t+\tau} h\{F(1, s_2; t, \tau - v)\}dG(v) + s_1 s_2\{1 - G(t + \tau)\},$$

$$(3.8)$$

with the initial condition $F(s_1, s_2; 0, 0) = s_1 s_2$ (Harris [8]). Let

$$\Psi(s_1, s_2; t, \tau) = E\{s_1^{Y(t)} s_2^{Y(t+\tau)} \mid Y(0) = 0\} \quad t, \tau \geq 0,$$

denote the bivariate p.g.f. associated with the process with immigration $Y(t)$. We have that

$$\Psi(s_1, s_2; t, \tau) = \exp\left\{-\int_0^t r(u)[1 - g\{F(s_1, s_2; t - u, \tau)\}]du\right.$$

$$\left. - \int_t^{t+\tau} r(v)[1 - g\{F(1, s_2; t, \tau - v)\}]dv\right\}. \quad (3.9)$$

The proof of (3.9) rests on a line of arguments similar to the one employed in establishing Eq. (3.5) (Yakovlev and Yanev [26, Theorem 1]), and it requires also a careful application of the *order statistics property* (Karlin and Taylor [18]) and of a random time change.

Introduce the second-order moments

$$A(t, \tau) = E\{Z(t)Z(t + \tau) \mid Z(0) = 1\} = \frac{\partial^2}{\partial s_1 \partial s_2} F(s_1, s_2; t, \tau) \mid_{s_1 = s_2 = 1},$$

and

$$M(t, \tau) = E\{Y(t)Y(t + \tau) \mid Y(0) = 0\} = \frac{\partial^2}{\partial s_1 \partial s_2} \Psi(s_1, s_2; t, \tau) \mid_{s_1 = s_2 = 1}.$$

Using appropriate conditioning argument and decomposition of the integrals, we obtain the following equations:

$$A(t, \tau) = m \int_0^t A(t - u, \tau)dG(u) + m_2 \int_0^t A(t - v)A(t + \tau - v)dG(v)$$

$$+ m \int_t^{t+\tau} A(t + \tau - v)dG(v) + 1 - G(t + \tau), \tag{3.10}$$

and

$$M(t, \tau) = \int_0^t r(u)[\gamma A(t - u, \tau) + \gamma_2 A^2(t - u)]du$$

$$+ \gamma^2 \left[\int_0^t r(u)A(t - u)du \right]^2, \tag{3.11}$$

with the relevant initial conditions $A(0, \tau) = A(\tau)$ and $M(0, \tau) = 0$.

Comment 1 By applying integration by parts to (3.6), one obtains

$$R(t) + \int_0^t R(t - u)dA(u) = M(t)/\gamma, \tag{3.12}$$

which is a Volterra equation of the second kind in $R(t)$. This equation is known for admitting a unique solution under mild regularity conditions (e.g., Linz [20]). Recall that in the Markov case $A(t) = \exp(\alpha t)$. Then it is not difficult to show that the solution of (3.12) is

$$R(t) = \left\{ M(t) - \alpha \int_0^t M(x)dx \right\} /\gamma. \tag{3.13}$$

This equation offers a means to estimate the immigration rate when α and γ are known.

4 Asymptotics for First- and Second-Order Moments

The asymptotic behavior of the process is primarily governed by the Malthus parameter, denoted as α, and defined as the root to the equation

$$m \int_0^\infty \exp(-\alpha x)dG(x) = 1. \tag{4.1}$$

In what follows we shall proceed from the assumption that α exists. The process $Z(t)$ is classified as subcritical if $\alpha < 0$ ($m < 1$), critical if $\alpha = 0$ ($m = 1$), and supercritical if $\alpha > 0$ ($m > 1$) (Harris [8]; Athreya and Ney [1]). It is well-known that $A(t) \equiv 1$ in the critical case, whereas the expectation of the process behaves as an exponential in both the sub- and supercritical cases; that is,

$$A(t) \sim C \exp(\alpha t) \tag{4.2}$$

whenever $\alpha \neq 0$, where the constant C has the expression

$$C = \frac{m-1}{\alpha m \widetilde{\mu}}, \tag{4.3}$$

with

$$\widetilde{\mu} = m \int_0^\infty x \exp(-\alpha x) dG(x) = \int_0^\infty x d\widetilde{G}(x), \tag{4.4}$$

and where the notation $A(t) \sim a(t)$ means $A(t)/a(t) \to 1$ as $t \to \infty$, as usual. We shall assume that $C < \infty$. Clearly, the constant $\widetilde{\mu}$ is finite in the critical and supercritical cases; we shall assume that it remains finite in the subcritical case too. It is worth noting that

$$\widetilde{G}(t) = m \int_0^t \exp(-\alpha x) dG(x)$$

is a c.d.f. in view of Eq. (4.1), so that $\widetilde{G}(t) \uparrow 1$ as $t \to \infty$.

The asymptotic behavior of the expectation of the process with immigration is now deduced by combining Eqs. (3.6) and (4.2). Firstly, in the critical case ($\alpha = 0$), we have that

$$M(t) = \gamma R(t), \quad t \geq 0. \tag{4.5}$$

In both the sub- and supercritical cases ($\alpha \neq 0$), we obtain that

$$M(t) = \gamma \exp(\alpha t) \int_0^t \overline{A}(t-u) d\widehat{R}_\alpha(u),$$

where $\overline{A}(t) = A(t) \exp(-\alpha t) \to C$ as $t \to \infty$, and where

$$\widehat{R}_\alpha(t) = \int_0^t \exp(-\alpha x) dR(x) = \int_0^t \exp(-\alpha x) r(x) dx. \tag{4.6}$$

We finally deduce, when $\alpha \neq 0$, that

$$M(t) \sim \gamma C \exp(\alpha t) \widehat{R}_\alpha(t). \tag{4.7}$$

Comment 2 (Homogeneous Poisson Process) When the immigration process is homogeneous, that is, when $R(t) = rt$, the asymptotic behavior of the expectation reduces to

$$M(t) \sim \begin{cases} r\gamma C/(-\alpha) & \text{if } \alpha < 0 \\ r\gamma t & \text{if } \alpha = 0 \\ r\gamma C \exp(\alpha t)/\alpha & \text{if } \alpha > 0. \end{cases} \tag{4.8}$$

Comment 3 By comparing Eqs. (4.5) and (4.7) to Eq. (4.8), it is not difficult to see that the asymptotic behavior of $M(t)$ may be quite different in the non-homogeneous and in the homogeneous cases as this behavior is strongly influenced by the rate $r(t)$ of the Poisson process. Cases of practical interest for $R(t)$ include, but are not limited to,

$$R(t) \sim \begin{cases} D \exp(\beta t) & 0 < \beta < \infty \\ D t^\rho & 0 < \rho < \infty \\ D \log^\theta (1+t) & 0 < \theta < \infty, \end{cases} \tag{4.9}$$

where D denotes a positive constant. When $R(t) = D \exp(\beta t)$ a little algebra gives

$$M(t) \sim \begin{cases} \dfrac{\gamma C D \beta}{\beta - \alpha} \exp(\beta t) & \text{if } \alpha < 0 \\ D\gamma \exp(\beta t) & \text{if } \alpha = 0 \\ \dfrac{\gamma C D \beta}{|\alpha - \beta|} \exp(\max(\alpha, \beta)t) & \text{if } \alpha > 0. \end{cases}$$

Let us now investigate the asymptotic behavior of the variance of the Bellman–Harris process.

Proposition 1 *Assume that m_2 and $\widetilde{\mu}$ are finite, and that $G(x)$ is non-lattice. Then the asymptotic behavior of $V(t)$ as $t \to \infty$ is given by*

$$V(t) \sim \begin{cases} \overline{V}_1 \exp(\alpha t) & \text{if } \alpha < 0 \\ m_2 t/\mu & \text{if } \alpha = 0 \\ \overline{V}_2 \exp(2\alpha t) & \text{if } \alpha > 0, \end{cases} \tag{4.10}$$

where

$$\overline{V}_1 = \frac{1 - m - \alpha m_2 \int_0^\infty \exp(-\alpha x) A^2(x) dx}{(-\alpha) m \widetilde{\mu}}, \tag{4.11}$$

$$\overline{V}_2 = \frac{(m-1)^2 [\widehat{G}(2\alpha)(m_2 + m) - 1]}{m^2 \alpha^2 \widetilde{\mu}^2 [1 - m\widehat{G}(2\alpha)]}, \tag{4.12}$$

and where

$$\widehat{G}(z) = \int_0^\infty \exp(-zx)dG(x).$$

Proof Since $V(t) = B(t) + A(t)[1 - A(t)]$, it suffices to use the asymptotic approximation to $A(t)$ (see Eq. (4.2)) and to obtain that of $B(t)$ (see below).

(i) *Critical case:* If $\alpha = 0$, by replacing $A(t)$ by its asymptotic approximation in Eq. (3.3) yields the renewal-type equation

$$B(t) = \int_0^t B(t - u)dG(u) + m_2 G(t),$$

from which we deduce that $B(t) \sim m_2 t/\mu$ by using a renewal theorem (Feller [5]).

(ii) *Subcritical case:* If $\alpha < 0$, by proceeding similarly to previously in (i), we obtain the renewal-type equation

$$\overline{B}(t) = \int_0^t \overline{B}(t - u)d\widetilde{G}(u) + f(t),$$

where $\overline{B}(t) = B(t) \exp(-\alpha t)$, and where $f(t) = \exp(-\alpha t)m_2 \int_0^t A^2(t - u)dG(u)$. Since

$$\int_0^\infty f(t)dt < \infty,$$

the *key renewal theorem* (Feller [5]) yields

$$B(t) \sim \overline{B}_1 \exp(\alpha t),$$

where $\overline{B}_1 = m_2/m\widetilde{\mu} \int_0^\infty \exp(-\alpha x)A^2(x)dx < \infty$.

(iii) *Supercritical case:* If $\alpha > 0$ we obtain the following renewal-type equation:

$$\overline{B}(t) = \kappa \int_0^t \overline{B}(t - u)d\overline{G}(u) + f(t),$$

where $\overline{B}(t) = B(t) \exp(-2\alpha t)$, $\kappa = m\widehat{G}(2\alpha) < 1$, $\overline{G}(t) = (m/\kappa) \int_0^t \exp(-2\alpha x)dG(x)$, and $f(t) = (m_2 \kappa/m) \int_0^t \overline{A}^2(t - u)d\overline{G}(u)$, with $\overline{A}(t) = \exp(-\alpha t)A(t)$. Since $f(t) \to C^2 m_2 \kappa/m$ as $t \to \infty$, by applying renewal theory (Feller [5]) we obtain that

$$B(t) \sim \overline{B}_2 \exp(2\alpha t),$$

where $\overline{B}_2 = C^2 m_2 \kappa/m(1 - \kappa)$.

$\square$

Comment 4 The asymptotic behavior of $B(t)$ in the supercritical case can be found in Harris [8] and in Athreya and Ney [1]. To the best of our knowledge, Proposition 1 is the first that provides exact expressions for the coefficients that appear in the limiting form of the variance in both the critical and subcritical cases.

Let us now investigate the asymptotic behavior of the variance of the process with immigration, $W(t)$. Since $W(t) = M_2(t) + M(t)[1 - M(t)]$, it follows from Eqs. (3.6) and (3.7) that

$$W(t) = M(t) + \gamma \int_0^t r(t-u)B(u)du + \gamma_2 \int_0^t r(t-u)A^2(u)du. \qquad (4.13)$$

When $\alpha = 0$, the above identity can be further developed to give

$$W(t) \sim \frac{\gamma m_2}{\mu}[tR(t) - \overline{R}(t)] + (\gamma_2 + \gamma)R(t), \qquad (4.14)$$

where $\overline{R}(t) = \int_0^t xr(x)dx$. When $\alpha > 0$, using Eqs. (4.2), (4.7), and Proposition 1, one obtains

$$W(t) \sim \exp(2\alpha t)\widehat{R}_{2\alpha}(t)\frac{C^2}{1 - m\widehat{G}(2\alpha)}[\widehat{G}(2\alpha)(\gamma m_2 - \gamma_2 m) + \gamma_2]. \qquad (4.15)$$

Similarly, when $\alpha < 0$, Eq. (4.13) gives

$$W(t) \sim \gamma C \exp(\alpha t)\widehat{R}_\alpha(t) + \exp(2\alpha t)\widehat{R}_{2\alpha}(t)[\gamma_2 C^2 + \gamma \overline{B}_1]. \qquad (4.16)$$

Comment 5 (Homogeneous Poisson Process) In the case where the immigration process is a homogeneous Poisson process (that is, when $R(t) = rt$) the asymptotic results for the variance simplifies to:

$$W(t) \sim \begin{cases} \overline{W}_1 & \text{if } \alpha < 0 \\ \dfrac{r\gamma m_2}{2\mu}t^2 + r(\gamma + \gamma_2)t & \text{if } \alpha = 0 \\ \overline{W}_2 \exp(2\alpha t) & \text{if } \alpha > 0, \end{cases} \qquad (4.17)$$

where

$$\overline{W}_1 = \frac{rC}{2m\widetilde{\mu}\alpha^2}\{-\alpha m\widetilde{\mu}(2\gamma + rC\gamma_2) + 2\gamma Cm_2\},$$

and

$$\overline{W}_2 = \frac{r\gamma C^2}{2\alpha\{1 - m\widehat{G}(2\alpha)\}}\{\widehat{G}(2\alpha)(\gamma m_2 - \gamma_2 m) + \gamma_2\}.$$

□

We finally focus on the limiting behavior of the covariance function, starting with the process without immigration.

Proposition 2 *Assume that both m_2 and $\widetilde{\mu}$ are finite, and that $G(x)$ is non-lattice. Then, as $t \to \infty$, we have that*

$$A(t, \tau) \sim \begin{cases} \overline{A}_1(\tau) \exp(\alpha(t+\tau)) & \text{if } \alpha < 0 \\ m_2 t / \mu & \text{if } \alpha = 0 \\ \overline{A}_2 \exp(\alpha(2t+\tau)) & \text{if } \alpha > 0, \end{cases} \quad (4.18)$$

where

$$\overline{A}_1(\tau) = \frac{m_2}{m\widetilde{\mu}} \int_0^\infty \exp(-\alpha x) A(x) A(x+\tau) dx$$

$$+ \frac{1}{\widetilde{\mu}} \left\{ \frac{1-m}{m(-\alpha)} - \int_0^\tau \exp(-\alpha x)[1 - G(x)] dx + \frac{1}{m} \int_0^\tau \exp(-\alpha x) A(x) dx \right.$$

$$\left. - \int_0^\tau \exp(-\alpha v) \int_0^{\tau-v} \exp(-\alpha x) A(x) dx \, dG(v) \right\}, \quad (4.19)$$

and where

$$\overline{A}_2 = \frac{m_2 C^2 \widehat{G}(2\alpha)}{1 - m\widehat{G}(2\alpha)}.$$

Notice that $\lim_{\tau \to 0} \overline{A}_1(\tau) = (m_2 C^2 + 1 - m)/m(-\alpha)\widetilde{\mu} = \overline{A}_1(0)$.

Proof

(i) *Critical case:* Consider first the case where $\alpha = 0$. Equation (3.10) yields the renewal equation

$$A(t, \tau) = \int_0^t A(t - u, \tau) dG(u) + f(t),$$

where $f(t) = m_2 G(t) + 1 - G(t)$, which converges to m_2 as $t \to \infty$. Hence, applying renewal theory (Feller [5]), we find that $A(t, \tau) \sim m_2 t / \mu$.

(ii) *Supercritical case:* If $\alpha > 0$, Eq. (3.10) gives the renewal-type equation

$$\overline{A}(t, \tau) = \kappa \int_0^t \overline{A}(t - u, \tau) d\overline{G}(u) + f(t, \tau),$$

where $\overline{A}(t, \tau) = A(t, \tau) \exp(-2\alpha t)$, $\kappa = m\widehat{G}(2\alpha) < 1$, $\overline{G}(t) = (m/\kappa) \int_0^t \exp(-2\alpha x) dG(x)$, $\overline{A}(t) = \exp(-\alpha t) A(t)$, and

$$f(t, \tau) = \exp(\alpha\tau)(m_2\kappa/m) \int_0^t \overline{A}(t-u)\overline{A}(t+\tau-u)d\overline{G}(u)$$

$$+ \exp(-2\alpha t)\{m \int_t^{t+\tau} A(t+\tau-v)dG(v) + 1 - G(t+\tau)\}.$$

Notice that

$$0 \leq \int_t^{t+\tau} A(t+\tau-v)dG(v) \leq \max_{0 \leq x \leq \tau} A(x)[G(t+\tau) - G(t)].$$

Since the right-hand side of the above inequality converges to 0 as $t \to \infty$, we conclude that

$$f(t, \tau) \sim \exp(\alpha\tau)C^2 m_2\kappa/m,$$

and, using renewal theory (Feller [5]), we deduce that

$$\overline{A}(t, \tau) \sim \exp(\alpha\tau)C^2 m_2\kappa/m(1 - \kappa)$$

uniformly in $\tau \geq 0$, which completes the proof.

(iii) *Subcritical case:* When $\alpha < 0$ we obtain the renewal-type equation

$$\overline{A}(t, \tau) = \int_0^t \overline{A}(t-u, \tau)d\widetilde{G}(u) + f(t, \tau),$$

where $\overline{A}(t, \tau) = A(t, \tau)\exp(-\alpha t)$, and

$$f(t, \tau) = \exp(-\alpha t)\left\{m_2 \int_0^t A(t-u)A(t+\tau-v)dG(v)\right.$$

$$\left. + m \int_t^{t+\tau} A(t+\tau-v)dG(v) + 1 - G(t+\tau)\right\}.$$

Firstly, a little algebra gives .

$$\int_0^\infty \exp(-\alpha t)[1 - G(t+\tau)]dt$$

$$= \exp(\alpha\tau)\left[(1-m)/m(-\alpha) - \int_0^\tau \exp(-\alpha x)(1 - G(x))dx\right].$$

Secondly, one can easily check that

$$\int_0^\infty \exp(-\alpha t) \int_0^t A(t-v)A(t+\tau-v)dG(v)dt$$

$$= \frac{1}{m} \int_0^\infty \exp(-\alpha t) A(t) A(t+\tau) dt.$$

Lastly,

$$\int_0^\infty \exp(-\alpha x) \int_x^{x+\tau} A(x+\tau - v) dG(v) dx$$

$$= \int_0^\tau \int_0^v \exp(-\alpha x) A(x+\tau - v) dx dG(v)$$

$$+ \int_\tau^\infty \int_{v-\tau}^v \exp(-\alpha x) A(x+\tau - v) dx dG(v)$$

$$= \exp(\alpha \tau)[m^{-1} \int_0^\tau \exp(-\alpha x) A(x) dx$$

$$- \int_0^\tau \exp(-\alpha v) \int_0^{\tau - v} \exp(-\alpha x) A(x) dx dG(v)].$$

We finally deduce that $\int_0^\infty f(t, \tau) dt < \infty$, and (4.18) and (4.19) follows from the *key renewal theorem* (Feller [5]). □

Comment 6 The asymptotic behavior of $A(t, \tau)$ was given in Harris [8] in the supercritical case only; the results of Proposition 2 appear to be new in the context of critical and subcritical cases.

Finally, the asymptotic behavior of the second-order factorial moment $M(t, \tau)$ as $t \to \infty$, for fixed $\tau \geq 0$, can be deduced from Eq. (3.11) and by applying Proposition 2. Specifically, we obtain

Corollary 1 *Assume that both m_2 and $\tilde{\mu}$ are finite, and that $G(x)$ is non-lattice. Then, as $t \to \infty$, we have that*

$$M(t, \tau) \sim$$
$$\begin{cases} e^{\alpha t} \{\gamma \overline{A}_1(\tau) e^{\alpha \tau} \widehat{R}_\alpha(t) + \gamma_2 C^2 e^{\alpha t} \widehat{R}_{2\alpha}(t) + \gamma^2 C^2 e^{\alpha t} \widehat{R}_\alpha^2(t)\} & \text{if } \alpha < 0 \\ \frac{\gamma m_2}{\mu} \{t R(t) - \overline{R}(t)\} + \gamma^2 R^2(t) + \gamma_2 R(t) & \text{if } \alpha = 0 \\ e^{2\alpha t} C^2 \left\{\gamma^2 \widehat{R}_\alpha^2(t) + \left[\gamma_2 + e^{\alpha \tau} \frac{\gamma m_2 \widehat{G}(2\alpha)}{1 - m\widehat{G}(2\alpha)}\right] \widehat{R}_{2\alpha}(t)\right\} & \text{if } \alpha > 0, \end{cases} \quad (4.20)$$

where $\overline{A}_1(\tau)$ is determined as in Eq. (4.19).

Comment 7 In the case of a homogeneous Poisson process for the immigration the following asymptotic results hold as $t \to \infty$:

$$M(t,\tau) \sim \begin{cases} \dfrac{r}{2\alpha^2}[C^2\{2\gamma^2 r - \gamma_2\alpha\} - 2\alpha\gamma\overline{A}_1(\tau)e^{\alpha\tau}] & \text{if } \alpha < 0 \\[2mm] \dfrac{\gamma r}{2\mu}(m_2 + 2\mu\gamma r)t^2 & \text{if } \alpha = 0 \\[2mm] \dfrac{rC^2}{\alpha}e^{2\alpha t}\left\{\dfrac{\gamma^2 r}{\alpha} + \dfrac{1}{2}\left[\gamma_2 + e^{\alpha\tau}\dfrac{\gamma m_2\widehat{G}(2\alpha)}{1 - m\widehat{G}(2\alpha)}\right]\right\} & \text{if } \alpha > 0. \end{cases}$$

We conclude this section with comments about the autocorrelation function of the process.

Corollary 2 *For every $t, \tau \geq 0$, let*

$$\rho(t,\tau) = Corr\{Z(t), Z(t+\tau)\} = [A(t,\tau) - A(t)A(t+\tau)]/\sqrt{V(t)V(t+\tau)}$$

denote the correlation coefficient for the Bellman–Harris process. Then, as $t \to \infty$, we have that

$$\rho(t,\tau) \sim \begin{cases} \exp(\alpha\tau/2) & \text{if } \alpha < 0 \\ 1/\sqrt{1+\tau/t} & \text{if } \alpha = 0 \\ \exp(-\alpha\tau/2) & \text{if } \alpha > 0. \end{cases}$$

The proof is immediate from the definition of $\rho(t,\tau)$.

Comment 8 One can check that for $\tau \geq 0$ the correlation in the process with immigration $\rho_1(t,\tau) = [M(t,\tau) - M(t)M(t+\tau)]/\sqrt{W(t)W(t+\tau)}$ has the same asymptotic behavior as in the process without immigration $\rho(t,\tau)$, $t \to \infty$.

Comment 9 From (3.12) and (4.2) we have the asymptotic solution

$$R(t) = \{M(t) - \alpha\int_0^t M(x)dx\}/\gamma C,$$

where C is determined by (4.3) for $\alpha \neq 0$, and $R(t) = M(t)/\gamma$ for $\alpha = 0$. This formula can be used to estimate the rate of immigration $R(t)$ if one has estimators for the mean $M(t)$ and for the constants α, γ, and C. In addition, to check whether the immigration process is homogeneous or not, one does not need to estimate the parameters α, γ, and C.

5 Statistical Inference

Motivated by our biological examples, we consider estimation when multiple cell populations are observed at several discrete time points. Define the vector $\mathbf{Y}_i = (Y_{i1}, \ldots, Y_{im})'$, where Y_{ij} denotes the size of the ith population at time t_j, with

$j = 1, \ldots, m$ and $i = 1, \ldots, n$. Write $\mathbf{t} = (t_1, \ldots, t_m)$ for the vector of ordered time points at which the ith population is observed. Let $\overline{\mathbf{Y}} = \sum_{i=1}^{n} \mathbf{Y}_i / n$, and, for every $i = 1, \ldots, m$, define the vector $\mathbf{Y}_i^{(2)} = \mathrm{diag}\{(\mathbf{Y}_i - \overline{\mathbf{Y}})'(\mathbf{Y}_i - \overline{\mathbf{Y}})\}$, and let $\overline{\mathbf{Y}}^{(2)} = \sum_{i=1}^{n} \mathbf{Y}_i^{(2)} / n$. Let

$$\widehat{\Omega}^{(2)} = \frac{1}{n-1} \sum_{i=1}^{n} \{\mathbf{Y}_i^{(2)} - \overline{\mathbf{Y}}^{(2)}\}' \{\mathbf{Y}_i^{(2)} - \overline{\mathbf{Y}}^{(2)}\}$$

and

$$\widehat{\Omega}^{(12)} = \frac{1}{n-1} \sum_{i=1}^{n} \{\mathbf{Y}_i - \overline{\mathbf{Y}}\}' \{\mathbf{Y}_i^{(2)} - \overline{\mathbf{Y}}^{(2)}\},$$

denote empirical estimators for the variance–covariance matrices of $\mathbf{Y}_i$ and $\mathbf{Y}_i^{(2)}$, and their covariance matrices.

Introduce the observed processes $\mathbf{Y}(\mathbf{t}) = \{Y(t_1), \ldots, Y(t_m)\}'$, $Y^{(2)}(t) = [Y(t) - E\{Y(t)\}]^2$, $\mathbf{Y}^{(2)}(\mathbf{t}) = \{Y^{(2)}(t_1), \ldots, Y^{(2)}(t_m)\}'$, and define the associated moments $\mathbf{m}(\mathbf{t}; \theta) = E_\theta\{\mathbf{Y}(\mathbf{t})\}$, $\mathbf{m}^{(2)}(\mathbf{t}; \theta) = E_\theta\{\mathbf{Y}^{(2)}(\mathbf{t})\}$, $\mathbf{v}(\mathbf{t}; \theta) = \mathrm{Var}_\theta\{\mathbf{Y}(\mathbf{t})\}$, $\mathbf{v}^{(2)}(\mathbf{t}; \theta) = \mathrm{Var}_\theta\{\mathbf{Y}^{(2)}(\mathbf{t})\}$, and $\mathbf{c}(\mathbf{t}; \theta) = \mathrm{Cov}_\theta\{\mathbf{Y}(\mathbf{t}), \mathbf{Y}^{(2)}(\mathbf{t})\}$. We assume that $\{\mathbf{Y}_i\}_{i=1,\ldots,n}$ is a collection of n i.i.d. copies of the process $\mathbf{Y}(\mathbf{t})$. In the above notation, θ is a vector that parameterizes the offspring and the lifespan distributions, as well as the rate of the Poisson immigration process. Further, it is easy to check that $\mathbf{m}^{(2)}(\mathbf{t}; \theta) = \mathrm{diag}\{v(\mathbf{t}; \theta)\}$.

To estimate θ from the observed sample $\{\mathbf{Y}_1, \ldots, \mathbf{Y}_n\}$, we consider two classes of estimating functions. The first one is derived from the pseudo-likelihood function

$$P(\theta) = - \sum_{i=1}^{n} [\{\mathbf{Y}_i - \mathbf{m}(\mathbf{t}; \theta)\}' \mathbf{v}(\mathbf{t}; \theta)^{-1} \{\mathbf{Y}_i - \mathbf{m}(\mathbf{t}; \theta)\} + \log \det \mathbf{v}(\mathbf{t}; \theta)]$$

used by Hyrien et al. [14, 15] and Hyrien [11] in the context of multi-type age-dependent branching processes without immigration. Our pseudo-likelihood estimator is defined as the vector $\hat{\theta}_{PL}$ that maximizes the approximation to $P(\theta)$ obtained by replacing $\mathbf{m}(\mathbf{t}; \theta)$ and $\mathbf{v}(\mathbf{t}; \theta)$ by their asymptotic approximations derived in Sect. 4. A somewhat simplified estimator is obtained by ignoring the correlation between observations measured at different time points, and based on the following estimation function:

$$P^m(\theta) = - \sum_{i=1}^{n} \sum_{j=1}^{m} \left[\{\mathbf{Y}_{ij} - m(t_j; \theta)\}' v(t_j; \theta)^{-1} \{\mathbf{Y}_{ij} - m(t_j; \theta)\} + \log \det v(t_j; \theta) \right].$$

Because of the similarity with marginal likelihoods, we shall refer to $P^m(\theta)$ as a marginal pseudo-likelihood function. It can be shown that $P^m(\theta)$ satisfies the inequality

$$E_{\theta_0}\{P^m(\theta)\} \leq E_{\theta_0}\{P^m(\theta_0)\},$$

for every vector θ, which, together with additional regularity conditions, will ensure consistency of the estimator that maximizes $P^m(\theta)$ when $n \to \infty$ and $m < \infty$.

Our second estimator is a quasi-likelihood estimator $\hat{\theta}_{QL}$ defined as the solution to the quasi-score equation $Q(\theta) = 0$, where

$$Q(\theta) = \sum_{i=1}^{n} \begin{pmatrix} \partial \mathbf{m}(t;\theta)/\partial\theta' \\ \partial \mathbf{m}^{(2)}(t;\theta)/\partial\theta' \end{pmatrix} \mathbf{\Psi}(\theta) \begin{pmatrix} \mathbf{Y}_i - \mathbf{m}(t;\theta) \\ \mathbf{Y}_i^{(2)} - \mathbf{m}^{(2)}(t;\theta) \end{pmatrix},$$

where $\mathbf{\Omega}(\theta)$ denotes a $2m \times 2m$ positive definite matrix that may (but does not have to) depend on θ. The explicit expressions for the expectations $\mathbf{m}(t;\theta)$ and $\mathbf{m}^{(2)}(t;\theta)$ being unknown, we consider replacing them by their asymptotic approximations developed in Sect. 4. A possible choice for $\mathbf{\Psi}(\theta)$ includes the identity matrix, in which case $\mathbf{\Psi}(\theta)$ is a generalized estimation equation (GEE) estimator with working correlation matrix the identity matrix. It is well-known that an optimal choice for $\hat{\theta}_{QL}$ is the inverse of the variance–covariance matrix of $(\mathbf{Y}_i, \mathbf{Y}_i^{(2)})$; that is, $\mathbf{\Psi}(\theta) = \mathbf{\Omega}(\theta)^{-1}$, where

$$\mathbf{\Omega}(\theta) = \begin{pmatrix} \mathbf{v}(t;\theta) & \mathbf{c}(t;\theta)' \\ \mathbf{c}(t;\theta) & \mathbf{v}^{(2)}(t;\theta) \end{pmatrix}$$

(Heyde [9]). The variance–covariance matrix $\mathbf{v}(t;\theta)$ can be approximated using the asymptotic approximations derived in Sect. 4, as done for the expectations. We have no such approximations for the matrices $\mathbf{v}^{(2)}(t;\theta)$ and $\mathbf{c}(t;\theta)$, but they can be replaced by empirical estimators such as $\hat{\Omega}^{(2)}$ and $\hat{\Omega}^{(12)}$. Under correct specifications of $\mathbf{m}(t, \theta)$ and $\mathbf{m}^{(2)}(t, \theta)$, the pseudo- and quasi-likelihood estimators would be consistent and asymptotically Gaussian when (e.g.) the sample size n increases (m remaining fixed) under additional mild regularity conditions. When m, rather than the sample size n, increases to infinity, the large sample properties of these estimators are no longer as simple and need to be determined on a case by case basis.

6 Examples and Applications

6.1 An Age-Dependent Branching Process with Homogeneous Poisson Immigration

Consider again the process formulated in Sect. 2, for which the p.g.f. of the offspring distribution is given by $h(s) = 1 - p + ps^2$, so that $m = m_2 = 2p$. Assume that the lifespan of any cell follows a gamma distribution with c.d.f. parameterized as

$$G(x) = \frac{\beta^v}{\Gamma(v)} \int_0^x u^{v-1} \exp(-\beta u) du.$$

The mean and variance of the lifespan are $\mu = v/\beta$ and $\sigma^2 = v/\beta^2$, respectively. Assume further that a single immigrant arrives at each immigration time, so that $g(s) = s$, which implies that $\gamma = 1$ and $\gamma_2 = 0$. The Malthus parameter for this process is $\alpha = \beta(m^{1/v} - 1)$, and a little algebra gives

$$\tilde{\mu} = \frac{v}{\beta m^{1/v}}, \quad \widehat{G}(2\alpha) = \frac{1}{(2m^{1/v} - 1)^v}, \quad C = \frac{(m-1)m^{1/v-1}}{v(m^{1/v} - 1)},$$

$$\overline{B}_1 = \frac{C^2 m^{1/v}}{v(1 - m^{1/v})}, \quad \text{and} \quad \overline{B}_2 = \frac{C^2 m}{(2m^{1/v} - 1)^v - m}.$$

The asymptotic approximations to the moments are deduced by plugging the above quantities into the appropriate equations. For instance, the approximations to the expectation and variance–covariance functions of the number of cells for the process with immigration become

$$M(t) \sim \begin{cases} rt & \text{if } m = 1 \\ \dfrac{r(m-1)m^{1/v-1}}{v\beta(m^{1/v} - 1)^2} e^{\beta(m^{1/v}-1)t} & \text{if } m > 1, \end{cases} \qquad (6.1)$$

$$W(t) \sim \begin{cases} \dfrac{r\beta m t^2}{2v} + rt & \text{if } m = 1 \\ \dfrac{r}{2\beta} \dfrac{(m-1)^2 m^{2/v-1}}{(m^{1/v} - 1)^3 v^2 \{(2m^{1/v} - 1)^v - m\}} e^{2\beta(m^{1/v}-1)t} & \text{if } m > 1, \end{cases} \qquad (6.2)$$

and

$$M(t, \tau) \sim \begin{cases} \dfrac{r\beta}{2v}\left\{m - \dfrac{2r\tau}{\beta t}\right\} t^2 & \text{if } m = 1 \\ \dfrac{r(m-1)^2 m^{2/v-2}}{\beta v^2(m^{1/v} - 1)^3}\left[\dfrac{r\{1 - e^{\beta(m^{1/v}-1)\tau}\}}{\beta(m^{1/v} - 1)} \\ \qquad\qquad + \dfrac{m e^{\beta(m^{1/v}-1)\tau}}{2(2m^{1/v} - 1)^v - m}\right] e^{2\beta(m^{1/v}-1)t} & \text{if } m > 1, \end{cases}$$

where

$$\overline{A}_1(\tau) = \frac{\beta m^{1/v}}{v}\left\{-\frac{m(m-1)^2 m^{2/v-2} + 1 - m}{m\beta v^2(m^{1/v} - 1)^3}\right.$$

$$+ \frac{1}{\alpha}(e^{-\alpha\tau} - 1) + \frac{\Gamma_{v,\beta+\alpha}(t)}{\beta m(m^{1/v} - 1)} - \frac{e^{-\beta(m^{1/v}-1)}}{\beta(m^{1/v} - 1)}G(t)$$

$$+ \int_0^\tau e^{-\alpha x}A(x)dx - m\int_0^\tau e^{-\alpha v}\int_0^{\tau-v} e^{-\alpha x}A(x)dxdG(v)\Bigg\},$$

and where $\Gamma_{v,\beta+\alpha}(t)$ denotes the c.d.f. of a gamma distribution with parameters v and $\alpha + \beta$.

We assessed the accuracy of these approximations by comparing their values to those obtained by simulating the process many times and by computing the corresponding empirical moments. Overall, the asymptotic approximations were found to be quite accurate. Figure 2 displays the examples for two sets of parameter values (see figure legend for detail).

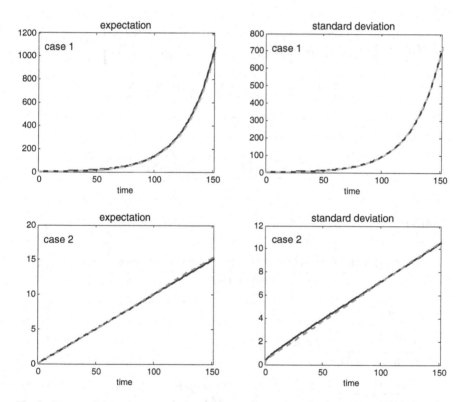

Fig. 2 Accuracy of the asymptotic approximations (solid line) vs. simulation-based approximations (dashed line) for two sets of parameters: $v = 1, \beta = 20, p = 0.9$, and $\rho = 0.1$ (case 1); $v = 4, \beta = 3, p = 0.5$, and $\rho = 0.1$ (case 2). The left panels show the expectation of the process as a function of time ($M(t)$), and the right panels show the standard deviation of the process ($\sqrt{W(t)}$)

6.2 A Simulation Study

We performed a simulation study to investigate estimation in the context of the process of Sect. 6.1 when cells do not die, that is when $p = 1$, implying that $m = 2$. Such an assumption would be relevant to study the proliferation of leukemia blast cells, in which cell death is believed to be negligible. The vector of unknown model parameters reduces to $\theta = (r, \beta, v)'$ for this simplified process. The Malthus parameter is given by $\alpha = \beta(2^{1/v} - 1)$. The process is supercritical ($\alpha > 0$) whenever $\beta > 0$ and $v > 0$, and the asymptotic expectation, variance, and covariance functions are

$$M(t) = \frac{r2^{1/v-1}}{v\beta(2^{1/v}-1)^2}e^{\beta(2^{1/v}-1)t}, \qquad W(t) = \frac{r}{\beta}\frac{2^{2/v-2}}{h_4(v)}e^{2\beta(2^{1/v}-1)t},$$

and

$$M(t,\tau) = \frac{r2^{2/v-2}}{\beta v^2(2^{1/v}-1)^3}\left\{\frac{r\{1-e^{\beta(2^{1/v}-1)\tau}\}}{\beta(2^{1/v}-1)} + \frac{e^{\beta(2^{1/v}-1)\tau}}{\beta(2^{1/v}-1)}\right\}e^{2\beta(2^{1/v}-1)t}.$$

The first-order partial derivatives of $M(t)$ w.r.t. to r, β, and v are

$$\frac{\partial M(t)}{\partial r} = \frac{2^{1/v-1}}{v\beta(2^{1/v}-1)^2}e^{\beta(2^{1/v}-1)t},$$

$$\frac{\partial M(t)}{\partial \beta} = \frac{r2^{1/v-1}}{v\beta(2^{1/v}-1)}\left\{t - \frac{1}{\beta(2^{1/v}-1)}\right\}e^{\beta(2^{1/v}-1)t},$$

$$\frac{\partial M(t)}{\partial v} =$$
$$\frac{r}{2\beta}\frac{h_1(v)\{1+h_1(v)\beta t\}v(2^{1/v}-1)-2^{1/v}\{(2^{1/v}-1)+2vh_1(v)\}}{v^2(2^{1/v}-1)^3}e^{\beta(2^{1/v}-1)t},$$

and those of $W(t)$ are

$$\frac{\partial W(t)}{\partial r} = \frac{2^{2/v-2}}{\beta h_4(v)}e^{\beta(2^{1/v+1}-2)t},$$

$$\frac{\partial W(t)}{\partial \beta} = \frac{r2^{2/v-2}}{h_4(v)}\left\{2(2^{1/v}-1)t - \frac{1}{\beta^2}\right\}e^{\beta(2^{1/v+1}-2)t},$$

$$\frac{\partial W(t)}{\partial v} = \frac{r}{\beta h_4(v)}e^{2\beta(2^{1/v}-1)t}\left\{\beta h_1(v)2^{2/v-1}t\right.$$
$$\left. + \frac{1}{4}\left[h_2(v) - 2^{2/v}\left(\frac{3h_1(v)}{2^{1/v}-1} + \frac{2}{v} + \frac{h_3(v)}{(2^{1/v}-1)v-2}\right)\right]\right\},$$

Table 1 Results of a simulation study on estimation in an age-dependent branching process with homogeneous Poisson immigration

Sample size (n)	Design	Parameter (true value)											
		μ (= 20)			σ (= 10)			r (= 0.5)			α (= 0.0378)		
		A	B	C	A	B	C	A	B	C	A	B	C
25	Average	20.42	19.44	19.83	17.6	8.03	8.71	0.248	0.483	0.502	0.044	0.0383	0.0378
	Bias	−0.42	0.56	0.17	−7.6	1.96	1.29	0.252	0.017	−0.002	−0.007	-5.10^{-4}	-3.10^{-5}
	Std. error	1.10	1.22	1.17	2.56	4.46	4.27	0.016	0.048	0.064	0.0003	0.0006	0.0009
100	Average	20.51	19.17	19.55	17.8	7.95	8.86	0.25	0.473	0.483	0.044	0.0386	0.0383
	Bias	−0.51	0.83	0.45	−7.81	2.05	1.14	0.25	0.027	0.017	−0.007	-7.10^{-4}	-4.10^{-4}
	Std. error	0.60	0.69	0.75	1.34	2.65	2.64	0.007	0.023	0.026	0.0002	0.0002	0.0002

where

$$h_1(v) = -2^{1/v}\frac{\log 2}{v^2},$$

$$h_2(v) = -2^{2/v+1}\frac{\log 2}{v^2},$$

$$h_3(v) = \frac{\partial\{(2^{1/v} - 1)^v\}}{\partial v}$$

$$= \left\{\log(2^{1/v+1} - 1) + \frac{2vh_1(v)}{(2^{1/v+1} - 1)}\right\}(2^{1/v+1} - 1)^v, \text{ and}$$

$$h_4(v) = v^2(2^{1/v} - 1)^3\{(2^{1/v+1} - 1)^v - 2\}.$$

These derivatives are used to compute quasi-likelihood estimators.

For the purpose of this study, data were simulated under three experimental designs. Under *Design A*, the sizes of n independent populations were longitudinally recorded every 6 h starting from $t = 0$ and up to 120 h; *Designs B* and *C* were identical to *Design A* in every respect, except that observations prior to 80 and to 100 h, respectively, were discarded. Table 1 reports the results for the marginal pseudo-likelihood estimator for two sample sizes: $n = 25$ and 100. The simulations suggested that the mean lifespan is properly estimated under all three designs. The estimates for the variance of the lifespan and for the rate of the immigration process, however, were clearly biased under *Design A*. By discarding the earliest observations, where the asymptotic approximations are less accurate, the bias decreased substantially, so that, under *Design C*, the estimators of all three parameters were essentially unbiased.

7 Dedication

This paper is dedicated to the memory of our friends and colleagues Dr. Andrei Yakovlev and Dr. Alexander Zorin. We will always remember our friendship, and continue to develop the ideas that germinated during stimulating discussions with both of them.

Acknowledgements The paper is supported by NIH R01 grant AI129518 (Hyrien) and grant KP-6-H22/3 from the NSF of the Ministry of Education and Science of Bulgaria.

This research is supported by NIH grants AI129518, NS039511, CA134839, and AI069351 (Hyrien) and grant KP6-H22/3 from the NSF of the Ministry of Education and Science of Bulgaria. The authors are grateful to their colleagues Drs. Mark Noble, Margot Mayer-Pröschel, and Craig Jordan for valuable discussions about the biological examples that motivated this work.

References

1. Athreya, K. B., & Ney, P. E. (1972). *Branching processes.* New York: Springer.
2. Boucher, K., Zorin, A. V., Yakovlev, A. Y., Mayer-Proschel, M., & Noble, M. (2001). An alternative stochastic model of generation of oligodendrocytes in cell culture. *Journal of Mathematical Biology, 43*, 22–36.
3. Chen, R., & Hyrien, O. (2011). Quasi-and pseudo-maximum likelihood estimators for discretely observed continuous-time Markov branching processes. *Journal of Statistical Planning and Inference, 141*, 2209–2227.
4. Chen, R., Hyrien, O., Noble, M., & Mayer-Pröschel, M. (2010). A composite likelihood approach to the analysis of longitudinal clonal data on multitype cellular systems under an age-dependent branching process. *Biostatistics, 12*, 173–191.
5. Feller, W. (1971). *An introduction to probability theory and its applications* (Vol. 2). New York: Wiley.
6. Guttorp, P. (1991). *Statistical inference for branching processes.* New York: Wiley.
7. Haccou, P., Jagers, P., & Vatutin, V. A. (2005). *Branching processes: Variation, growth and extinction of populations.* Cambridge: Cambridge University Press.
8. Harris, T. E. (1963). *The theory of branching processes.* Berlin: Springer.
9. Heyde, C. C. (1997). *Quasi-likelihood and its applications: A general approach to optimal parameter estimation.* New York: Springer.
10. Hoel, D., & Crump, K. (1974). Estimating the generation-time of an age-dependent branching process. *Biometrics, 30*, 125–235.
11. Hyrien, O. (2007). A pseudo-maximum likelihood estimator for discretely observed multitype Bellman–Harris branching processes. *Journal of Statistical Planning and Inference, 137*, 1375–1388.
12. Hyrien, O., Ambescovich, I., Mayer-Pröschel, M., Noble, M., & Yakovlev, A. Y. (2006). Stochastic modeling of oligodendrocyte generation in cell culture: model validation with time-lapse data. *Theoretical Biology & Medical Modelling, 3*, 21.
13. Hyrien, O., Chen, R., Mayer-Pröschel, M., & Noble, M. (2010). Saddlepoint approximations to the moments of multitype age-dependent branching processes, with applications. *Biometrics, 66*, 567–577.
14. Hyrien, O., Mayer-Pröschel, M., Noble, M., & Yakovlev, A. Y. (2005). A stochastic model to analyze clonal data on multi-type cell populations. *Biometrics, 61*, 199–207.
15. Hyrien, O., Mayer-Pröschel, M., Noble, M., & Yakovlev, A. Y. (2005). Estimating the lifespan of oligodendrocytes from clonal data on their development in cell culture. *Mathematical Biosciences, 193*, 255–274.
16. Hyrien, O., & Zand, M. S. (2008). A mixture model with dependent observations for the analysis of CFSE-labeling experiments. *Journal of the American Statistical Association, 103*, 222–239.
17. Jagers, P. (1975). *Branching processes with biological applications.* London: Wiley.
18. Karlin, S., & Taylor, H. M. (1981). *A second course in stochastic processes.* San Diego: Academic Press.
19. Kimmel, M., & Axelrod, D. E. (2002). *Branching processes in biology.* New York: Springer.
20. Linz, P. (1985). *Analytical and numerical methods for Volterra equations.* Philadelphia: SIAM.
21. Nedleman, J., Downs, H., & Pharr, P. (1987). Inference for an age-dependent, multitype branching-process model of mast cells. *Journal of Mathematical Biology, 25*, 203–226.
22. Yakovlev, A. Y., Boucher, K., Mayer-Proschel, M., & Noble, M. (1998). Quantitative insight into proliferation and differentiation of oligodendrocyte-type 2 astrocyte progenitor cells *in vitro. Proceedings of the National Academy of Sciences of the United States of America, 95*, 14164–14167.

23. Yakovlev, A. Y., Stoimenova, V. K., & Yanev, N. M. (2008). Branching processes as models of progenitor cell populations and estimation of the offspring distributions. *Journal of the American Statistical Association, 103*, 1357–1366.
24. Yakovlev, A. Y., & Yanev, N. M. (1989). *Transient processes in cell proliferation kinetics.* Berlin: Springer.
25. Yakovlev, A. Y., & Yanev, N. M. (2006). Branching stochastic processes with immigration in analysis of renewing cell populations. *Mathematical Biosciences, 203*, 37–63.
26. Yakovlev, A. Y., & Yanev, N. M. (2007). Age and residual lifetime distributions for branching processes. *Statistics & Probability Letters, 77*, 503–513.
27. Yanev, N. M. (2008). Statistical inference for branching processes. In M. Ahsanullah & G. P. Yanev (Eds.), *Records and branching processes* (pp. 143–168). New York: NOVA Science Publishers.
28. Yanev, N. M., Jordan, C. T., Catlin, S., & Yakovlev, A. (2005). Two-type Markov branching processes with immigration as a model of leukemia cell kinetics. *Comptes Rendus de l'Académie Bulgare des Sciences, 58*, 1025–1032.
29. Zorin, A. A., Yakovlev, A. Y., Mayer-Proschel, M., & Noble, M. (2000). Estimation problems associated with stochastic modeling of proliferation and differentiation of O-2A progenitor cells *in vitro. Mathematical Biosciences, 167* 109–121.

A Study of the Correlation Structure of Microarray Gene Expression Data Based on Mechanistic Modeling of Cell Population Kinetics

Linlin Chen, Lev Klebanov, Anthony Almudevar, Christoph Proschel, and Andrei Yakovlev

Abstract Sample correlations between gene pairs within expression profiles are potentially informative regarding gene regulatory pathway structure. However, as is the case with other statistical summaries, observed correlation may be induced or suppressed by factors which are unrelated to gene functionality. In this paper, we consider the effect of heterogeneity on observed correlations, both at the tissue and subject level. Using gene expression profiles from highly enriched samples of three distinct embryonic glial cell types of the rodent neural tube, the effect of tissue heterogeneity on correlations is directly estimated for a simple two component model. Then, a stochastic model of cell population kinetics is used to assess correlation effects for more complex mixtures. Finally, a mathematical model for correlation effects of subject-level heterogeneity is developed. Although decomposition of correlation into functional and nonfunctional sources will generally not be possible,

On February 27, 2008, Dr. Andrei Yakovlev tragically passed away. We deeply grieve the loss of our colleague, advisor, and friend who was a source of inspiration for all around him.

L. Chen (✉)
School of Mathematical Sciences, Rochester Institute of Technology, Rochester, NY, USA
e-mail: lxcsma@rit.edu

L. Klebanov
Department of Probability and Statistics, Charles University, Prague, Czech Republic
e-mail: Lev.Klebanov@mff.cuni.cz

A. Almudevar · A. Yakovlev
Department of Biostatistics and Computational Biology, University of Rochester, Rochester, NY, USA
e-mail: anthony_almudevar@urmc.rochester.edu

C. Proschel
Department of Biomedical Genetics, University of Rochester, Rochester, NY, USA
e-mail: chris_proschel@urmc.rochester.edu

© Springer Nature Switzerland AG 2020
A. Almudevar et al. (eds.), *Statistical Modeling for Biological Systems*,
https://doi.org/10.1007/978-3-030-34675-1_3

47

since this depends on nonobservable parameters, reasonable bounds on the size of
such effects can be made using the methods proposed here.

Keywords Gene expression data · Cellular kinetics

1 Introduction

Analyses of gene expression profiles often make use of gene pair sample correlations to investigate differential gene expression patterns [2] or to infer gene
regulatory pathway structure [7, 22, 23]. The idea of correlation as a measure
of gene interaction appears intuitive and natural, but it is important to establish
whether or not correlation has significant sources beyond the functional pathway
level. As is well known, what is observed is the correlation coefficient between
the aggregation of the gene expression of a large number of cells. It may fail
to accurately measure the true correlation among genes, leaving the functional
analysis groundless. Therefore, it is necessary to be cautious utilizing the correlation
structures of gene expression data in this type of analysis.

A number of sources of correlation beyond pathway interactions can be identified
[3], ranging from large-scale negative regulation by microRNAs to correlation
induced by commonly used gene profile normalization procedures [16]. Furthermore, these sources may induce correlation structures which are either short-range
(localized) or long-range (involving thousands of genes) [10, 11]. Of particular
concern is any slide-associated *random effect*, which refers to any source of variation
affecting all gene expressions within a slide in a similar manner, inducing positive
correlation. These may have both biological and technical origin.

In this paper, we consider in particular the effect of heterogeneity on observed
correlations, both at the tissue and subject level. In Sect. 2, using gene expression
profiles from highly enriched samples of three distinct embryonic glial cell types
of the rodent neural tube, the effect of tissue heterogeneity on correlation is
directly estimated for a simple two component model. Then, in Sect. 3 a stochastic
model of cell population kinetics is used to assess correlation effects for more
complex mixtures. Finally, in Sect. 4 a mathematical model for correlation effects
of subject-level heterogeneity is developed. Although decomposition of correlation
into functional and nonfunctional sources will generally not be possible, since this
depends on nonobservable parameters, reasonable bounds on the size of such effects
can be made using the methods proposed here.

2 The Effect of Cell Mixtures on Observed Correlations

We first develop a model for the effect of tissue heterogeneity on pairwise sample
correlation coefficients. The analysis will be based on three samples of highly

enriched tissue cell types which might be expected to appear commonly in a single tissue sample. This will permit us to directly predict the effect of tissue heterogeneity on the sample properties of correlation coefficients.

We will first consider a simple model involving two cell types. Suppose a gene expression profile is obtained from such a mixture. We denote the observed gene expression levels of the jth gene in group k by Z_{kj}, $k = 1, 2$. The gene expression level of the jth gene in the mixture, denoted Z_j, is then given by

$$Z_j = \theta Z_{1j} + (1 - \theta) Z_{2j}, \tag{2.1}$$

where θ is a mixture proportion in $[0, 1]$. Since groups 1 and 2 represent distinct sets of cells, we will assume covariances $cov(Z_{1i}, Z_{2j}) = cov(Z_{1j}, Z_{2i}) = 0$ for any genes i, j (although nonzero covariances may emerge from other sources). We otherwise denote $\mu_{ki} = E[Z_{ki}]$ and $\sigma_{kij} = cov(Z_{ki}, Z_{kj})$. We also define correlation coefficients $\rho_{kij} = \sigma_{kij}/(\sigma_{kii}\sigma_{kjj})^{1/2}$. Let $\bar{\theta}$ and τ_θ be the mean and variance of θ. The correlation coefficient of Z_i and Z_j then becomes

$$corr(Z_i, Z_j) = \frac{\eta_1\sigma_{1ij} + \eta_2\sigma_{2ij} + \tau_\theta(\mu_{1i} - \mu_{2i})(\mu_{1j} - \mu_{2j})}{[\eta_1\sigma_{1ii} + \eta_2\sigma_{2ii}]^{1/2}[\eta_1\sigma_{1jj} + \eta_2\sigma_{2jj}]^{1/2}}, \text{ where}$$

$$\eta_1 = \left(\bar{\theta}^2 + \tau_\theta\right),$$

$$\eta_2 = \left((1 - \bar{\theta})^2 + \tau_\theta\right). \tag{2.2}$$

To fix ideas, suppose we have equal variances $\sigma_{kii} = \sigma_{kjj} = \sigma^2$, $k = 1, 2$. Then expression (2.2) may be written

$$corr(Z_i, Z_j) = \frac{\eta_1\rho_{1ij} + \eta_2\rho_{2ij}}{\eta_1 + \eta_2} + \frac{\tau_\theta(\mu_{1i} - \mu_{2i})(\mu_{1j} - \mu_{2j})}{\sigma^2(\eta_1 + \eta_2)}. \tag{2.3}$$

The correlation consists of two terms. The first is, as would be expected, a linear, convex combination of the group specific correlations ρ_{1ij} and ρ_{2ij}. The second term will play a role when there is sufficient variability of mixture parameter θ, as well as systematic differences in mean gene expression levels between tissue groups. If expression levels of one group are consistently higher, this will result in a positive bias, otherwise, this effect will increase variability of the observed correlations. Either case will have implications for the statistical detection of nonzero correlations.

The effect will be illustrated using a set of gene expression profiles obtained from a mRNA expression analysis of three distinct and highly enriched cell populations that were derived from embryonic glial precursors of the rodent neural tube. The first population represents the glial-restricted precursor (GRP) cells that can be isolated from the 13-day-old embryonic rat neural tube using immuno-affinity based methods directed at surface antigens recognized by the A2B5 and 5A1 antibodies [17]. The second and third populations are two distinct population of astrocytes that can be generated from GRP cells by in vitro application of bone morphogenetic protein-4 (BMP4) or ciliary neurotrophic factor (CNTF) [4]. These highly enriched (>99% pure) astrocyte populations are referred to as GDABMP and GDACNTF, respectively. Although these three cell populations are derived from the same lineage, in vitro and transplant studies reveal that these glial populations have very distinct phenotypic and functional properties. The ability to generate highly enriched cultures of these cell types in vitro makes them an attractive target for microarray studies aimed at identifying the transcriptionally unique determinants of these three glial cell populations. Each cell population was analyzed using six biologically independent replicates. For simplicity, we will refer to the three cell populations here as group 1 (GRP), group 2 (GDABMP), and group 3 (GDACNTF).

Analysis of the potential mixture effects of the three groups was reduced to 3 pairwise comparisons. Each group pair was analyzed in the following way. Since each group represents a nearly homogeneous cell sample, a gene expression profile of a mixture can be simulated by randomly selecting pairs of profiles, one from each group, then constructing gene expression mixtures of the form (2.1) for a given mixture parameter θ. Each group contains six profiles, so a new simulated mixture sample can be created by first permuting the sample indices of one group, creating profile pairs by matching the new profile indices, then constructing the gene expression mixture using mixture parameters sampled independently for each profile pair. The mixture parameter was sampled from a beta density with parameters selected to yield $\tau_\theta = \bar{\theta}(1 - \bar{\theta})/51$, and then allowing $\bar{\theta} = 0, 0.1, \ldots, 0.9, 1$. In particular, $\bar{\theta} = 0, 1$ represent homogeneous samples, and for $\bar{\theta} = 0.5$ the standard deviation of θ is $\tau_\theta^{1/2} \approx 0.016$. For a given group pair, the same analysis was performed on a control group pair, created by exchanging three profiles between the two original groups. To control the size of the correlation matrix, only 1000 test genes were used in the study. In our study, 1000 replications were simulated, and for each all pairwise correlations from the test genes were calculated and summarized by the mean (CM) standard deviation (CSD) and variance (CV).

The second term of (2.3) will be influential in the presence of significant differential expression between groups, as can be anticipated for distinct cell types. To gauge this effect in our own study, a two-sample t-test was performed for each gene within each group pair and within their respective control group pairs. An estimate of the false discovery rate (FDR) was calculated using the Benjamini–Hochberg procedure. All genes ($n = 31042$) were used for this analysis. For group pairs 1–2, 1–3, and 2–3, we found that 486,368, and 818 genes, respectively, were differentially expressed with FDR $= 0.05$, and 6654, 4123, and 4772 were differentially expressed with FDR $= 0.25$ (Table 1). No genes in any of the control group pairs were differentially expressed at FDR $= 0.25$.

In addition, using sample means and variances of the gene expressions, an estimate of the quantity $K_{ij} = (\mu_{1i} - \mu_{2i})(\mu_{1j} - \mu_{2j})/\sigma^2$ was calculated for all pairs of test genes. We may then interpret the quantity K_{IJ} as a randomly selected value of K_{ij}. The means and variances of K_{IJ} are given in Table 1. Note that the

Table 1 Summary statistics for mixture sample pairs, with respective controls (ctl)

Pairs	# FDR ≤ 0.05	# FDR ≤ 0.25	$E[K_{IJ}]$	$var(K_{IJ})$	Obs R var	Interp R var	Diff
1,2	486	6654	0.052	2.389	0.246	0.234	0.0116
1,2 ctl	0	0	0.038	0.049	0.249	0.250	−0.00132
1,3	368	4123	0.084	3.852	0.250	0.237	0.01309
1,3 ctl	0	0	0.018	0.008	0.257	0.257	0.00001
2,3	818	4772	0.000	2.321	0.243	0.239	0.00392
2,3 ctl	0	0	0.001	0.014	0.247	0.249	−0.00190

Columns 1 and 2 give the number of positive tests for differential expression controlled at a FDR rate of 0.05 and 0.25, estimated by the Benjamini–Hochberg procedure. Columns 3 and 4 give the mean and variances of K_{IJ}, as described in the Sect. 2. Columns 5 and 6 give the observed and interpolated CV, the difference appearing in column 7

variance of K_{IJ} for the group pairs is considerably larger than for the control pairs, which is anticipated by the prevalence of differential expression observed.

The results of the study are summarized in Fig. 1. Each boxplot summarizes the 1000 replications of CM or CSD for each combination of group pair and mixture parameter mean $\bar{\theta}$. The span of the boxplots shows considerable variations of these quantities, as would be expected for the relatively small sample size of 6 per group. Despite this, a clear trend can be seen in the boxplot medians. The expected value of the mixture correlations can be approximated using formula (2.3) based on the correlation means of the homogeneous groups. These are superimposed on the boxplot in the first two columns of Fig. 1, and closely agree with the sample medians obtained by the simulations. Thus, as predicted, the correlation of a mixture is obtainable as a convex combination of the homogeneous correlations.

The situation for the CSD measurements is complicated by the variability of the sample correlation coefficient, which is not modeled in Eq. (2.3). To illustrate the point, a simple parametric simulation study similar to our experimental design was performed. Two groups of $n = 6, 24, 100$ gene expression profiles of 1000 genes were simulated. Within each sample, genes were normally distributed with unit variance and a mutual correlation of $\rho = 0.5$. The mean expression levels μ_{ki} were then randomly sampled from a normal distribution with mean 0 and variance $\eta = 0, 1, 2$. A mixture sample was then created by combining the two groups using Eq. (2.1) as described above. The mixture parameter had a beta distribution with mean $\bar{\theta} = 0.5$ and standard deviations $\tau_{\theta}^{1/2} \approx 0.022, 0.050, 0.070, 0.109$. The simulation used 1000 replications, and the quantity CV was calculated for each. Table 2 gives the mean CV for one of the homogeneous samples, and the ratio of the mean CV of the mixture and homogeneous samples, values larger than 1 signifying variance inflation induced by mixture effects. It can be clearly seen that variance inflation occurs as parameters τ_{θ} and η increase, as predicted by Eq. (2.3). Interestingly, variance inflation also increases with sample size n. Presumably, the reduction in sampling variation attenuates the effect of the mixture parameters, while for smaller sample sizes, sampling variation is large relative to these effects.

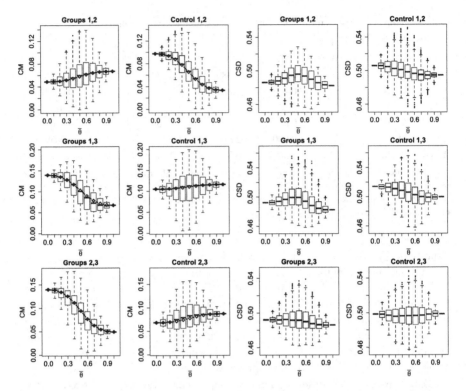

Fig. 1 Boxplots of CM and CSD measurements obtained from simulated mixture samples ($n =$ 1000 replications). All pairs of groups, with their controls, are represented, with varying mixture parameter $\bar{\theta}$ represented within each plot. The predicted value of CM obtained from equation (2.3) is superimposed in the boxplots of columns 1 and 2

Table 2 Summary of parametric simulation study of Sect. 2

		CV for homogeneous sample $\tau_\theta^{1/2} =$				Ratio of CV (mixture/homog) $\tau_\theta^{1/2} =$			
n	η	0.022	0.050	0.070	0.109	0.022	0.050	0.070	0.109
6	0	0.1110	0.1111	0.1127	0.1132	1.00	1.00	0.99	0.99
6	1	0.1136	0.1134	0.1108	0.1126	1.00	1.02	1.04	1.07
6	2	0.1111	0.1107	0.1142	0.1112	0.99	1.07	1.11	1.26
24	0	0.0201	0.0202	0.0205	0.0201	1.01	1.00	0.99	1.01
24	1	0.0204	0.0203	0.0201	0.0203	1.00	1.02	1.05	1.18
24	2	0.0200	0.0203	0.0201	0.0204	1.02	1.14	1.34	2.04
100	0	0.0045	0.0045	0.0045	0.0045	1.00	1.00	1.00	1.01
100	1	0.0045	0.0045	0.0045	0.0045	1.01	1.03	1.12	1.52
100	2	0.0045	0.0045	0.0045	0.0045	1.03	1.41	2.19	5.36

For each parameter set, the mean CV for the 1000 homogeneous samples is reported in columns 1–4, and the ratio of CV for the mixture versus homogeneous samples is reported in columns 5–8. Values of this ratio greater than 1 signify variance inflation attributable to mixture effects

The CSD values of mixture sample simulations based on the tissue samples are shown as boxplot summaries in columns 3 and 4 of Fig. 1. For the control samples generated from pairs 1, 2 and 1, 3 the mixture CSDs are a convex combination of the CSD of the homogeneous samples, while for pair 2, 3 the mixture CSDs are smaller than for the homogeneous groups. However, when examining all 3 mixture samples it can be seen that correlation variance is inflated as $\bar{\theta}$ approaches 0.5 in comparison to the respective control samples. To provide a rough comparison, for each mixture sample pair the mean CV at $\bar{\theta} = 0.5$ is given in Table 1, along with the average CV of the two homogeneous groups, equivalent to the interpolated variance at $\bar{\theta} = 0.5$, which would be expected under a convex combination model. Examining the differences (also given) it can be seen that for the mixture groups the observed CV exceeds that predicted by the convex combination model, in contrast with the control groups. Thus, tissue heterogeneity appears to result in some amount of correlation variance inflation, and, perhaps counterintuitively, our parametric simulation suggests that this effect can be expected to become more dominant with increased sample size.

3   A Comprehensive Study of Correlation Based on Mechanistic Modeling of Cell Population Kinetics

The statistical effect of tissue nonhomogeneity on the correlation between two genes was examined in the previous section. We now consider the broader effects of nonhomogeneity on a specific tissue culture, using a model for cellular heterogeneity arising from cell differentiation. We will use a probabilistic model described in Hyrien et al. [9] in which nonidentical time-to-transformation distributions are incorporated into a multi-type Bellman–Harris branching process. In our model, the cell division involves three types of cells: stem cells, progenitor cells, and differentiated cells. Originating from a single stem cell, the population eventually grows to a pre-defined number. In this model (Fig. 2), the following rules will be used:

1. A single stem cell is always divided into two daughter cells: one stem cell and one progenitor cell.
2. When a progenitor cell is divided into two cells, each of the daughter cells have probability p of being the same type as the parent cell and probability $1 - p$ of being a differentiated cell, which will expire after a period of time (the length of the duration time is defined in the table below).
3. During the cell cycle, the stem cell has three stages G_1, S and $G_2 + M$, denoted as $stage_1$, $stage_2$, $stage_3$. The stem cell will enter each stage in order before the cell division occurs.
4. The progenitor has three similar stages, denoted differently as $stage_4$, $stage_5$, $stage_6$. Like stem cells, the progenitor cell will enter each stage in order before the cell is divided.

Fig. 2 Cell division stage
transitions

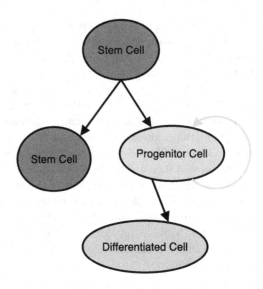

Table 3 Gamma parameters for the cell division simulation

Cell type	Stage index	Stage name	τ	CoV	α	β
Stem cell	1	G_1	5	0.5	4	1.25
	2	S	10	0.2	25	0.4
	3	$G_2 + M$	4	0.3	11.1	0.36
	4	G_1	5	0.5	4	1.25
Progenitor cell	5	S	10	0.2	25	0.4
	6	$G_2 + M$	4	0.3	11.1	0.36
Differentiated cell	7	–	20	0.1	100	0.2

5. To simplify the problem, there will be only one stage for the differentiated cell
   type ($stage_7$).
6. The duration of each stage of different cell types is assumed to be random and
   follows a gamma distribution with shape parameter α and scale parameter β. The
   parameters differ across the stages. In our study, we manually set the parameter
   values as shown in Table 3, where $\tau = \alpha\beta$ defined as the mean duration time,
   and CoV as the variation coefficient.

   Table 4 summarizes the mixture proportions obtained from 10 experimental
replications, using an initial population of 100 stem cells. As can be seen, the
proportions of the progenitor cells at each stage and the differentiated cells are
stable, with estimated standard deviations ranging from 0.008 to 0.019. This
compares with a standard deviation of $\tau_\theta^{1/2} \approx 0.016$ for $\bar{\theta} = 0.5$ used in the tissue
example of Sect. 2.

Table 4 Cell type mixtures (as percentages) at different stages for the 10 experimental replications

Experiment	Total time	Stem G_1	Stem S	Stem $M + G_2$	Prog G_1	Prog S	Prog $M + G_2$	Differentiated
1	196.24	0.07	0.25	0.01	25.02	41.02	13.29	20.33
2	202.51	0.12	0.22	0	22.39	43.19	12.94	21.16
3	208.98	0	0	0.33	26.55	37.95	13.72	21.45
4	219.52	0	0.33	0	25.03	41.79	14.24	18.61
5	208.35	0.15	0.13	0.05	22.96	42.11	13.49	21.11
6	209.97	0.2	0.13	0	26.16	38.32	12.43	22.76
7	203.86	0.02	0.32	0	25.13	41.19	13.71	19.64
8	205.20	0.19	0	0.14	23.16	42.82	12.25	21.44
9	234.13	0	0.31	0.02	23.46	42.41	12.39	21.42
10	211.22	0.01	0.19	0.14	23.88	43.12	11.76	20.90

3.1 Gene Expression Levels During the Cell Cycles

The observed gene expression levels will be the sum of the gene expression contributions from all cells, so the gene expression model may be based directly on the predicted cell type distributions. Let v be the total number of cells involved in the aggregation process of the final gene expression signal measurements, let $NStage$ be the total number of cell stages defined above, and let $NGene$ be the total number of genes in the study. The notation $Y_{j,k}(i)$ denotes the expression level for gene i in cell k at stage j, and v_j will be the number of cells at stage j, for $j = 1, \ldots, NStage; k = 1, \ldots, v_j; i = 1, \ldots, NGene$.

The gene expression signals are assumed to follow the normal distribution

$$Y_{j,k}(i) \sim N(\mu_{i,j}, \sigma), \qquad (3.1)$$

where $\mu_{i,j}$ is the mean expression level and, to simplify the analysis, σ is the common standard deviation for gene i at stage j. The final gene expression value for gene i is then

$$Y(i) = \sum_{j} \sum_{k=1}^{v_j} Y_{j,k}(i). \qquad (3.2)$$

We next consider the question of modeling gene expression homogeneity which, technically, involves the construction of a distribution for $\mu_{i,j}$. The analysis of the tissue samples of Sect. 2 suggests that $\mu_{i,j}$ may be modeled as a type of random effect, and estimates of the appropriate mean and variance are available in Table 1. To deal with the more complex mixture of this section, we will generate the random effect based on Table 5, to which an initial stage G_0 for stem cells ready to begin

Table 5 Proposed cell type heterogeneity pattern

| Group of genes | Stem cells | | | | Progenitors | | | |
	G_0	G_1	S	$G_2 + M$	G_1	S	$G_2 + M$	Diff cells
1	+	+	+	+	−	−	−	−
2	−	+	−	−	+	−	−	−
3	−	+	−	−	+	−	−	−
4	−	−	+	−	−	+	−	−
5	−	−	+	−	−	+	−	−
6	−	−	−	+	−	−	+	−
7	−	−	−	+	−	−	+	−
8	−	−	−	−	+	+	+	−
9	−	−	−	−	−	−	−	+
10	+	+	+	+	+	+	+	+

cell division has been added. It will be assumed that the 1000 genes are divided
evenly into 10 groups. Each row in the table defines a gene group for each stage
(column), which possesses a common mean $\mu_{i,j}$, determined by the table entry. In
particular, (−) corresponds to a baseline value $\mu_{i,j} = 0$, while (+) corresponds to
$\mu_{i,j} = \log(K)$ for scaling factors $K = 2, 5, 10$, which represent the variation of
the gene expression values across the cell cycles (logarithms will be in base 10).
Entries from Table 1 suggest that setting $\mu_{i,j}$ to be 1 or 2 standard deviations would
represent a plausible parameter choice. In addition, pairwise gene correlations may
be modeled. This will be done by setting a common exchangeable correlation ρ
within each group, which are otherwise independent.

3.2   Simulation Study

In each replication of our study, gene expression values of 1000 genes, conforming
to our model, are simulated, with 100 copies (or microarray slides). We use the
following steps:

1. Begin with the cell division process: for each copy, we start with 100 stem
   cells, and let them grow according to the rules of the simplified cell division
   model, until the number of cells reaches a pre-defined total number $N = 30,000$,
   within which a number of cells ν are randomly selected to contribute to the gene
   expression measurement.
2. Assign the gene expression values for the genes with respect to the cell types and
   cell stages, according to our model.
3. Sum the gene expression values for each gene across all stages and all cell types.
   This summation is regarded as one copy of the gene expression value for that
   particular gene.

4. Repeat the Steps 1–3 100 times to generate 100 copies of observations of the gene expression levels for the 1000 genes.

Each replication simulates a 1000×100 data matrix of gene expression measurements, with each row representing one gene, and each column representing one replicate (array). In order to assess a plausible range of effects, parameters were varied as follows (see Tables 6 and 7 for specific settings):

1. We will allow the total number of assayed cells ν to be constant or normally distributed.
2. The scaling factor K assumes values 2,5,10.
3. Group correlation is set to $\rho = 0, 0.5$.
4. Standard deviation is set to $\sigma = 0.01, 0.3$.

Because ν is a random quantity associated with a specific slide, it may be expected to act as a random effect, inducing spurious positive correlation between gene pairs. It will therefore be useful to examine the case of constant ν, to allow a direct observation of any heterogeneity effect, without confounding by ν. For each simulated replication the mean correlation summary CM is calculated. Tables 6 and 7 report the mean and variance of these summaries over all replications for each parameter set.

As shown in Table 7, when the number of cells ν is random, the correlations are considerably higher than when ν is constant. As anticipated, when ν varies across slides, it can be expected to function as a slide-specific random effect, inducing positive correlation. However, even when ν is constant, a smaller effect on aggregate correlation can be observed as scaling factor K is increased, conforming to our observations in Sect. 2. In Table 6, for $\rho = 0$, a positive correlation is clearly induced with increasing K for both values of σ. For both values of ρ, there is an increase in correlation variance with K for $\sigma = 0.01$ but not $\sigma = 0.3$. This is consistent with the finding reported in Table 2 that variance inflation increases with larger sample sizes,

Table 6 Mean and variance of correlation summary CM over all replications for each parameter set

ν	σ	K=2		K=5		K=10	
		$\rho = 0$	$\rho = 0.5$	$\rho = 0$	$\rho = 0.5$	$\rho = 0$	$\rho = 0.5$
20,000	0.01	0.0002	0.0537	0.0327	0.0367	0.0289	0.0359
		(0.0101)	(0.0307)	(0.2067)	(0.2081)	(0.2100)	(0.2158)
15,000	0.01	−0.0001	0.0534	0.0295	0.0361	0.0349	0.0414
		(0.0101)	(0.0314)	(0.2056)	(0.2078)	(0.2070)	(0.2134)
10,000	0.01	0.0002	0.0600	0.0356	0.0396	0.0424	0.0479
		(0.0101)	(0.0318)	(0.2065)	(0.2090)	(0.2109)	(0.2170)
10,000	0.3	−0.0001	0.0474	0.0018	0.0494	0.0043	0.0472
		(0.0101)	(0.0330)	(0.0104)	(0.0314)	(0.0116)	(0.0310)
20,000	0.3	0.0001	0.0449	0.0010	0.0540	0.0027	0.0426
		(0.0101)	(0.0322)	(0.0104)	(0.0313)	(0.0105)	(0.0330)

Total cell number ν is constant

Table 7 Mean and variance of average correlations over all replications for each parameter set

			K=2		K=5		K=10	
$V(v)$	$E(v)$	σ	$\rho = 0$	$\rho = 0.5$	$\rho = 0$	$\rho = 0.5$	$\rho = 0$	$\rho = 0.5$
$0.1 \times 20{,}000$	20,000	0.01	0.9994	0.9994	0.9295	0.9487	0.9023	0.9102
			(0.0000)	(0.0000)	(0.0028)	(0.0015)	(0.0039)	(0.0037)
$0.2 \times 20{,}000$	20,000	0.01	0.9994	0.9994	0.9295	0.9487	0.9023	0.9102
			(0.0000)	(0.0000)	(0.0028)	(0.0015)	(0.0039)	(0.0037)
$0.1 \times 20{,}000$	20,000	0.3	0.2738	0.3100	0.1427	0.1838	0.1131	0.1534
			(0.0081)	(0.0193)	(0.0121)	(0.0296)	(0.0112)	(0.0296)
$0.2 \times 20{,}000$	20,000	0.3	0.6953	0.7117	0.3258	0.3908	0.3149	0.3660
			(0.0018)	(0.0036)	(0.0134)	(0.0208)	(0.0123)	(0.0230)

Total cell number v is random

which would be equivalent to a reduction in σ. As a point of comparison with the mixtures effects reported in Table 1, we note that $(\log K)/\sigma = 1{,}100$ for $K = 10$ and $\sigma = 0.3, 0.01$, making the mixture nonhomogeneity effects relatively small or large, respectively, in comparison to the earlier study, while $(\log K)/\sigma = 2.33$ for $K = 5$ and $\sigma = 0.3$ is well within the observed range.

4 Heterogeneity of Subjects

Continuing our discussion of heterogeneity of biological samples, we will examine the effect of subject heterogeneity. Consider the simplest situation where only two types of subjects (patients, cell cultures, etc.) are present in the general population with their expected proportions being equal to π and $1 - \pi$. Let $X = (X_1, X_2)$ and $Y = (Y_1, Y_2)$ be two independent random vectors representing expression levels in a single pair of genes observed in the two types of subjects. Let ε_π be a Bernoulli random variable with parameter π. Assuming that ε_π is independent of the pair X, Y, consider the mixed random variable $Z = \varepsilon_\pi X + (1 - \varepsilon_\pi)Y$. For the covariance between the components of the vector $Z = (Z_1, Z_2)$ we have

$$cov(Z_1, Z_2) = \pi\, cov(X_1, X_2) + (1 - \pi)\, cov(Y_1, Y_2)$$
$$+ \pi(1 - \pi)(E[X_1] - E[Y_1])(E[X_2] - E[Y_2]). \tag{4.1}$$

Consider the case where the vectors X and Y have independent components. In this case, $cov(Z_1, Z_2) = 0$ if and only if either $E[X_1] = E[Y_1]$ or $E[X_2] = E[Y_2]$. It is easy to show that the same necessary and sufficient condition for stochastic independence (and not just correlation) of Z_1 and Z_2 is true with the expected values replaced with the respective characteristic functions. If none of the equalities holds for the mean values, the random variables Z_1 and Z_2 are correlated and the sign of their correlation coincides with that of the product $(E[X_1] - E[Y_1])(E[X_2] - E[Y_2])$. Therefore, by mixing vectors with different means one can induce either a positive or a negative correlation between their components. Does it mean that the observed positive correlation in microarray data should be attributed to this effect while assuming inherent independence of genes? The reason for a negative answer has to do with the fact that this condition is very restrictive. To induce spurious positive correlation in a given gene pair, it is required that both (presumably independent) genes undergo a unidirectional change in their expression levels across the two categories of subjects. Therefore, one can expect to see such correlations only in those gene pairs that are composed from a subset of the set of "differentially expressed" genes, the latter being typically a small proportion of the entire population of genes. The mixing effect does not extend to those numerous pairs of genes that include at least one gene with the same mean value for both types of subjects. Therefore, this effect is likely to induce only a short-range correlation. It should be noted in addition that the condition: $(E[X_1] \neq E[Y_1])$ *and* $(E[X_2] \neq E[Y_2])$ is untestable because it refers to a latent variable which is impossible to observe.

Giving up the unrealistic assumption $cov(X_1, X_2) = cov(Y_1, Y_2) = 0$, which was used only for the sake of simplicity, let us consider a typical gene pair with at least one of the genes satisfying the condition $E[X] = E[Y]$. This condition usually defines the null hypothesis when testing for differential expression and thus is expected to be met for the overwhelming majority of genes. Then it follows from formula (4.1) that

$$cov(Z_1, Z_2) = \pi\, cov(X_1, X_2) + (1 - \pi)\, cov(Y_1, Y_2),$$

i.e., $cov(Z_1, Z_2)$ is a convex combination of $cov(X_1, X_2)$ and $cov(Y_1, Y_2)$. Therefore,

$$\min\{cov(X_1, X_2), cov(Y_1, Y_2)\} \leq cov(Z_1, Z_2) \leq \max\{cov(X_1, X_2), cov(Y_1, Y_2)\},$$

indicating that the covariances are subject to averaging rather than to a proportional increase (or a decrease) in the majority of gene pairs. A similar analysis of correlation coefficients is less tractable but it does not seem to be necessary in order to illustrate the main point.

5 Discussion

In principle, evidence of functional gene pathway structure is available as observable correlation in gene expression profiles. However, these correlations are potentially confounded by effects which are unrelated to gene interactions. We have studied this phenomenon as it related to tissue and subject heterogeneity. The phenomenon proves to be quite complex, involving a number of parameters such as group heterogeneity of gene expression levels, variability in mixture proportions, and variability in the number of cells assayed. In addition, sample size and sample variation also plays an import role in these effects.

Decomposing sources of correlation explicitly will generally not be possible, since the components arising from heterogeneity are not identifiable when the heterogeneity itself is unobservable, as will usually be the case. However, there is nothing remarkable about this situation. The multiplicity of layers of correlation only reinforces the statement that stochastic dependence does not always reflect the basic processes of transcription regulation, which circumstance should not discourage the investigator from using correlation coefficients as a rich source of useful information on the functioning of the genome machinery. It is only a naive interpretation of correlation measures that is in doubt, and not the usefulness of such measures *per se*. It still makes sense to generate plausible hypotheses regarding between-gene interactions based on correlation analysis and test them in specially designed biological experiments.

Our discussion suggests that those causes of correlation discussed here are unlikely to exhaust all potential sources. This is particularly true for long-range correlations which do not conform to the usual gene pathway models, although all of these sources contribute to its specific features. We hypothesize that there is still a missing factor related to the processes of transcriptional regulation affecting extremely large sets of genes. This factor acts as a random effect, causing a high magnitude of inter-subject variability, but its biological origin has yet to be deciphered and quantitatively characterized. In addition to miRNA, there exists a formidable "dark matter" of ncRNAs, in particular long non-coding sequences known as macroRNA [14, 15, 18, 20]. While biological functions of macroRNAs remain largely unknown, their involvement in regulation of gene expression cannot be ruled out. In fact, the story of rethinking the role of miRNA that was long thought to be junk RNA ("transcriptional noise") makes this possibility a very real one. Similarly, the control of gene expression at the promoter level has been expanded beyond enhancer elements to include the epigenetic control of chromatin structure by DNA and histone modifications that may control the expression of large numbers of genes [1, 5, 8, 12, 19]. Higher order organization of chromatin, including silencers and isolators may further alter the expression of large numbers of genes at once [6, 13, 21].

In summary, there are multiple biological causes of correlations between gene expression signals, including regulatory processes, composite cellular make-up of tissues, and heterogeneity of subjects. We have examined various forms of heterogeneity, presenting methodologies with which the magnitude of their effects on correlation can be estimated. However, a complete understanding of the correlation structure observed in gene expression data will likely require the modeling of large-scale regulatory mechanisms of the type discussed here.

References

1. Berger, S. L. (2007). The complex language of chromatin regulation during transcription. *Nature, 447,* 407–412.

2. Braun, R., Cope, L., & Parmigiani, G. (2008). Identifying differential correlation in gene/pathway combinations. *BMC Bioinformatics, 9*, 1.
3. Chen, L., Almudevar, A., & Klebanov, L. (2013). Aggregation effect in microarray data analysis. In A. Y. Yakovlev, L. Klebanov, D. Gaile (Eds.), *Statistical methods for microarray data analysis*. Methods in Molecular Biology (Vol. 972, pp. 177–191). New York: Springer.
4. Davies, J., Proschel, C., Zhang, N., Mayer-Proschel, M., & Davies, S. (2008). Transplanted astrocytes derived from BMP- or CNTF-treated glial-restricted precursors have opposite effects on recovery and allodynia after spinal cord injury. *Brazilian Journal of Biology, 7*, 24.
5. Dulac, C. (2010). Brain function and chromatin plasticity. *Nature, 465*, 728–35.
6. Fraser, P., & Bickmore, W. (2007). Nuclear organization of the genome and the potential for gene regulation. *Nature, 447*, 413–417.
7. Friedman, N. (2004). Inferring cellular networks using probabilistic graphical models. *Science, 303*, 799–805.
8. Grewal, S. I., & Moazed, D. (2003). Heterochromatin and epigenetic control of gene expression. *Science, 301*, 798–802.
9. Hyrien, O., Mayer-Proschel, M., Noble, M., & Yakovlev, A. (2005). A stochastic model to analyze clonal data on multi-type cell populations. *Biometrics, 61*, 199–207.
10. Klebanov, L., Jordan, C., & Yakovlev, A. (2006). A new type of stochastic dependence revealed in gene expression data. *Statistical Applications in Genetics and Molecular Biology, 5*, Article 7.
11. Klebanov, L., & Yakovlev, A. (2007). Diverse correlation structures in microarray gene expression data and their utility in improving statistical inference. *The Annals of Applied Statistics, 1*, 538–559.
12. Li, E. (2002). Chromatin modification and epigenetic reprogramming in mammalian development. *Nature Reviews Genetics, 3*, 662–673.
13. Li, G., & Reinberg, D. (2011). Chromatin higher-order structures and gene regulation. *Current Opinion in Genetics & Development, 21*, 175–186.
14. Numata, K., Kanai, A., Saito, R., Kondo, S., Adachi, J., Wilming, L. G., et al. (2003). Identification of putative noncoding RNAs among the RIKEN mouse full-length cDNA collection. *Genome Research, 13*, 1301–1306.
15. Ponjavic, J., Ponting, C. P., & Lunter, G. (2007). Functionality or transcriptional noise? Evidence for selection within long noncoding RNAs. *Genome Research, 17*, 556–565.
16. Qiu, X., Brooks, A., Klebanov, L., & Yakovlev, A. (2005). The effects of normalization on the correlation structure of microarray data. *BMC Bioinformatics, 6*, 120.
17. Rao, M., Tanksale, A., Ghatge, M., & Deshpande, V. (1998). Molecular and biotechnological aspects of microbial proteases. *Microbiology and Molecular Biology Reviews, 62*, 597–635.
18. Ravasi, T., Suzuki, H., Pang, K. C., Katayama, S., Furuno, M., Okunishi, R., et al. (2006). Experimental validation of the regulated expression of large numbers of non-coding RNA from the mouse genome. *Genome Research, 16*, 11–19.
19. Saxena, A., & Carninci, P. (2011). Long non-coding RNA modifies chromatin: Epigenetic silencing by long non-coding RNAs. *Bioessays, 33*, 830–839.
20. Soares, L., & Valcarcel, J. (2006). The expanding transcriptome: The genome as the "Book of Sand". *The EMBO Journal, 25*, 923–931.
21. Vogelmann, J., Valeri, A., Guillou, E., Cuvier, O., & Nollmann, M. (2011). Roles of chromatin insulator proteins in higher-order chromatin organization and transcription regulation. *Nucleus, 2*, 358–369.
22. Werhli, A. V., Grzegorczyk, M., & Husmeier, D. (2006). Comparative evaluation of reverse engineering gene regulatory networks with relevance networks, graphical Gaussian models and Bayesian networks. *Bioinformatics, 22*, 2523–2531.
23. Zhou, X., Kao, M., & Wong, W. (2002). Transitive functional annotation by shortest-path analysis of gene expression data. *Proceedings of the National Academy of Sciences of the United States of America, 99*, 12783–12788.

Correlation Between the True and False Discoveries in a Positively Dependent Multiple Comparison Problem

Xing Qiu and Rui Hu

Abstract Testing multiple hypotheses when observations are positively correlated is very common in practice. The dependence between observations can induce dependence between test statistics and distort the joint distribution of the true and false positives. It has a profound impact on the performance of common multiple testing procedures. While the marginal statistical properties of the true and false discoveries such as their means and variances have been extensively studied in the past, their correlation remains unknown.

By conducting a thorough simulation study, we find that the true and false positives are likely to be negatively correlated if testing power is high and the opposite holds true—they are likely to be positively correlated if testing power is low. The fact that positive dependence between observations can induce negative correlation between the true and false discoveries may assist researchers in designing multiple testing procedures for dependent tests in the future.

Keywords Multiple testing · Correlation analysis · Microarray analysis

1 Background and Introduction

Testing statistical significance of a family of null hypotheses, denoted as $H_0^{(i)}$, $i = 1, \ldots, m$, is a main research field of modern inferential statistics. Here m is the number of null hypotheses to be tested. The multiplicity of the testing problems makes it a challenging task to control the per-family type I error without sacrificing too much testing power. For convenience, the definitions of several quantities that are useful for describing per-family errors and their relationship are provided in Table 1 (original source: Table 1 in [1]).

X. Qiu (✉) · R. Hu
Department of Biostatistics and Computational Biology, University of Rochester, Rochester, NY, USA
e-mail: Xing_Qiu@URMC.Rochester.edu; Rui_Hu@URMC.Rochester.edu

© Springer Nature Switzerland AG 2020
A. Almudevar et al. (eds.), *Statistical Modeling for Biological Systems*,
https://doi.org/10.1007/978-3-030-34675-1_4

Table 1 Number of errors
committed when testing m
null hypotheses

	Declared non-significant	Declared significant	Total
True null	U	V	m_0
True alternative	T	S	m_1
	$m - R$	R	m

Original source: Benjamini and Hochberg [1]

One straightforward method is to test each one of the hypotheses at a given significance level α without applying any multiple testing procedure (MTP). This method controls the per-comparison error rate (PCER): PCER $= E(V)/m$. Since m may be quite large, this method is likely to produce large number of false discoveries (V). For example, analyzing modern genomic data such as microarray data [4] often involves testing tens of thousands of hypotheses. Specifically, a GeneChip Human Genome U133 Plus 2.0 Array (Affymetrix Incorporated, Santa Clara, CA, U.S.A.) produces gene expression levels measured by over 47,000 probe sets. Testing hypotheses for all probe sets at significance level $\alpha = 0.05$ would *on average* falsely declare about 2350 differentially expressed genes. This high level of false positives is unacceptable by any reasonable standard. One solution to this problem is to apply one of the classical MTPs designed to control the familywise error rate (FWER), which is defined as the probability of making at least one type I error. This practice dates back to Bonferroni's seminal work published in 1935 [3]. Since then considerable efforts have been made to improve the power of the original Bonferroni procedure. Today, there exist many FWER controlling procedures, among which are the Šidák procedure [15], Holm procedure [9], Simes procedure [16], and Westfall–Young resampling procedure [20], and so on.

The stringency of the abovementioned procedures comes with a price, namely, they have much less statistical power than their unadjusted counterparts. As a compromise, Benjamini and Hochberg [1] invented a step-up procedure with a different principle. More specifically, the Benjamini–Hochberg procedure is designed to control the expected proportion of false discoveries among all rejections, i.e., the false discovery rate (FDR) defined as

$$\text{FDR} = E(Q), \quad \begin{cases} Q = \dfrac{V}{R} & \text{if } R > 0, \\ Q = 0 & \text{if } R = 0. \end{cases} \tag{1.1}$$

In addition to controlling for FWER or FDR, some multiple testing procedures, such as the extended Bonferroni procedure, are designed to control per-family error rate (PFER) [8], which is defined as the expected number of false positives, PFER $= E(V) = m \cdot \text{PCER}$ [4]. Since the extended Bonferroni procedure, the Benjamini–Hochberg procedure, and the "dummy" procedure without multiple testing adjustment all preserve the ranking of significance determined by unadjusted p-values, Gordon et al. [8] were able to show in their simulation study that the Benjamini–Hochberg procedure is approximately equivalent to an extended Bonferroni procedure with a matched PFER significance level.

Most classical multiple testing procedures are designed for testing independent hypotheses. In reality, hypotheses in one study are seldom independent of each other. A practical multiple testing procedure must take into account their dependencies.

There is one fundamental difference between a PCER/PFER controlling procedure and an FDR controlling procedure when tests are dependent: Both PCER and PFER depend only on the *marginal distributions* of testing statistics, but FDR depends on the *joint distribution* of them instead. Take PFER as an example, the dependence structure among test statistics is irrelevant in its definition:

$$\text{PFER} = E(V) = \sum_{i=1}^{m} P(t_i \in D), \tag{1.2}$$

where t_i is the statistic used in testing the ith null hypothesis, D is the rejection region of the testing statistic t_i at a pre-specified significance level. On the other hand, FDR is the mathematical expectation of a *nonlinear* function of two possibly dependent random variables V and R. Therefore its behavior cannot be determined solely by the marginal distributions of t_i.

The original Bonferroni procedure guarantees strong control of FWER for arbitrary dependence structure of t_i. Other procedures such as the Holm procedure [9] and the Šidák [15] procedure, which are more powerful than the Bonferroni procedure for independent tests, are not valid for arbitrary dependent tests [4].

In recent years, substantial efforts have been made to study the behavior of commonly used MTPs when tests are dependent. Benjamini and Yekutieli [2] proved that the Benjamini–Hochberg procedure controls FDR under positive dependence. They also provided a conservative FDR controlling procedure known as the Benjamini-Yekutieli procedure, which is valid for arbitrary dependence structure. Sarkar and Chang showed [14] that the Simes procedure is valid when the multivariate distributions of testing statistics exhibit a special type of positive dependence. The Šidák procedure controls the FWER for test statistics that satisfy the Šidák inequality (also known as the *positive orthant dependence property*, see [4, 5, 15] for more details). The Westfall and Young permutation procedure [20] can be considered as a permutation-based extension of the Šidák procedure. It controls FWER *weakly* for arbitrary dependent cases and controls FWER *strongly* as long as the unadjusted p-values have the *subset pivotality* property. Theoretical derivations of these properties can be found in [20].

These efforts have collectively made the classical multiple testing procedures more flexible and more applicable for real world applications such as microarray analysis. However, such flexibility usually comes with a price: low testing power. In order to design a good MTP, it is important to understand how false and true discoveries (V and S) behave with different levels of between-test dependence. Finner et al. [6, 7] studied the expected false discoveries of some FDR controlling procedures based on the independent test assumptions. Storey et al. [18, 19]

investigated the FDR estimating procedures in theory and simulation. Pavlidis and Noble [10] and Qiu et al. [12] evaluated the stability of several multiple testing procedures used in microarray analysis in the presence of intergene dependence and concluded that the reproducibility (in terms of the list of differentially expressed genes) of these procedures is very low. Furthermore, Qiu and Yakovlev [13] showed that the empirical estimators of FDR proposed in [18, 19] have high variance and are biased when the tests are correlated.

All these previous studies focus on the marginal statistical properties such as the mean and variance of V and S. In this paper, we intend to provide a thorough simulation study of the correlation between V and S. This simulation study is modeled after a common application of multiple testing in microarray analysis: selecting differentially expressed genes by running two-sample Student's t-test on data sampled from two biological conditions. Our main conclusions are:

1. V and S are independent when the tests are independent.
2. If the observations are positively dependent and the testing power is low, the correlation between V and S is positive.
3. If the observations are positively dependent and the testing power is high, the correlation between V and S is negative.

2 Methods

2.1 A Parametric Model with Two t-Tests

Consider testing two correlated hypotheses of which the alternative hypothesis is true for the first test and the null hypothesis is true for the second one. The false positive V and the true positive S are two correlated Bernoulli random variables.

More specifically, denote $\mathbf{X}$ and $\mathbf{Y}$ as two 2-dimensional random vectors with bivariate normal distributions:

$$\mathbf{X} = \begin{pmatrix} X_1 \\ X_2 \end{pmatrix} \sim N \left(\begin{pmatrix} \mu_{x,1} \\ \mu_{x,2} \end{pmatrix}, \begin{pmatrix} \sigma_1^2 & \rho\sigma_1\sigma_2 \\ \rho\sigma_1\sigma_2 & \sigma_2^2 \end{pmatrix} \right),$$

$$\mathbf{Y} = \begin{pmatrix} Y_1 \\ Y_2 \end{pmatrix} \sim N \left(\begin{pmatrix} \mu_{y,1} \\ \mu_{y,2} \end{pmatrix}, \begin{pmatrix} \sigma_1^2 & \rho\sigma_1\sigma_2 \\ \rho\sigma_1\sigma_2 & \sigma_2^2 \end{pmatrix} \right).$$

Here $\mu_{a,i}$ ($a = x, y, i = 1, 2$) represent the mean values of each distribution; σ_i^2 ($i = 1, 2$) represent the variances of these distributions. In addition $\rho = \text{corr}(X_1, X_2) = \text{corr}(Y_1, Y_2)$ represents the correlation coefficients between these random variables. Then

$$\{\mathbf{x}_j\} = \left\{ \begin{pmatrix} x_{1j} \\ x_{2j} \end{pmatrix} \right\}_{j=1}^{n_x} \text{ and } \{\mathbf{y}_k\} = \left\{ \begin{pmatrix} y_{1k} \\ y_{2k} \end{pmatrix} \right\}_{k=1}^{n_y}$$

are independent samples of size n_x and n_y drawn from $\mathbf{X}$ and $\mathbf{Y}$, respectively.

We conduct two separate two-sample Student's t-tests for testing the following (two-sided) hypotheses:

1. $H_0^{(1)}$: $\mu_{x,1} = \mu_{y,1}$ versus $H_1^{(1)}$: $\mu_{x,1} \neq \mu_{y,1}$.
2. $H_0^{(2)}$: $\mu_{x,2} = \mu_{y,2}$ versus $H_1^{(2)}$: $\mu_{x,2} \neq \mu_{y,2}$.

As mentioned earlier, we assume the alternative hypothesis, $H_1^{(1)}$, is true for the first test and the null hypothesis, $H_0^{(0)}$ is true for the second test. Since the two group t-test is invariant under translation and scaling of the observations, we can assume $\mu_{x,2} = \mu_{y,1} = \mu_{y,2} = 0$ and $\sigma_1^2 = \sigma_2^2 = 1$ without loss of generality. In other words, the joint distributions of $\mathbf{X}$ and $\mathbf{Y}$ are

$$\mathbf{X} = \begin{pmatrix} X_1 \\ X_2 \end{pmatrix} \sim N\left(\begin{pmatrix} d \\ 0 \end{pmatrix}, \begin{pmatrix} 1 & \rho \\ \rho & 1 \end{pmatrix} \right),$$

$$\mathbf{Y} = \begin{pmatrix} Y_1 \\ Y_2 \end{pmatrix} \sim N\left(\begin{pmatrix} 0 \\ 0 \end{pmatrix}, \begin{pmatrix} 1 & \rho \\ \rho & 1 \end{pmatrix} \right),$$

(2.1)

where constant d represents the standardized effect size of the first test.

Denote the two-sample t-statistics associated with these two tests as $\mathbf{t} = \begin{pmatrix} t_1 \\ t_2 \end{pmatrix}$. Under Model (2.1), the marginal distributions of t_1 and t_2 are

$$t_1 \sim T(\nu, \lambda_1) = T\left(n_x + n_y - 2, \, d\sqrt{\frac{n_x n_y}{n_x + n_y}} \right),$$

$$t_2 \sim T(\nu, \lambda_2) = T(n_x + n_y - 2, 0),$$

(2.2)

where $\nu = n_x + n_y - 2$ is the degree of freedom and $\lambda_1 = d\sqrt{n_x n_y/(n_x + n_y)}$, $\lambda_2 = 0$ are the noncentrality parameters, and $T(\nu, \lambda)$ denotes a noncentral Student's t-distribution with degree of freedom ν and noncentrality parameter λ.

2.2 Correlation Between the Two t-Statistics

The correlation between X_1, X_2 and Y_1, Y_2 induces certain correlation between t_1 and t_2 (denoted as ρ_t henceforth). The sample t-statistic is computed from the formula below (assuming a balanced design for simplicity):

$$t_i = \sqrt{\frac{n}{2}} \left(\frac{\bar{x}_i - \bar{y}_i}{\hat{\sigma}_p} \right),$$

(2.3)

where $x_i = \sum_j x_{ij}$, $y_i = \sum_k y_{ik}$ and $\hat{\sigma}_p$ is the pooled standard deviation. Since $\hat{\sigma}_p$ is an approximation of σ, the population standard deviation, $\hat{\sigma}_p \approx 1$. Therefore,

$$t_i \approx \sqrt{\frac{n}{2}} \, (\bar{x}_i - \bar{y}_i), \quad \sigma_t^2 = \text{var}(t_1) = \text{var}(t_2) \approx 1,$$

$$\text{cov}\,(t_1,\, t_2) \approx \frac{1}{2n} \, \text{cov} \left(\sum_{j=1}^{n} (x_{1j} - y_{1j}), \sum_{j=1}^{n} (x_{2j} - y_{2j}) \right) = \rho, \qquad (2.4)$$

$$\rho_t = \text{corr}\,(t_1,\, t_2) = \frac{\text{cov}\,(t_1,\, t_2)}{\sigma_t^2} \approx \rho.$$

A simulation study is carried out in order to show whether positive correlation between the observations (e.g., $\rho > 0$) induces positive correlation between the two observed t-statistics (e.g., $\rho_t > 0$). In this simulation, parameter d (the true difference between two groups, see Model (2.1)) takes two possible values: 0.5 and 1.0; ρ takes three possible values: 0.0, 0.4, and 0.8. We focus on the balanced design, i.e., $n_x = n_y = n$. There are five choices of n: 5, 10, 15, 20, and 25.

Table 2 summarizes the results of this simulation study. Numbers in this table are the estimated (Pearson) correlation coefficients (and their standard errors) between t_1 and t_2 under various simulation configurations. Each estimated value in the table is computed from a total of 100,000 i.i.d. repetitions grouped into 100 batches. More specifically, one Pearson correlation coefficient is computed from each batch (1000 t_1 and t_2), and the means and standard errors of these correlation coefficients computed from 100 batches are reported.

From the table, we can see that positive ρ induces positive ρ_t. We also notice that $\rho \approx \rho_t$. These observations are in accordance with the approximation results we derived above.

2.3   Correlation Between the True and False Positives

In a two-sided test with significance level α, the rejection region in terms of t-statistic is $D = D_c := (-\infty, -c) \cup (c, \infty) := D^- \cup D^+$, where the cutoff c is

Table 2 Mean and standard error (in parentheses) of Pearson correlation coefficients between t_1 and t_2

n	$d = 0.5$			$d = 1$		
	$\rho = 0$	$\rho = 0.4$	$\rho = 0.8$	$\rho = 0$	$\rho = 0.4$	$\rho = 0.8$
5	0.003 (0.003)	0.365 (0.002)	0.754 (0.002)	0.003 (0.003)	0.341 (0.003)	0.705 (0.002)
10	0.001 (0.003)	0.387 (0.003)	0.777 (0.001)	0.003 (0.003)	0.362 (0.003)	0.735 (0.001)
15	−0.001 (0.003)	0.385 (0.003)	0.782 (0.001)	0.001 (0.003)	0.363 (0.003)	0.745 (0.002)
20	0 (0.003)	0.388 (0.003)	0.784 (0.001)	−0.003 (0.003)	0.37 (0.003)	0.746 (0.001)
25	0.001 (0.003)	0.382 (0.003)	0.784 (0.001)	−0.005 (0.003)	0.366 (0.003)	0.747 (0.002)

Each correlation coefficient is computed from 1000 independent simulations. Means and standard errors of these correlation coefficients are computed from 100 batches

Table 3 Mean and standard error (in parentheses) of Pearson correlation coefficients between true and false positives

$d = 0.5$			$d = 1$		
n $\rho = 0$	$\rho = 0.4$	$\rho = 0.8$	$\rho = 0$	$\rho = 0.4$	$\rho = 0.8$
5 0.004 (0.003)	0.058 (0.003)	0.25 (0.004)	0.004 (0.003)	0.024 (0.003)	0.077 (0.003)
10 −0.008 (0.003)	0.052 (0.004)	0.172 (0.004)	0.002 (0.003)	−0.013 (0.003)	−0.048 (0.003)
15 0 (0.004)	0.031 (0.003)	0.113 (0.004)	0.004 (0.003)	−0.046 (0.004)	−0.127 (0.004)
20 −0.006 (0.003)	0.026 (0.003)	0.073 (0.003)	0.003 (0.003)	−0.052 (0.004)	−0.193 (0.006)
25 0.004 (0.003)	0.014 (0.003)	0.034 (0.003)	−0.002 (0.003)	−0.054 (0.004)	−0.242 (0.006)

Cutoff for t-statistic: $c = 1.96$. Each correlation coefficient is computed from the observed true and false positives of 1000 independent simulations. Means and standard errors of these correlation coefficients are computed from 100 batches

determined by this formula: $P(-c \leq t \leq c) = 1 - \alpha$. Clearly, $S = 1 \Longleftrightarrow t_1 \in D$ and $V = 1 \Longleftrightarrow t_2 \in D$. The pair (S, V) forms a bivariate Bernoulli random vector.

Table 3 summarizes the estimates of the Pearson correlation coefficients between the true and false positives computed from the simulation study described in Sect. 2.2. For simplicity and computational efficiency, null hypotheses are rejected based on fixed cutoffs of t-statistics ($c = 1.96$) rather than p-values. It is easy to show that $\alpha \approx 0.05$ when sample size is large.

From Table 3, we observe the following facts:

1. S and V are uncorrelated when $\rho = 0$.
2. S and V are positively correlated when the true effect is small and/or the sample size is small.
3. S and V are negatively correlated when the true effect is large and/or the sample size is large.
4. A larger value of ρ *amplifies* this correlation no matter whether it is positive or negative.

Some observations such as the first one are intuitive. The third one is rather surprising and deserves some discussion. The following proposition states that the sign of cov (S, V) is determined by $P(S = 1 \mid V = 1)$ and $P(S = 1)$.

Proposition 2.1 (Condition for Positive Correlation Between S and V) *We have* cov $(S, V) > 0$ *if and only if* $P(S = 1, V = 1) > P(S = 1) \cdot P(V = 1)$, *or equivalently,* $\Delta = P(S = 1 \mid V = 1) - P(S = 1) > 0$.

Proof See Appendix. □

It turns out that the sign of Δ is determined by λ, the noncentrality parameter. If λ is small (large), $P(S = 1 \mid V = 1)$ will be greater (less) than $P(S = 1)$. Consequently, the true and false positives are positively (negatively) correlated when λ is small (large) according to Proposition 2.1. This fact will be illustrated by Figs. 1, 2, and 3.

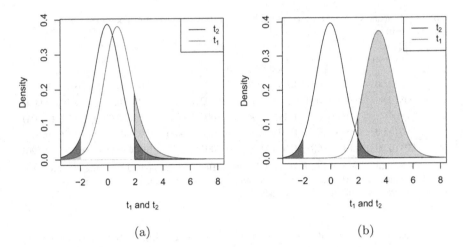

Fig. 1 Density plots of t_1 and t_2 with two sets of parameters. d is the true group difference, n is the sample size for one group, and λ is the noncentrality parameter of t_1. Shaded areas represent the rejection region. The cutoffs for t-statistics are set to be ± 1.96. (**a**) $d = 0.5, n = 5, \lambda = 0.791$. (**b**) $d = 1.0, n = 25, \lambda = 3.536$

Figure 1 shows the theoretical density functions of t_1 and t_2, denoted by $p(t_1)$ and $p(t_2)$, with two different sets of parameters. Figure 1a represents the small λ case (small true effect and small sample size), and Fig. 1b represents the large λ case.

For convenience, we denote the negative and positive branches of the rejection region as $D^- = (-\infty, -1.96)$ and $D^+ = (1.96, \infty)$. Also we denote

$$P_1^+ = P(t_1 \in D^+), \quad P_1^- = P(t_1 \in D^-);$$

$$P_1^{++} = P(t_1 \in D^+ \mid t_2 \in D^+), \quad P_1^{+-} = P(t_1 \in D^+ \mid t_2 \in D^-),$$

$$P_1^{-+} = P(t_1 \in D^- \mid t_2 \in D^+), \quad P_1^{--} = P(t_1 \in D^- \mid t_2 \in D^-).$$

By using these notations, we can express Δ as

$$\Delta = P(S = 1 \mid V = 1) - P(S = 1)$$

$$= \frac{(P_1^{+-} + P_1^{--})P(t_2 \in D^-) + (P_1^{++} + P_1^{-+})P(t_2 \in D^+)}{P_2(D)} - P_1^+ - P_1^-$$

$$= \frac{1}{2}\left((P_1^{++} - P_1^+) + (P_1^{+-} - P_1^+) + (P_1^{-+} - P_1^-) + (P_1^{--} - P_1^-)\right).$$

In both Fig. 1a and b, P_1^- is negligible compared to P_1^+ (theoretical values: $P_1^- = 0.007$ in the first case and $P_1^- = 3.6 \times 10^{-8}$ in the second case). Thus it is clear that

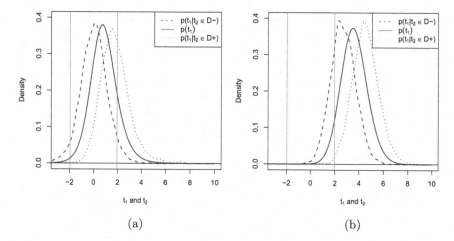

Fig. 2 Density plots. The curve in the middle is the unconditional density function of t_1. The other two curves represent conditional density functions of t_1 when $t_2 \in D^-$ and $t_2 \in D^+$. They are estimated from simulations. The correlation coefficient between observations: $\rho = 0.4$. (**a**) $d = 0.5, n = 5, \lambda = 0.791, s = 0.814$. (**b**) $d = 1.0, n = 25, \lambda = 3.536, s = 0.916$

$$P(S = 1) = P_1^+ + P_1^- \approx P_1^+.$$

As shown in Sect. 2.2, positive correlation between observations induces positive correlation between two t-statistics, so larger (smaller) values of t_2 imply larger (smaller) values of t_1. Consequently a "dragging effect" can be seen in Fig. 2. From these two figures we see that the conditional density function $p(t_1 \mid t_2 \in D^+)$ is approximately a copy of $p(t_1)$ shifted to the *right s* units, and $p(t_1 \mid t_2 \in D^-)$ is approximately a copy of $p(t_1)$ shifted to the *left s* units, i.e.,

$$p(t_1 \mid t_2 \in D^+) \approx p(t_1 - s), \quad p(t_1 \mid t_2 \in D^-) \approx p(t_1 + s), \qquad (2.5)$$

where the magnitude of this shifting effect

$$s = \frac{\text{median}(t_1 \mid t_2 \in D^+) - \text{median}(t_1 \mid t_2 \in D^-)}{2}$$

is controlled by the correlation coefficient ρ and cutoff c. So it is clear that

$$P_1^{-+} < P_1^-, \quad P_1^{--} > P_1^-, \quad P_1^{++} > P_1^+, \quad P_1^{+-} < P_1^+.$$

Since P_1^- is negligible, so is P_1^{-+}. By using this approximation and Eq. (2.5), we conclude that

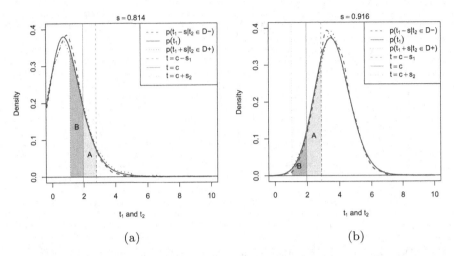

(a)   (b)

Fig. 3 Aligned density plots. The curve in the middle is $p(t_1)$, the unconditional density function of t_1. The other two curves are estimates of $p(t_1 - s \mid t_2 \in D^-)$ and $p(t_1 + s \mid t_2 \in D^+)$; the conditional density functions of t_1 (on $t_2 \in D^-$ and $t_2 \in D^+$) shifted in such a way that the medians of the three curves are the same. The correlation coefficient between observations: $\rho = 0.4$. The light shadowed area (area A) represents $\int_c^{c+s} p(t_1)dt_1$; the dark shadowed area (area B) represents $\int_{c-s}^c p(t_1)dt_1$. (**a**) $d = 0.5, n = 5, \lambda = 0.791$. (**b**) $d = 1.0, n = 25, \lambda = 3.536$

$$2\Delta \approx (P_1^{++} - P_1^+) + (P_1^{+-} - P_1^+) + P_1^{--}$$

$$\approx \left(\int_c^\infty p(t_1 - s) - p(t_1)dt_1 \right) + \left(\int_c^\infty p(t_1 + s) - p(t_1)dt_1 \right) + P_1^{--}$$

$$= \int_{c-s}^c p(t_1)dt_1 - \int_c^{c+s} p(t_1)dt_1 + P_1^{--}.$$

When λ is small Fig. 2a, it suffices to show that $\int_{c-s}^c p(t_1)dt_1 - \int_c^{c+s} p(t_1)dt_1 > 0$. When λ is large Fig. 2b, $P_1^{--} \approx \int_{-\infty}^{-c} p(t_1+s)dt_1 = \int_{-\infty}^{-c+s} p(t_1)dt \approx 0$. So we only need to focus on the right branch and show that $\int_{c-s}^c p(t_1)dt_1 - \int_c^{c+s} p(t_1)dt_1 < 0$.

It is apparent from Fig. 3 that $\int_c^{c+s} p(t_1)dt_1 = P(A) < (>) P(B) = \int_{c-s}^c p(t_1)dt_1$ when λ is small (large). By Proposition 2.1, these observations lead to our main conclusions:

- Small noncentrality parameters imply corr $(S, V) > 0$;
- Large noncentrality parameters imply corr $(S, V) < 0$,

which also explains what we have observed in Table 3.

2.4 A General Multiple Testing Design Motivated by Microarray Analysis

In microarray analysis, tens of thousands of gene expressions need to be tested simultaneously for differentiation across different phenotypes or biological conditions. Typically, observations are divided into two groups based on their phenotypes with n_x and n_y samples. Each observation has m gene expressions. The observations in these two groups can be viewed as n_x and n_y independent realizations of two m-dimensional random vectors.

We will focus on a parametric model based on multivariate normal distribution. In this model, n_x and n_y observations are sampled from two m-dimensional random vectors $\mathbf{X}$ and $\mathbf{Y}$, respectively. $\mathbf{X}$ and $\mathbf{Y}$ have the following multivariate normal distributions:

$$
\mathbf{X} \sim N(\boldsymbol{\mu}_x, \Sigma), \quad \mathbf{Y} \sim N(\boldsymbol{\mu}_y, \Sigma);
$$

$$
\boldsymbol{\mu}_x = (\mu_{x,1}, \mu_{x,2}, \dots, \mu_{x,m}), \quad \boldsymbol{\mu}_y = (\mu_{y,1}, \mu_{y,2}, \dots, \mu_{y,m});
$$

$$
\Sigma = \{\Sigma_{ij}\}_{i,j=1,\dots,m}; \quad \Sigma_{ij} = \begin{cases} \sigma_i^2 & i = j; \\ \rho_{ij}\sigma_i\sigma_j & i \neq j. \end{cases}
$$

(2.6)

Here $\boldsymbol{\mu}_x$ and $\boldsymbol{\mu}_y$ represent the population means for the first and second groups, Σ represents the covariance matrix for both groups, and σ_i^2 represents the marginal variance for the ith variable in both groups. This implies that the homoscedasticity condition is met for each individual test. ρ_{ij} is the correlation coefficient between the ith and the jth gene expressions in both conditions. Observations $\{\mathbf{x}_j\}$ are sampled from $\mathbf{X}$ where $j = 1, 2, \dots, n_x$ and $\mathbf{x}_j = (x_{1j}, \dots, x_{mj})$. Similarly, observations $\{\mathbf{y}_j\}$ are sampled from $\mathbf{Y}$ where $k = 1, 2, \dots, n_y$ and $\mathbf{y}_k = (y_{1k}, \dots, y_{mk})$.

The following hypotheses will be tested by two-sample Student's t-tests ($i = 1, 2, \dots, m$):

$$
H_0^{(i)}: \ \mu_{x,i} = \mu_{y,i};
$$
$$
H_1^{(i)}: \ \mu_{x,i} \neq \mu_{y,i}.
$$

Let M_0 and M_1 be the index sets of true null and alternative hypotheses, respectively. Denote $m_0 = \#(M_0)$ as the number of non-differentially expressed genes (NDEGs) and $m_1 = \#(M_1)$ as the number of differentially expressed genes (DEGs). The random numbers of true and false discoveries are

$$
S = \sum_{i \in M_1} S_i; \quad V = \sum_{j \in M_0} V_j,
$$

where $S_i, i = 1, 2, \dots, m_1$ and $V_j, j = 1, 2, \dots, m_0$ are the indicator functions of true and false positives, respectively.

Empirical evidences show that positive dependence is dominant for microarray data [11]. For simplicity, we assume $\rho_{ij} \geqslant 0$ for all i and j. Below are some other applications which may fit well with the above positive correlation model:

1. comparing two cohorts of patients with multiple clinical endpoints that are known to be positively correlated;
2. assessing the academic performance of two schools by means of exams in different but related subjects for each student.

Under the positive correlation model, the accumulative covariance between S and V can be written as a summation of covariances between individual true and false positives.

Proposition 2.2 (Summation Formula for Covariance) *The covariance between S and V is*

$$\text{cov}(S, V) = \sum_{i \in M_1} \sum_{j \in M_0} \text{cov}(S_i, V_j).$$

Therefore, if the individual $\text{cov}(S_i, V_j)$ are all positive (negative), the summary covariance $\text{cov}(S, V)$ must be positive (negative) as well. Based on the discussion in Sect. 2.3, we make the following predictions:

- If the majority of genes in M_1 are barely differentially expressed and/or the sample size is small such that the overall testing power is poor, we will find $\text{corr}(S, V) > 0$.
- If the majority of genes in M_1 are highly differentially expressed and/or the sample size is large such that the overall testing power is good, we will find $\text{corr}(S, V) < 0$.

3 Simulation Results

We conduct a simulation study to test our predictions. The configurations of this study are described as follows:

- The number of genes is $m = 1000$, of which $m_1 = 200$ are DEGs, $m_0 = 800$ are NDEGs. Their indexes are $M_1 = \{1, \ldots, 200\}$ and $M_0 = \{201, \ldots, 1000\}$, respectively.
- Sample size n (number of slides in one group) takes values of 5, 10, 15, 20, and 25. Again, we focus on the balanced design.
- Gene expression levels for each slide are modeled as a multivariate normal random variable described in Model (2.6). $\sigma_i = 0.36$ for $1 \leqslant i \leqslant 1000$. For NDEGs, $\mu_{x,j} = \mu_{y,j} = 9.0$ where $j \in M_0$. For DEGs, $\mu_{x,k} = 9.36$, $\mu_{y,k} = 9.0$ where $k \in M_1$. These values are estimated empirically from a large childhood leukemia dataset [17]. In other words, an effect size of one

standard deviation is imposed upon each DEG to produce differences between two biological conditions.

- The correlation coefficient ρ has three choices: $\rho = 0, 0.4$, and 0.8.
- two-sample t-test is used to test $H_0^{(i)}$ ($1 \leqslant i \leqslant 1000$). Three different significance levels (for unadjusted p-values) are used in these tests: $\alpha = 0.001, 0.005, 0.01$. The corresponding per-family error rate (PFER) levels are: 1, 5, and 10.
- For each testing data we compute and record the true and false positives according to the corresponding significance levels. We repeat this procedure 100,000 times. These repetitions are divided into 100 batches, each containing 1000 repeats. One Pearson correlation coefficient between S and V is computed in each batch.

We report the means and standard errors of these estimates in Table 4.[1] The following conclusions can be made based on this table, and they fit our predictions well:

1. S and V are uncorrelated when $\rho = 0$.
2. Both positive and negative dependence between S and V are possible.
3. For the dependent cases, corr (S, V) is positive when the sample size is small and/or the significance level α is small.
4. For the dependent cases, corr (S, V) is negative when the sample size is large and/or the significance level α is large.

4 Discussion

While the "curse of dimension" is a major challenge of testing multiple hypotheses and receives great attention from statisticians, the impact of between-test dependence is also a very important factor that cannot be ignored.

To overcome the "curse of dimension," a pertinent multiple testing procedure should be applied to control variant statistical error rates related to the true and false discoveries (S and V) such as FWER, PFER, and FDR. Knowing the dependence structure between S and V for a particular application may facilitate designing more powerful MTPs.

In this paper, we conduct a simulation study motivated by microarray analysis to investigate the correlation between S and V when the observations (expression levels) are positively correlated. We find that S and V can still be either positively or negatively correlated. More specifically, the sign of correlation coefficient between S and V depends on the average testing power: when testing power is low, corr (S, V) is likely to be positive; the increase of testing power by raising sample

---

[1] In order to achieve a better formatting effect, Table 4 only includes simulation results with $\rho = 0.0$ and 0.4. Table 5 shows the results with $\rho = 0.8$.

Table 4 Mean and standard error (in parentheses) of Pearson correlation coefficients between true and false positives

n	$\alpha = 0.001$		$\alpha = 0.005$		$\alpha = 0.01$	
	$\rho = 0$	$\rho = 0.4$	$\rho = 0$	$\rho = 0.4$	$\rho = 0$	$\rho = 0.4$
5	0.001 (0.003)	0.438 (0.007)	0.001 (0.003)	0.445 (0.007)	0.004 (0.003)	0.416 (0.006)
10	−0.002 (0.003)	0.357 (0.007)	0.004 (0.003)	0.271 (0.006)	−0.001 (0.003)	0.205 (0.006)
15	−0.005 (0.003)	0.233 (0.006)	−0.003 (0.003)	0.111 (0.006)	−0.002 (0.003)	0.032 (0.006)
20	−0.001 (0.003)	0.103 (0.006)	−0.004 (0.003)	−0.046 (0.006)	−0.002 (0.003)	−0.133 (0.006)
25	0.006 (0.003)	0.012 (0.006)	−0.001 (0.003)	−0.156 (0.006)	−0.004 (0.003)	−0.245 (0.006)

Each correlation coefficient is computed from the observed true and false positives of 1000 independent simulations. Means and standard errors of these correlation coefficients are computed from 100 batches

Table 5 Mean and standard error (in parentheses) of Pearson correlation coefficients between true and false positives

n	$\alpha = 0.001$	$\alpha = 0.005$	$\alpha = 0.01$
5	0.5 (0.011)	0.441 (0.008)	0.391 (0.007)
10	0.302 (0.008)	0.185 (0.005)	0.133 (0.005)
15	0.145 (0.005)	0.057 (0.004)	0.006 (0.004)
20	0.054 (0.004)	−0.036 (0.004)	−0.098 (0.004)
25	−0.011 (0.004)	−0.121 (0.005)	−0.199 (0.005)

The intergene correlation coefficient is $\rho = 0.8$. Each correlation coefficient is computed from the observed true and false positives of 1000 independent simulations. Means and standard errors of these correlation coefficients are computed from 100 batches

size or effect size in general is associated with the decrease of corr (S, V). In reality, it is not unusual that a research project using microarray data is constrained by budget so the sample size is limited and the average testing power is expected to be low. Based on our observations, we predict that S and V will be positively correlated in this case.

On the other hand, a well-designed study usually requires a power analysis in advance, and the targeted power level is typically very high (0.8 or 0.9). We believe S and V are likely to be *negatively* correlated in such a study. Therefore, it is possible in theory that increasing the true positives (higher testing power) could be accompanied by decreasing false positives (lower type I error). This is a rather surprising result which highlights the intricacy of the impact of between-test dependence on the statistical analysis. More robust and more powerful MTPs could be developed by exploiting this result. Another implication of our finding is that if researchers intend to design a particular MTP to work with positive correlation between S and V, they should be aware of the behavior of the correlation between S and V when testing multiple hypotheses with known positive dependence between observations/tests.

Acknowledgements This research is supported by NIH Grant GM079259 (X. Qiu). We would like to express our deepest gratitude to our late advisor and friend, Dr. Andrei Yakovlev, whose vision and encouragement are critical for this paper. We also want to thank Ms. Jing Che for her proofreading efforts.

Appendix: Proof of Proposition 2.1

Suppose $B = (S, V)$ is a 2-dimensional Bernoulli random variable with the following probability table:

$$\begin{pmatrix} p_{00} & p_{01} \\ p_{10} & p_{11} \end{pmatrix}.$$

Then

$$S \sim \text{Bernoulli}(p_{10} + p_{11}), \quad V \sim \text{Bernoulli}(p_{01} + p_{11}),$$

$$E(B) = (p_{10} + p_{11}, \ p_{01} + p_{11}),$$

$$\text{var}(S) = p_{10}p_{00} + p_{10}p_{01} + p_{11}p_{00} + p_{01},$$

$$\text{var}(V) = p_{01}p_{00} + p_{01}p_{10} + p_{11}p_{00} + p_{10},$$

$$\text{cov}(S, V) = E(S \cdot V) - E(S)E(V) = P(S = 1, V = 1) - P(S = 1)P(V = 1)$$

$$= p_{11} - (p_{10} + p_{11})(p_{01} + p_{11})$$

$$= p_{00} - (p_{00} + p_{01})(p_{00} + p_{10})$$

$$= P(S = 0, V = 0) - P(S = 0)P(V = 0).$$

$$(A.1)$$

The last statement shows that:

Proposition A.1 *The two components of a bivariate Bernoulli random vector, denoted as S and V, are positively correlated if and only if*

$$p_{11} \geqslant P(S = 1, V = 1) - P(S = 1)P(V = 1) = (p_{10} + p_{11})(p_{01} + p_{11}).$$

Furthermore, if S and V are positively correlated, they are positive quadrant dependent (PQD).

Proof We only need to prove the PQD part. Since (S, V) is a bivariate Bernoulli vector, it suffices to show that

$$P(S = 0, V = 0) = p_{00} \geqslant P(S = 0)P(V = 0).$$

It follows from

$$p_{11} \geqslant P(S = 1)P(V = 1)$$

$$p_{11} \geqslant 1 - P(S = 0)$$

$$- P(V = 0) + P(S = 0)P(V = 0)$$

$$p_{11} + P(S = 0) + P(V = 0) - 1 \geqslant P(S = 0)P(V = 0)$$

$$p_{00} \geqslant P(S = 0)P(V = 0).$$

$$\square$$

References

1. Benjamini, Y., & Hochberg, Y. (1995). Controlling the false discovery rate: A practical and powerful approach to multiple testing. *Journal of the Royal Statistical Society, Series B (Methodology), 57*, 289–300.

2. Benjamini, Y., & Yekutieli, D. (2001). The control of the false discovery rate in multiple testing under dependency. *Annals of Statistics, 29*, 1165–1188.
3. Bonferroni, C. E. (1935). Il calcolo delle assicurazioni su gruppi di teste. In *Studi in onore del professore Salvatore Ortu Carboni*, Carboni, S. O. (Ed.) (pp. 13–60). Bardi: Tipografia del Senato.
4. Dudoit, S., Shaffer, J. P., & Boldrick, J. C. (2003). Multiple hypothesis testing in microarray experiments. *Statistical Science, 18*, 71–103.
5. Dunn, O. J. (1958). Estimation of the means of dependent variables. *Annals of Mathematical Statistics, 29*, 1095–1111.
6. Finner, H., & Roters, M. (2001). On the false discovery rate and expected type 1 errors. *Biometrical Journal, 43*, 985–1005.
7. Finner, H., & Roters, M. (2002). Multiple hypotheses testing and expected number of type 1 errors. *Annals of Statistics, 30*, 220–238.
8. Gordon, A., Glazko, G., Qiu, X., & Yakovlev, A. (2007). Control of the mean number of false discoveries, Bonferroni, and stability of multiple testing. *Annals of Applied Statistics, 1*, 179–190.
9. Holm, S. (1979). A simple sequentially rejective multiple test procedure. *Scandinavian Journal of Statistics, 6*, 65–70.
10. Pavlidis, P., Li, Q., & Noble, W. (2003). The effect of replication on gene expression microarray experiments. *Bioinformatics, 19*, 1620–1627.
11. Qiu, X., Brooks, A. I., Klebanov, L., & Yakovlev, A. (2005). The effects of normalization on the correlation structure of microarray data. *BMC Bioinformatics, 6*, 120.
12. Qiu, X., Xiao, Y., Gordon, A., & Yakovlev, A. (2006). Assessing stability of gene selection in microarray data analysis. *BMC Bioinformatics, 7*, 50.
13. Qiu, X., & Yakovlev, A. (2006). Some comments on instability of false discovery rate estimation. *Journal of Bioinformatics and Computational Biology, 4*, 1057–1068.
14. Sarkar, S. K., & Chang, C. K. (1997). The Simes method for multiple hypothesis testing with positively dependent test statistics. *Journal of the American Statistical Association, 92*, 1601–1608.
15. Sidak, Z. (1967). Rectangular confidence regions for the means of multivariate normal distributions. *Journal of the American Statistical Association, 62*, 626–633.
16. Simes, R. (1986). An improved Bonferroni procedure for multiple tests of significance. *Biometrika, 73*, 751.
17. St. Jude Children's Research Hospital (SJCRH) Database on Childhood Leukemia. http://www.stjuderesearch.org/data/ALL1/
18. Storey, J. D. (2002). A direct approach to false discovery rates. *Journal of the Royal Statistical Society, Series B (Methodology), 64*, 479–498.
19. Storey, J. D., Taylor, J. E., & Siegmund, D. (2003). Strong control, conservative point estimation and simultaneous conservative consistency of false discovery rates: a unified approach. *Journal of the Royal Statistical Society, Series B Methodology, 66*, 187–205.
20. Westfall, P. H., & Young, S. (1993). *Resampling-based multiple testing*. New York: Wiley.

Multiple Testing Procedures: Monotonicity and Some of Its Implications

Alexander Y. Gordon

Abstract We review some results concerning the levels at which multiple testing procedures (MTPs) control certain type I error rates under a general and unknown dependence structure of the p-values on which the MTP is based. The type I error rates we deal with are (1) the classical family-wise error rate (FWER); (2) its immediate generalization: the probability of k or more false rejections (the generalized FWER); (3) the per-family error rate—the expected number of false rejections (PFER). The procedures considered are those satisfying the condition of *monotonicity*: reduction in some (or all) of the p-values used as input for the MTP can only increase the number of rejected hypotheses. It turns out that this natural condition, either by itself or combined with a property of being a *step-down* or *step-up* MTP (where the terms "step-down" and "step-up" are understood in their most general sense), has powerful consequences. Those include optimality results, inequalities, and identities involving different numerical characteristics of a procedure, and computational formulas.

Keywords Multiple testing procedure · Monotone procedure · Step-down procedure · Step-up procedure · Family-wise error rate · Generalized family-wise error rate · Per-family error rate

1 Introduction

Multiple testing procedures (MTPs) are widely used in biological and medical studies, such as DNA microarray experiments, clinical trials with multiple end-

---

Dedicated with deep gratitude and admiration to the memory of Andrei Yakovlev, who always inspired and encouraged those around him.

A. Y. Gordon (deceased)
Professor Gordon wrote this chapter while at University of North Carolina at Charlotte, Charlotte, NC, USA

A. Almudevar et al. (eds.), *Statistical Modeling for Biological Systems*,
https://doi.org/10.1007/978-3-030-34675-1_5

points, and functional neuroimaging. The development of the theory and methods of multiple hypothesis testing in the last 25 years was largely motivated by the needs of those applications. For a brief introduction to MTPs in the context of microarray experiments, see [3]. More information can be found in [4, 14, 22, 28].

The problem of multiple testing arises when we want to simultaneously test several hypotheses about the probability distribution from which the data are drawn. Based on the data, we want to select some of these hypotheses for rejection. Any rule providing such a selection is called a *multiple testing procedure*.

For example, in the problem of detection of the genes differentially expressed between two tissues, represented by two sets of microarrays (see [3]), the ith hypothesis may state that the expression levels of the ith gene in the two tissues have the same probability distribution. Rejection of this hypothesis can then be interpreted as a claim that the ith gene is differentially expressed between the two tissues.

We want an MTP to reject as many hypotheses as possible (to maximize the procedure's ability to reject false hypotheses), but at the same time the occurrence of false rejections (type I errors, which in many cases can be interpreted as *false discoveries*) should be somehow controlled. One commonly used quantity to be controlled is the *family-wise error rate* (FWER) defined as the probability of rejecting at least one true hypothesis. Its immediate generalization is the probability of k or more false rejections, called the *generalized family-wise error rate (of order k)*. Another important quantity is the expected number of false rejections, called the *per-family error rate* (PFER). In these terms, the word "family" refers to the set of hypotheses tested simultaneously based on a certain data set.

The MTPs that we consider are of the most common type: they use as input p-values associated with the hypotheses being tested. (There are other types of MTPs; see, e.g., [4].) In the present work we review some results concerning the levels at which MTPs control the aforementioned type I error rates under a general and unknown dependence structure of the p-values. All the results reviewed below pertain to MTPs satisfying the condition of *monotonicity*: reduction in some (or all) of the p-values used as input for the MTP can only increase the number of rejected hypotheses.

In some of the results reviewed below, the condition of monotonicity is combined with the property of being a *step-down* or *step-up* MTP. Here the terms "step-down" and "step-up" are understood in their most general sense (defined below in Sect. 2.3). The well-known and commonly used classes of what we call *threshold step-down* (TSD) and *threshold step-up* (TSU) procedures (exemplified by the Holm procedure and the Benjamini–Hochberg procedure, respectively) are parametric subclasses of the classes of monotone step-down and monotone step-up procedures, respectively, so that our conclusions hold in those subclasses automatically.

The rest of the paper is organized as follows. In Sect. 2, the necessary notions and notation are introduced; Sect. 3 contains the results.

2 Basic Notions

2.1 Uninformed MTPs

An MTP is a decision rule that, based on randomly generated data, selects for rejection a subset of the given set of hypotheses about the probability distribution from which the data are drawn.

We assume that there are in total m hypotheses $H_1, H_2, \ldots, H_m$, and associated with them are p-values $P_1, P_2, \ldots, P_m$.

The p-value P_i is a random variable (determined by the data) such that

1. $0 \le P_i \le 1$;

2. if the hypothesis H_i is true, then

$$\mathrm{pr}\{P_i \le x\} \le x \quad \text{for all } x \ (0 \le x \le 1) \tag{2.1}$$

(with an equality for all $x \in I = [0, 1]$ if P_i is based on a test statistic with a continuous distribution).

The p-value P_i measures the strength of the evidence against the hypothesis H_i provided by the data: the smaller P_i, the stronger the evidence. For a more detailed discussion of p-values, see, for example, [19, pp. 63–64].

From now on we assume that the hypothesis H_i is true if and only if (2.1) holds.

The MTPs that we consider are *marginal-p-value-only based*, or *uninformed*. Such a procedure, $\mathcal{M}$, takes as input a vector of observed p-values $\mathbf{p} = (p_1, p_2, \ldots, p_m) \in [0, 1]^m = I^m$ (*p-vector*) and transforms it into a set $\mathcal{M}(\mathbf{p}) \subseteq \mathbf{N}_m = \{1, 2, \ldots, m\}$. If $i \in \mathcal{M}(\mathbf{p})$, we say that, given p-vector $\mathbf{p}$, procedure $\mathcal{M}$ *rejects hypothesis* H_i, or equivalently, the corresponding p-value p_i is $\mathcal{M}$-*significant*. If, on the contrary, $i \notin \mathcal{M}(\mathbf{p})$, we say that, given p-vector $\mathbf{p}$, the procedure $\mathcal{M}$ *fails to reject* (or *retains*) the hypothesis H_i, or equivalently, the corresponding p-value p_i is $\mathcal{M}$-*insignificant*.

The classical Bonferroni procedure $\mathcal{M} = Bonf^\alpha (0 < \alpha < 1)$ is an example:

$$Bonf^\alpha(\mathbf{p}) = \{i \in \mathbf{N}_m : \ p_i \le \alpha/m\}.$$

Other examples will appear in the context of specific classes of MTPs.

Furthermore, we assume that an MTP has two basic properties: it is

(a) *symmetric* (permuting components of a *p*-vector does not affect which of them will be found significant), and
(b) *cutting* (significant components are smaller than insignificant ones).

We denote the set of all such procedures by Proc^m, where m is the number of hypotheses being tested.

In the sequel, by saying "$\mathcal{M}$ is an MTP" we will mean that $\mathcal{M} \in$ Proc^m.

2.2 Monotonicity

We call an MTP $\mathcal{M} \in$ Proc^m *monotone* if for any pair $\mathbf{p}, \mathbf{p}' \in I^m$ such that $\mathbf{p}' \leq \mathbf{p}$ (that is, $p_i' \leq p_i$ for all i) we have

$$|\mathcal{M}(\mathbf{p}')| \geq |\mathcal{M}(\mathbf{p})|. \tag{2.2}$$

In words, monotonicity means that if some *p*-values are replaced by smaller ones (while the others remain the same), the number of rejected hypotheses can only increase.

We note that in order to establish monotonicity of an MTP, it suffices to verify this property for *p*-vectors that belong to the *m*-dimensional simplex Simp^m defined as follows:

$$\text{Simp}^m = \{\mathbf{p} \in \mathbf{R}^m : 0 \leq p_1 \leq p_2 \leq \ldots \leq p_m \leq 1\}.$$

Proposition 2.1 ([8]) *An MTP $\mathcal{M}$ is monotone if and only if (2.2) holds for any pair of p-vectors $\mathbf{p}, \mathbf{p}' \in$ Simp^m such that $\mathbf{p}' \leq \mathbf{p}$.*

This fact greatly simplifies the verification of monotonicity.

On the other hand, as is shown in [11], the definition of monotonicity can also be given in the following (seemingly stronger, but actually equivalent) form: an MTP $\mathcal{M}$ is monotone if the relation $\mathbf{p}' \leq \mathbf{p}$ ($\mathbf{p}, \mathbf{p}' \in I^m$) implies that $\mathcal{M}(\mathbf{p}') \supseteq \mathcal{M}(\mathbf{p})$. That is, if some *p*-values are reduced (while the others remain unchanged), the hypotheses that were rejected still remain rejected. We note that in [23] the latter property is called "*p*-value monotonicity."

2.3 Step-Down and Step-Up Procedures

We present a generalized version of the concepts of step-down and step-up MTPs.

We call an MTP a *step-down* procedure [6] if its decision to reject or not reject hypothesis $H_{(i)}$ depends only on the *p*-values $p_{(j)}$, $1 \leq j \leq i$. Here $p_{(j)}$ denotes the jth ordered p-value (so that $p_{(1)} \le p_{(2)} \le \ldots \le p_{(m)}$) and $H_{(j)}$ denotes the corresponding hypothesis.

Informally speaking, a step-down procedure rejects hypotheses sequentially, starting from those with the smallest p-values. When deciding whether to reject the currently considered hypothesis, it takes into account only the corresponding p-value and the p-values encountered before. If some hypothesis is retained, all remaining hypotheses are retained too, and the procedure stops. (The name *step-down procedure* is due to the fact that in most cases smaller p-values correspond to higher values of the original test statistic.)

The most important subclass of step-down MTPs consists of procedures determined by a sequence of thresholds (critical values).

Let $\mathbf{u} = (u_1, u_2, \ldots, u_m)$ be a sequence of real numbers such that $0 \le u_1 \le u_2 \le \ldots \le u_m \le 1$.

The *threshold step-down procedure determined by* $\mathbf{u}$, denoted by $TSD_{\mathbf{u}}$, is defined as follows. Given p-vector $\mathbf{p}$, the ith ordered p-value $p_{(i)}$ is significant if and only if

$$p_{(j)} \le u_j \text{ for all } j, \ 1 \le j \le i.$$

This class of procedures is well-known; in the literature, MTPs of this type are called "step-down procedures," but we use this unrestricted term in the more general sense indicated above.

Example 2.1 The Bonferroni procedure $Bonf^{\alpha}$ is $TSD_{\mathbf{u}}$ with

$$\mathbf{u} = \left(\frac{\alpha}{m}, \frac{\alpha}{m}, \ldots, \frac{\alpha}{m} \right).$$

Example 2.2 The [15] procedure $Holm^{\alpha}$ is $TSD_{\mathbf{u}}$ with

$$\mathbf{u} = \left(\frac{\alpha}{m}, \frac{\alpha}{m-1}, \ldots, \frac{\alpha}{2}, \frac{\alpha}{1} \right).$$

It is readily seen that any TSD procedure is a monotone step-down MTP.

In parallel with step-down procedures, we consider their step-up counterparts.

An MTP is a *step-up* procedure [8] if its decision to reject or not reject hypothesis $H_{(i)}$ depends only on the p-values $p_{(j)}, i \le j \le m$.

Informally speaking, a step-up procedure *retains* hypotheses sequentially, starting from those with the largest p-values. When deciding whether to retain or reject the currently considered hypothesis, it takes into account only the corresponding p-value and the p-values encountered before. If some hypothesis is rejected, all remaining ones are rejected too, and the procedure stops.

For any sequence $\mathbf{u}$ of real numbers u_i such that $0 \le u_1 \le u_2 \le \ldots \le u_m \le 1$, we define the corresponding *threshold step-up procedure*, denoted by $TSU_{\mathbf{u}}$, as follows.

Given p-vector $\mathbf{p}$, the ith ordered p-value $p_{(i)}$ is significant if and only if

$$p_{(j)} \leq u_j \text{ for some } j, \ i \leq j \leq m.$$

Again, this class of MTPs, under the name of *step-up procedures*, is well-known; however, we use the unrestricted term in the more general sense described above.

Example 2.3 The Bonferroni procedure $Bonf^\alpha$ is $TSU_\mathbf{u}$, where $u_i = \alpha/m$, $i = 1, 2, \ldots, m$.

Example 2.4 The Hochberg procedure [13] $Hoch^\alpha$ is $TSU_\mathbf{u}$, where $u_i = \alpha/(m - i + 1)$, $i = 1, 2, \ldots, m$ (the same critical values as those of $Holm^\alpha$).

Example 2.5 The Benjamini-Hochberg procedure [1] BH^α is $TSU_\mathbf{u}$, where

$$\mathbf{u} = \left(\frac{\alpha}{m}, \frac{2\alpha}{m}, \ldots, \frac{m\alpha}{m} \right).$$

Again, any TSU procedure is a monotone step-up MTP.

2.4 Threshold Step-Up-Down Procedures

The TSD and TSU procedures are special cases of what we call *threshold step-up-down* (TSUD) procedures. This class of MTPs, introduced by Tamhane et al. [24] under the name of generalized step-up-down procedures, interpolates between the TSD and TSU classes.

A *threshold step-up-down procedure* $\mathcal{M}$, denoted below as $\text{TSUD}_\mathbf{u}^r$, is defined by a vector $\mathbf{u} \in Simp^m$ and an integer r $(1 \leq r \leq m)$. Given a p-vector $\mathbf{t} \in Simp^m$, it acts as follows. For $i \leq r$, $\mathcal{M}$ declares t_i significant if and only if $t_j \leq u_j$ for at least one j such that $i \leq j \leq r$. For $i \geq r$, $\mathcal{M}$ declares t_i significant if and only if $t_j \leq u_j$ for all j such that $r \leq j \leq i$.

Clearly, $\text{TSUD}_\mathbf{u}^1 = \text{TSD}_\mathbf{u}$ and $\text{TSUD}_\mathbf{u}^m = \text{TSU}_\mathbf{u}$. It is also obvious that a TSUD procedure is monotone.

2.5 Generalized Family-Wise Error Rates

The kth generalized family-wise error rate (we denote it by FWER_k) is the probability of k or more false rejections. This concept was first introduced by Victor [27] and further explored by Hommel and Hoffman [16], Korn et al. [17], van der Laan et al. [26], and Lehmann and Romano [18].

Let $\mathcal{M}$ be an MTP, $\mathcal{P}$ a probability distribution on the unit cube I^m, and $1 \leq k \leq m$. Let $\mathbf{P} = (P_1, \ldots, P_m)$ be a random vector with distribution $\mathcal{P}$. We put

$$\text{FWER}_k(\mathcal{M}, \mathcal{P}) := \text{pr}\{\mathcal{M}, \text{ given } \mathbf{P}, \text{ rejects} \geq k \text{ true hypotheses } H_i\}.$$

The procedure $\mathcal{M}$ is said to *strongly control* FWER_k *at level* α if $\text{FWER}_k(\mathcal{M}, \mathcal{P}) \leq \alpha$ for all probability distributions $\mathcal{P}$ on the unit cube I^m.

The quantity

$$\text{FWER}_k(\mathcal{M}) := \sup_{\mathcal{P}} \text{FWER}_k(\mathcal{M}, \mathcal{P}),$$

therefore, represents the exact level at which $\mathcal{M}$ strongly controls FWER_k.

Furthermore, the procedure $\mathcal{M}$ *weakly controls* FWER_k *at level* α if $\text{FWER}_k(\mathcal{M}, \mathcal{P}) \leq \alpha$ for all probability distributions $\mathcal{P}$ on the unit cube I^m satisfying all hypotheses H_i $(i = 1, \ldots, m)$.

Hence, the quantity

$$\text{wFWER}_k(\mathcal{M}) := \sup_{\mathcal{P}: \text{ all } H_i \text{ are true}} \text{FWER}_k(\mathcal{M}, \mathcal{P})$$

represents the exact level at which $\mathcal{M}$ weakly controls FWER_k.

If we put $k = 1$, then FWER_k becomes the traditional family-wise error rate (FWER) introduced by Tukey [25]. In this case we will write $\text{FWER}(\mathcal{M})$ and $\text{wFWER}(\mathcal{M})$ instead of $\text{FWER}_1(\mathcal{M})$ and $\text{wFWER}_1(\mathcal{M})$, respectively.

2.6 Per-Family Error Rate

The per-family error rate (PFER) [25] is the expected number of false rejections.

Let $\mathcal{M}$ be an MTP, $\mathcal{P}$ a probability distribution on the unit cube I^m, and $\mathbf{P} = (P_1, \ldots, P_m)$ a random vector with distribution $\mathcal{P}$. We put

$$\text{PFER}(\mathcal{M}, \mathcal{P}) := \mathbf{E}(|\mathcal{M}(\mathbf{P}) \cap \mathcal{T}_{\mathcal{P}}|),$$

where $\mathcal{T}_{\mathcal{P}} = \{i \in \mathbf{N}_m: H_i \text{ is true for the distribution } \mathcal{P}\}$. That is, $\text{PFER}(\mathcal{M}, \mathcal{P})$ is the expected number of true hypotheses falsely rejected by $\mathcal{M}$, given a random vector with distribution $\mathcal{P}$. The procedure $\mathcal{M}$ is said to *strongly control* PFER *at level* γ if $\text{PFER}(\mathcal{M}, \mathcal{P}) \leq \gamma$ for any probability distribution $\mathcal{P}$ on the unit cube I^m. The number

$$\text{PFER}(\mathcal{M}) := \sup_{\mathcal{P}} \text{PFER}(\mathcal{M}, \mathcal{P})$$

is, therefore, the exact level at which $\mathcal{M}$ strongly controls the PFER.

Furthermore, the procedure $\mathcal{M}$ *weakly controls* PFER *at level* γ if $\mathrm{PFER}(\mathcal{M}, \mathcal{P}) \leq \gamma$ for all probability distributions $\mathcal{P}$ on the unit cube I^m satisfying all the hypotheses H_i ($i = 1, \ldots, m$). Hence, the number

$$\mathrm{wPFER}(\mathcal{M}) := \sup_{\mathcal{P}:\, \mathcal{T}_{\mathcal{P}}=\mathbf{N}_m} \mathrm{PFER}(\mathcal{M}, \mathcal{P})$$

is the exact level at which $\mathcal{M}$ weakly controls the PFER.

2.7 Comparison of Procedures

Following [20], we say that a multiple testing procedure $\mathcal{M}$ *dominates* a procedure $\mathcal{M}'$, if for all $\mathbf{p} \in I^m$ we have $\mathcal{M}(\mathbf{p}) \supseteq \mathcal{M}'(\mathbf{p})$, i.e., $\mathcal{M}$ rejects all hypotheses H_i rejected by $\mathcal{M}'$ (and maybe some others); in this case we write $\mathcal{M} \succeq \mathcal{M}'$ and call $\mathcal{M}$ an *extension* of $\mathcal{M}'$. The extension is *nontrivial* if, in addition, $\mathcal{M} \neq \mathcal{M}'$ (that is, $\mathcal{M}(\mathbf{p}) \neq \mathcal{M}'(\mathbf{p})$ for at least one vector $\mathbf{p} \in I^m$).

Let C be a class of procedures, and let $\mathcal{M} \in C$; $\mathcal{M}$ is *the most rejective* (or *optimal*) in the class C if $\mathcal{M} \succeq \mathcal{M}'$ for all $\mathcal{M}' \in C$; $\mathcal{M}$ is *unimprovable* (or *weakly optimal*) in the class C if the relations $\mathcal{M}' \in C$ and $\mathcal{M}' \succeq \mathcal{M}$ imply that $\mathcal{M}' = \mathcal{M}$.

Note that the partial order $\succeq$ between MTPs is not linear: procedures $\mathcal{M}$ and $\mathcal{M}'$ may be incomparable, i.e., both relations $\mathcal{M}' \succeq \mathcal{M}$ and $\mathcal{M} \succeq \mathcal{M}'$ may be false. In particular, a class may contain more than one unimprovable MTP. (Of course, the most rejective MTP in the class, if it exists, is unique.)

Example 2.6 Given two vectors $\mathbf{u}, \mathbf{v} \in \mathrm{Simp}^m$, we have $TSD_\mathbf{u} \succeq TSD_\mathbf{v}$ if and only if $\mathbf{u} \succeq \mathbf{v}$ (that is, $u_i \geq v_i$ for all $i \in \mathbf{N}_m$). Similarly, $TSU_\mathbf{u} \succeq TSU_\mathbf{v}$ if and only if $\mathbf{u} \succeq \mathbf{v}$.

Example 2.7 For any $\mathbf{u} \in \mathrm{Simp}^m$, we have $TSU_\mathbf{u} \succeq TSD_\mathbf{u}$.

3 Some Implications of Monotonicity

In this section we present a review of some results that concern monotone MTPs and their control of the FWER, its generalization FWER_k, and the PFER.

3.1 Optimality of the Holm Procedure

As is well-known, the Bonferroni procedure $Bonf^\alpha$ strongly controls the FWER at (exact) level α and can be extended to the Holm step-down procedure $Holm^\alpha$, which still strongly controls the FWER at the same level α.

It turns out that the Holm procedure has the following optimality property.

Theorem 3.1 ([12]) *Let $\mathcal{M}$ be a monotone step-down MTP strongly controlling the FWER at level $\alpha < 1$. Then $\mathcal{M} \preceq Holm^\alpha$.*

In other words, the class of all monotone step-down procedures that strongly control the FWER at level $\alpha < 1$ contains a procedure that dominates all others—namely, the Holm procedure $Holm^\alpha$.

A special case of this theorem pertaining to threshold step-down (TSD) procedures, formulated in terms of their thresholds, was obtained earlier ([19], Chapter 9):

If $\mathcal{M} = TSD_\mathbf{u}$ and $\mathrm{FWER}(\mathcal{M}) \leq \alpha < 1$, then $u_j \leq \alpha/(m - j + 1)$ for all $j \in \mathbf{N}_m$.

Theorem 3.1 extends this result to its natural generality.

3.2 Extensions of the Holm Procedure

Theorem 3.1 implies, in particular, that in the class of monotone step-down MTPs, the Holm procedure $Holm^\alpha$ cannot be further nontrivially extended without weakening the control of the FWER:

If $\mathcal{M}$ is a monotone step-down procedure, $\mathcal{M} \succeq Holm^\alpha$ ($\alpha < 1$) and $\mathcal{M} \neq Holm^\alpha$, then $\mathrm{FWER}(\mathcal{M}) > \alpha$.

What if we look for extensions of the Holm procedure beyond the class of step-down procedures? Is it possible to nontrivially extend the Holm procedure in the class of all monotone MTPs without weakening the control of the FWER? If so, what are those extensions?

The answer is given by the following theorem:

Theorem 3.2 ([10]) *If $\mathcal{M}$ is a monotone procedure controlling the FWER at level $\alpha < 1$ and such that $\mathcal{M} \succeq Holm^\alpha$, then $\mathcal{M} = Holm^\alpha$.*

In other words, the Holm procedure $Holm^\alpha$ is *unimprovable*, or *weakly optimal*, in the class of monotone MTPs strongly controlling the FWER at level α.

We note that, as is shown in [10], this class of MTPs does not contain an optimal (most rejective) procedure.

Remark The result of Theorem 3.2 can be viewed as somewhat surprising: whereas in the setting of Theorem 3.1 (or its special case pertaining to TSD procedures) the procedure $Holm^\alpha$ is being compared with other step-down MTPs, nothing in the "step-down" design of the Holm procedure indicates that it may have any optimality properties in any class of MTPs not entirely consisting of step-down procedures.

3.3 Extensions of the Bonferroni Procedure

As was said, the Holm procedure $Holm^\alpha$ was introduced as a TSD extension of $Bonf^\alpha$ still controlling the FWER at the same level α. Theorems 3.1 and 3.2 establish certain optimality properties of this extension.

Can the Bonferroni procedure $Bonf^\alpha$ be similarly extended in the class of all monotone *step-up* procedures? What is the most rejective of such extensions, if it exists?

The answer is given by the following theorem, which shows that the situation in the class of monotone step-up MTPs is quite different from that in the class of monotone step-down procedures.

Theorem 3.3 ([8]) *Let $\mathcal{M}$ be a monotone step-up procedure such that*

$$\text{FWER}(\mathcal{M}) \le \alpha < 1 \text{ and } \mathcal{M} \succeq Bonf^\alpha.$$

Then $\mathcal{M} = Bonf^\alpha$.

In other words, in the class of monotone step-up MTPs, the Bonferroni procedure cannot be nontrivially extended without weakening the control of the FWER.

3.4 Quasi-Thresholds

Let $\mathcal{M}$ be a monotone MTP. Define its *ith quasi-threshold* $U_i(\mathcal{M})$ ($i = 1, 2, \ldots, m$) as the supremum of such values of τ that, given the p-vector $\mathbf{t} = (0, \ldots, 0, \tau, 1, \ldots, 1)$, its ith component, equal to τ, is $\mathcal{M}$-significant.

The quasi-thresholds can be easily calculated for any explicitly defined MTP. In particular, it is obvious that

if $\mathcal{M} = TSD_\mathbf{u}$ or $\mathcal{M} = TSU_\mathbf{u}$, then $U_i(\mathcal{M}) = u_i$, $i = 1, 2, \ldots, m$.

Therefore, the quasi-thresholds are a generalization of the thresholds (critical values) of TSD and TSU procedures.

The following result shows that in the case of a monotone step-down MTP its quasi-thresholds determine the exact level at which the procedure strongly controls the FWER.

Theorem 3.4 ([7]) *If $\mathcal{M}$ is a monotone step-down procedure, then*

$$\text{FWER}(\mathcal{M}) = \left(\max_{0 \le w \le m-1} (m - w) \, U_{w+1}(\mathcal{M}) \right) \wedge 1$$

(we use the standard notation $a \wedge b = \min(a, b)$, $a \vee b = \max(a, b)$).

Example 3.1 Benjamini and Liu [2] introduced a step-down procedure $\mathcal{M} = TSD_{\mathbf{u}}$ where the thresholds u_i are

$$u_i = \left(\frac{mq}{(m-i+1)^2} \right) \wedge 1, \quad i = 1, 2, \ldots, m \quad (0 < q < 1).$$

According to [2], this procedure controls the FDR (false discovery rate, or the proportion of false rejections among all rejections, see [1]) at level q. At what level does $\mathcal{M}$ control the FWER?

It can be easily derived from Theorem 3.4 (see [9]) that

$$\text{FWER}(\mathcal{M}) = mq \wedge 1.$$

3.5 Some Sharp Inequalities

Let $\mathcal{M}$ be an MTP; $\text{FWER}_k(\mathcal{M})$, the exact level at which $\mathcal{M}$ strongly controls FWER_k (the probability of $\geq k$ false rejections), where $2 \leq k \leq m$, satisfies an obvious inequality $\text{FWER}_k(\mathcal{M}) \leq \text{FWER}(\mathcal{M})$. For a monotone step-down MTP, this inequality can be strengthened.

Theorem 3.5 ([12]) *Let $\mathcal{M}$ be a monotone step-down procedure such that $\text{FWER}(\mathcal{M}) < 1$. Then for all $k = 2, 3, \ldots, m$*

$$\text{FWER}_k(\mathcal{M}) \leq C_k \text{FWER}(\mathcal{M}),$$

where

$$C_k = \begin{cases} 4k/(k+1)^2 & \text{if } k \text{ is odd;} \\ 4/(k+2) & \text{if } k \text{ is even.} \end{cases}$$

The inequalities are sharp.

Remark In the class of monotone *step-up* MTPs, the inequalities $\text{FWER}_k(\mathcal{M}) \leq \text{FWER}(\mathcal{M})$ $(2 \leq k \leq m)$ cannot be strengthened. Indeed, let $\mathcal{M} = TSU_{\mathbf{u}}$, where $u_1 = \ldots = u_{m-1} = 0$ and $0 < u_m < 1$. Let, furthermore, $\mathcal{P}$ be any distribution on I^m. Then, given a random vector $\mathbf{P} \sim \mathcal{P}$, the procedure $\mathcal{M}$ rejects either all hypotheses H_i, if $\max_{1 \leq i \leq m} P_i \leq u_m$, or only those for which $P_i = 0$ (a zero probability event if H_i is true) otherwise. Therefore, given any distribution $\mathcal{P}$ on I^m for which all hypotheses H_i are true, for any k $(1 \leq k \leq m)$ we have

$$\text{FWER}_k(\mathcal{M}, \mathcal{P}) = \text{pr}\{ \max_{1 \leq i \leq m} P_i \leq u_m \} \leq \text{pr}\{ P_1 \leq u_m \} \leq u_m,$$

so that $\text{FWER}_k(\mathcal{M}) \leq u_m$. In the special case where $P_1 = \ldots = P_m \sim U[0, 1]$, we have $\text{FWER}_k(\mathcal{M}, \mathcal{P}) = u_m$ for any k $(1 \leq k \leq m)$. Therefore, $\text{FWER}_k(\mathcal{M}) = \text{FWER}(\mathcal{M}) = u_m$.

Here again we see a dramatic difference between the classes of monotone step-down and monotone step-up procedures.

3.6   Bounds on Generalized Family-Wise Error Rates

Let us fix an arbitrary MTP $\mathcal{M}$ and consider the sequence $w_k = \text{wFWER}_k(\mathcal{M})$ of exact levels at which the procedure $\mathcal{M}$ *weakly* controls the generalized family-wise error rates of orders $k = 1, 2, \ldots, m$.

Obviously, this sequence is non-increasing:

$$1 \geq w_1 \geq w_2 \geq \ldots \geq w_m \geq 0.$$

It turns out that it cannot decrease arbitrarily fast if the procedure $\mathcal{M}$ is monotone.

Theorem 3.6 ([9]) *Let $\mathcal{M}$ be a monotone MTP. Then the finite sequence $w_k = \text{wFWER}_k(\mathcal{M})$ satisfies inequalities $w_{k+1} \geq (k/(k+1))w_k$ $(k = 1, 2, \ldots, m-1)$ or, equivalently,*

$$w_1 \leq 2w_2 \leq 3w_3 \leq \ldots \leq mw_m.$$

Theorem 3.6 can be reversed.

Theorem 3.7 ([9]) *Let $w_k \in [0, 1]$, $k = 1, 2, \ldots, m$, be such a sequence that*

$$w_1 \geq w_2 \geq \ldots \geq w_m \tag{3.1}$$

and

$$w_1 \leq 2w_2 \leq 3w_3 \leq \ldots \leq mw_m. \tag{3.2}$$

Then there exists a monotone MTP $\mathcal{M}$ for which $\text{wFWER}_k(\mathcal{M}) = w_k$ for all $k = 1, 2, \ldots, m$. Moreover, such an MTP can be found in the class of threshold step-down (TSD) procedures.

Remark The proof of Theorem 3.7 presented in [9], in the extreme cases $w_k \equiv \alpha$ and $kw_k \equiv \alpha$ (the slowest and the fastest decay of w_k allowed by (3.1) and (3.2)), gives the procedures $TSD_{\mathbf{u}}$ with $u_j = j\alpha/m$ (see [21] and [5]) and $u_j \equiv \alpha/m$ (the Bonferroni procedure), respectively. It follows, in particular, that among all monotone MTPs $\mathcal{M}$ with $\text{wFWER}(\mathcal{M}) = \alpha$, the Bonferroni procedure has the smallest value of $\text{wFWER}_k(\mathcal{M})$ for each $k = 2, 3, \ldots, m$.

The statement similar to Theorem 3.6, where $\text{wFWER}_k(\mathcal{M})$ is replaced by $\text{FWER}_k(\mathcal{M})$, is not true. However, Theorem 3.6 and its inverse can be generalized to the case where a limited number of hypotheses may be false.

Given a distribution $\mathcal{P}$ on I^m, let $T(\mathcal{P})$ be the number of hypotheses H_i which $\mathcal{P}$ satisfies.

Theorem 3.8 ([9]) *For a given monotone procedure $\mathcal{M}$ and a given integer $\tau(1 \leq \tau \leq m)$, let*

$$z_k := \text{FWER}_k^{T(\mathcal{P}) \geq \tau}(\mathcal{M}) \equiv \sup_{\mathcal{P}: \, T(\mathcal{P}) \geq \tau} \text{FWER}_k(\mathcal{M}, \mathcal{P}).$$

Then the non-increasing sequence z_k satisfies the inequalities

$$z_1 \leq 2z_2 \leq \ldots \leq \tau z_\tau.$$

Conversely, if a sequence $z_1, z_2, \ldots, z_\tau$ is such that

$$1 \geq z_1 \geq z_2 \geq \ldots \geq z_\tau \geq 0$$

and

$$z_1 \leq 2z_2 \leq \ldots \leq \tau z_\tau,$$

then there exists a TSD procedure $\mathcal{M}$ for which

$$\text{FWER}_k^{T(\mathcal{P}) \geq \tau}(\mathcal{M}) = z_k, \quad k = 1, 2, \ldots, \tau.$$

Interpretation of the inequalities $w_1 \leq 2w_2 \leq \ldots \leq mw_m$ established in Theorem 3.6. The quantity $\text{FWER}_k(\mathcal{M}, \mathcal{P})$ is the probability of the "bad event": k or more false rejections, when the p-values have the joint distribution $\mathcal{P}$. The outcomes with less than k false rejections are "invisible" to FWER_k, so that k characterizes the *resolution* of this type I error rate: the smaller k, the higher the resolution.

On the other hand, $w_k = \text{wFWER}_k(\mathcal{M})$ characterizes the degree of *certainty* that the above "bad event" does not occur (assuming the joint distribution of p-values satisfies all hypotheses H_i): the smaller w_k, the higher the certainty.

Therefore, the numbers k^{-1} and w_k^{-1} may be considered as measures of the resolution and certainty, respectively.

The inequality $kw_k \leq k'w_{k'}$ $(k < k')$ can be re-written (unless $w_{k'} = 0$) as

$$\frac{w_k}{w_{k'}} \leq \frac{k'}{k} \quad \text{if } k < k'. \tag{3.3}$$

Both sides are ≥ 1. The left-hand side (denote it by g) equals $w_{k'}^{-1}/w_k^{-1}$—the gain in certainty as we switch from k to k' ($k' > k$).

The right-hand side, equal to $k^{-1}/(k')^{-1}$, represents the loss in resolution.

The inequality (3.3) can be interpreted as follows: *in order to gain, by increasing the parameter k, a g-fold increase in certainty, we have to give up at least a g-fold loss in resolution.*

This bears a certain resemblance to the *golden rule of mechanics*: if, using a simple machine (such as a lever or inclined plane) we gain g-fold in force, then we lose g-fold in displacement.

3.7   An "All-or-Nothing" Theorem

By the definition, PFER($\mathcal{M}$) is the exact level at which a given procedure $\mathcal{M}$ controls the PFER under a general and unknown dependence structure of the p-values. Roughly speaking, this is the expected number of false rejections (falsely rejected true hypotheses) for the least favorable joint distribution of the p-values.

The following statement (the All-or-Nothing Theorem) says that, under the additional assumption that all the hypotheses H_i are true, the search for such a least favorable distribution can be restricted to those distributions that have the following property: given a random vector with such distribution, the procedure almost surely rejects either all hypotheses or none.

Theorem 3.9 (All-or-Nothing Theorem [11])   *If $\mathcal{M} \in \mathrm{Proc}^m$ is a monotone multiple testing procedure, then*

$$\mathrm{wPFER}(\mathcal{M}) = m \cdot \mathrm{wFWER}_m(\mathcal{M}). \tag{3.4}$$

Theorem 3.9 allows one to explicitly calculate the exact level of control of the PFER (both with and without the above assumption) for the commonly used classes of stepwise procedures.

Theorem 3.10 ([11]) *For any* $\mathbf{u} \in Simp^m$

(a) $\mathrm{wPFER}(\mathrm{TSD_u}) = m^2 \cdot \min_{1 \leq j \leq m} \left(\frac{u_j}{j} \right)$;

(b) $\mathrm{PFER}(\mathrm{TSD_u}) = \max_{0 \leq l \leq m-1} \left[(m-l)^2 \cdot \min_{1 \leq j \leq m-l} \left(\frac{u_{l+j}}{j} \right) \right]$.

Remark The above formulas (a) and (b) allow the following graphical interpretation. Let L be a line with a nonnegative slope in the Cartesian x, y plane, intersecting the x-axis at a point $(l, 0)$ with an integer l ($0 \leq l < m$) and such that all the points (i, u_i), $i = 1, \ldots, m$, lie on or above L. This line, together with the lines $y = 0$ and $x = m$, makes a right triangle. The maximum of the double areas of all such triangles equals $\mathrm{PFER}(\mathrm{TSD_u})$; if we fix $l = 0$, the restricted maximum equals $\mathrm{wPFER}(\mathrm{TSD_u})$.

Theorem 3.11 ([11]) $\text{wPFER}(\text{TSU}_\mathbf{u}) = \text{PFER}(\text{TSU}_\mathbf{u}) = mu_m$.

Both theorems 3.10 and 3.11 are special cases of the following statement.

Theorem 3.12 ([11]) *For any* $\mathbf{u} \in Simp^m$ *and* $r \in \{1, \ldots, m\}$,

(a) $\text{wPFER}(\text{TSUD}_\mathbf{u}^r) = m^2 \cdot \min_{r \leq i \leq m} \left(\frac{u_i}{i}\right)$;

(b) $\text{PFER}(\text{TSUD}_\mathbf{u}^r) = \max_{0 \leq l \leq m-1} \left[(m-l)^2 \cdot \min_{1 \vee (r-l) \leq j \leq m-l} \left(\frac{u_{l+j}}{j}\right)\right]$.

3.8 It is obvious that for any procedure $\mathcal{M} \in \text{Proc}^m$ we have $\text{PFER}(\mathcal{M}) \leq m \cdot \text{FWER}(\mathcal{M})$. It turns out that for monotone step-down MTPs this inequality can be significantly strengthened.

Theorem 3.13 ([11]) *Let* $\mathcal{M} \in \text{Proc}^m$ *be a monotone step-down procedure. If* $\text{FWER}(\mathcal{M}) = \alpha < 1$, *then* $\text{PFER}(\mathcal{M}) \leq (4 \wedge m)\alpha$.

Remark Nothing similar holds for monotone step-up procedures. For them, the coefficient m in the inequality $\text{PFER}(\mathcal{M}) \leq m \cdot \text{FWER}(\mathcal{M})$ cannot be reduced. Indeed, let $\mathcal{M}$ and $\mathcal{P}$ be the TSU procedure and the probability distribution introduced in the remark following Theorem 3.5. Then $m\text{FWER}(\mathcal{M}) \geq \text{PFER}(\mathcal{M}) \geq \text{PFER}(\mathcal{M}, \mathcal{P}) = mu_m = m\text{FWER}(\mathcal{M})$, so that $\text{PFER}(\mathcal{M}) = m\text{FWER}(\mathcal{M})$.

Alex Gordon died on May 13, 2019. We thank Anna Mitina for permission to include his paper in this book.

References

1. Benjamini, Y., & Hochberg, Y. (1995). Controlling the false discovery rate: A practical and powerful approach to multiple testing. *Journal of the Royal Statistical Society, Series B: Statistical Methodology, 57*, 289–300.
2. Benjamini, Y., & Liu, W. (1999). *A distribution-free multiple test procedure that controls the false discovery rate*. Technical Report. RP-SOR-99-3, Department of Statistics and Operations Research, Tel Aviv University.
3. Dudoit, S., Shaffer, J. P., & Boldrich, J. C. (2003). Multiple hypothesis testing in microarray experiments. *Statistical Science, 18*, 71–103.
4. Dudoit, S., & van der Laan, M. J. (2008). *Multiple testing procedures with applications to genomics*. New York, NY: Springer.
5. Finner, H., & Roters, M. (2002). Multiple hypothesis testing and expected number of type I errors. *The Annals of Statistics, 30*, 220–238.
6. Gordon, A. Y. (2007a). Explicit formulas for generalized family-wise error rates and unimprovable step-down multiple testing procedures. *Journal of Statistical Planning and Inference, 137*, 3497–3512.
7. Gordon, A. Y. (2007b). Family-wise error rate of a step-down procedure. *Random Operators and Stochastic Equations, 15*, 399–408.
8. Gordon, A. Y. (2007c). Unimprovability of the Bonferroni procedure in the class of general step-up multiple testing procedures. *Statistics & Probability Letters, 77*, 117–122.
9. Gordon, A. Y. (2009). Inequalities between generalized familywise error rates of a multiple testing procedure. *Statistics & Probability Letters, 79*, 1996–2004.
10. Gordon, A. Y. (2011). A new optimality property of the Holm step-down procedure. *Statistical Methodology, 8*, 129–135.

11. Gordon, A. Y. (2012). A sharp upper bound for the expected number of false rejections. *Statistics & Probability Letters, 82*, 1507–1514.
12. Gordon, A. Y., & Salzman, P. (2008). Optimality of the Holm procedure among general step-down multiple testing procedures. *Statistics & Probability Letters, 78*, 1878–1884.
13. Hochberg, Y. (1988). A sharper Bonferroni procedure for multiple tests of significance. *Biometrika, 75*, 800–802.
14. Hochberg, Y., & Tamhane, A. C. (1987). *Multiple comparison procedures*. New York, NY: Wiley.
15. Holm, S. (1979). A simple sequentially rejective multiple test procedure. *Scandinavian Journal of Statistics, 6*, 65–70.
16. Hommel, G., & Hoffman, T. (1988). Controlled uncertainty. In P. Bauer, G. Hommel, & E. Sonnemann (Eds.), *Multiple hypothesis testing* (pp. 154–161). Heidelberg: Springer.
17. Korn, E. L., Troendle, J. F., McShane, L. M., & Simon, R. (2004). Controlling the number of false discoveries: application to high-dimensional genomic data. *Journal of Statistical Planning and Inference, 124*, 379–398.
18. Lehmann, E. L., & Romano, J. P. (2005a). Generalizations of the familywise error rate. *The Annals of Statistics, 33*, 1138–1154.
19. Lehmann, E. L., & Romano, J. P. (2005b). *Testing statistical hypotheses* (3rd ed.). New York, NY: Springer.
20. Liu, W. (1996). Multiple tests of a non-hierarchical family of hypotheses. *Journal of the Royal Statistical Society, Series B: Statistical Methodology, 58*, 455–461.
21. Sarkar, S. K. (2002). Some results on false discovery rate in stepwise multiple testing procedures. *The Annals of Statistics, 30*, 239–257.
22. Shaffer, J. (1995). Multiple hypothesis testing: A review. *Annual Review of Psychology, 46*, 561–584.
23. Tamhane, A. C., & Liu, W. (1995). On weighted Hochberg procedures. *Biometrika, 95*, 279–294.
24. Tamhane, A. C., Liu, W., & Dunnett, C. W. (1998). A generalized step-up-down multiple test procedure. *The Canadian Journal of Statistics, 26*, 353–363.
25. Tukey, J. W. (1953). The problem of multiple comparison. Unpublished manuscript. In *The collected works of John W. Tukey VIII. Multiple comparisons: 1948–1983* (pp. 1–300). New York, NY: Chapman and Hall.
26. van der Laan, M. J., Dudoit, S., & Pollard, K. S. (2004). Augmentation procedures for control of the generalized family-wise error rate and tail probabilities for the proportion of false positives. *Statistical Applications in Genetics and Molecular Biology, 3*, Article 15.
27. Victor, N., (1982). Exploratory data analysis and clinical research. *Methods of Information in Medicine, 21*, 53–54.
28. Yang, H. Y., & Speed, T. (2003). Design and analysis of comparative microarray experiments. In T. Speed (ed.), *Statistical analysis of gene expression microarray data* (pp. 35–92). Boca Raton, FL: Chapman and Hall.

Applications of Sequential Methods in Multiple Hypothesis Testing

Anthony Almudevar

Abstract One of the main computational burdens in genome-wide statistical applications is the evaluation of large scale multiple hypothesis tests. Such tests are often implemented using replication-based methods, such as the permutation test or bootstrap procedure. While such methods are widely applicable, they place a practical limit on the computational complexity of the underlying test procedure. In such cases it would seem natural to apply sequential procedures. For example, suppose we observe the first ten replications of an upper-tailed statistic under a null distribution generated by random permutations, and of those ten, five exceed the observed value. It would seem reasonable to conclude that the P-value will not be small enough to be of interest, and further replications should not be needed.

While such methods have been proposed in the literature, for example by Hall in 1983, by Besag and Clifford in 1991 and by Lock in 1991, they have not been widely applied in multiple testing applications generated by high dimensional data sets, where they would likely be of some benefit. In this article related methods will first be reviewed. It will then be shown how commonly used multiple testing procedures may be modified so as to introduce sequential procedures while preserving the validity of reported error rates. A number of examples will show how such procedures can reduce computation time by an order of magnitude with little loss in power.

Keywords Multiple hypothesis testing · Sequential hypothesis testing · Gene expression analysis

A. Almudevar (✉)
Department of Biostatistics and Computational Biology, University of Rochester, Rochester, NY, USA
e-mail: anthony_almudevar@urmc.rochester.edu

© Springer Nature Switzerland AG 2020
A. Almudevar et al. (eds.), *Statistical Modeling for Biological Systems*,
https://doi.org/10.1007/978-3-030-34675-1_6

1 Introduction

Sequential hypothesis tests form a class of methods intended as an alternative to the fixed sample size design, particularly where sampling carries a high cost. A sample size N may be chosen in the design stage to achieve to specified power and significance level, but in practice a decision may be reachable well before the complete sample is observed. If the data are, or can be, observed sequentially in a controllable way, it may be possible to devise a sampling stopping time resulting in a decision which preserves the original significance level and power. The obvious benefit is a reduction in cost due to smaller samples (at least on average) or, in clinical trials, an earlier decision regarding treatment efficacy. The general theory for these techniques was originally developed in [17], largely concerning the *sequential probability ratio test* (SPRT) which will be discussed below (see, for example, [15] for a general treatment of the subject).

In this article we will consider a more recent application. Significance levels are often assessed using simulated replications, usually either a permutation procedure or a bootstrap. In cases were replications are computationally expensive, there is a clear motivation to terminate sampling when a statistically valid decision can be reached. The problem is compounded in multiple testing applications commonly used in the analysis of high-throughput data, such as gene expression profiles obtained from microarrays.

Sequential procedures were applied to permutation tests in [9, 11] based on the SPRT, and were demonstrated to reduce computation time considerably with little loss of power. A sequential estimate of a significance level was proposed in [5], applicable to independent replications, as well as to dependent replications generated by Markov chain sampling as described in [4]. This methodology has been recently applied to estimate simulation-based empirical P-values in high dimensional linkage analysis. See [12] for a discussion. A more recent survey of these methods can be found in [8].

2 The Empirical Hypothesis Test as Stopped Binary Process

We first introduce the basic stochastic model underlying the remaining discussion. Define, for $\theta \in (0, 1)$,

$$\tilde{U} = (U_1, U_2, \ldots), \text{ where } U_i \sim Bernoulli(\theta)$$

$$Z_n = \sum_{i=1}^{n} U_i, n \geq 1, \tag{2.1}$$

and suppose T is a stopping time for $\tilde{U}$ (formally, T is a stopping time if the event $\{T = t\}$ depends only on $U_1, \ldots, U_t$). We refer to the pair (T, Z_T) as a *stopped binary process* (SBP) with parameter θ. Note that if $P(T = M) = 1$, then (M, Z_M) is equivalent to a summary of a fixed sized Bernoulli sample. In addition, if $P(T \leq M) = 1$ and $P(T = M) > 0$, we say the SBP is *truncated* at M, in which case the stopping time will generally be denoted T_M. The sequence $\tilde{U}$ is usually independent, but important exceptions exist, for example, [4, 5].

2.1 Monte Carlo Hypothesis Tests as SBPs

Suppose data X_{obs} from distribution F_X is sampled, and we wish to test against a null hypothesis H_0 using statistic $S(X)$. Assume we may simulate a sample $X_1^*, \ldots, X_n^*$ for any n from null distribution P_0 when H_0 is simple, or from a conditional distribution $P_0(X \mid U(X))$ for some suitable sufficient statistic $U(X)$. For convenience, we will assume that H_0 is rejected for large values of $S(X)$, although this convention is easily adjusted. The observed significance level is then

$$\alpha_{obs} = P_0(S(X) \geq S(X_{obs}) \mid U(X) = U(X_{obs})).$$

In this case, we define a SBP (T, Z_T) by setting in (2.1)

$$U_i = I\{S(X_i^*) \geq S(X_{obs})\}, \quad i \geq 1,$$
$$\theta = \alpha_{obs} \tag{2.2}$$

coupled with a stopping time T. For any fixed n, $\hat{\alpha}_{obs} = Z_n/n$ is an unbiased estimate of α_{obs}. We advocate the practice of including the original data X_{obs} among the replications, in effect setting $\hat{\alpha}_{obs} = (Z_n + 1)/(n + 1)$, but it will be more convenient to use the unbiased form throughout the following discussion. The recommended adjustment can then be appended to the algorithms proposed here where relevant.

Possibly, the simulation of X_i^* or the calculation of $S(X_i^*)$ may be computationally expensive, and in addition the test may have to be performed multiple times. In this case the objective is to select stopping time T which will be in some sense small, while maintaining the required error rates for the hypothesis test. Intuitively, if after the first 10 replications we have a current estimate $\hat{\alpha}_{obs} = 1/2$, it seems reasonable to predict that a small P-value is unlikely, and further simulation is unnecessary. The role of sequential analysis is to formalize this idea.

In practice, stopped samples are often truncated at M, that is, the stopping time is not allowed to exceed M. In such cases it may be useful to regard the stopped sample as being embedded in a fixed sample design, that is, a sample of size M which can be terminated early. This gives natural criteria for the design of a sequential procedure as being approximately equal to the fixed sample design in accuracy, but with a significantly reduced expected sample size.

An important distinction to make is between those tests for which an estimate of α_{obs} is desired, and those for which a fixed significance level is specified. We discuss each briefly.

2.2 Estimation of Significance Level

When an estimate of α_{obs} in needed, one approach is to adopt the following stopping rule, often referred to as *inverse sampling*. For some fixed K, let T_K be the minimum replication number for which $Z_{T_K} = K$. If the replications are independent (possibly conditionally), then T_K has a negative binomial distribution which depends on α_{obs}. Furthermore, $E[T_K]$ is smaller for larger α_{obs}, which is a desirable property for this application. This type of procedure is discussed in [5] in greater generality, in which truncated stopping rules are considered, as well as dependent replications generated by Markov chain sampling as described in [4].

2.3 Fixed Level Tests

If the test is performed with a nominal fixed significance level α, then we reject H_0 when $\alpha_{obs} \leq \alpha$. Since α_{obs} is itself being estimated, we have a *secondary test* involving hypothesis $H_0^* : \alpha_{obs} \leq \alpha$. We then refer to the test concerning H_0 as the *primary test*. Note that it is inappropriate to treat H_0^* as a null hypothesis in the usual sense. If we set a small Type 1 error for H_0^*, then the probability of falsely rejecting H_0 when α_{obs} is marginally larger than α will be high.

We will briefly formalize some concepts associated with the replacement of α_{obs} with a simulated estimate. We assume we may generate a SBP (T_M, Z_{T_M}) with parameter $\theta = \alpha_{obs}$ and truncated stopping time T_M. We have an acceptance region E for H_0^* based on (T_M, Z_{T_M}). The properties of the secondary test can be summarized by the *acceptance curve* $A(\theta \mid E) = P_\theta(E)$ (this has been referred to as an *operating characteristic curve* in earlier literature [17]). We define the three quantities:

$$Q(E) = \int_0^1 A(\theta \mid E)d\theta$$

$$K(E) = \int_0^1 E_\theta[T_M]d\theta$$

$$D(E \mid \alpha) = \int_0^1 |I\{\theta \leq \alpha\} - A(\theta \mid E)|d\theta, \qquad (2.3)$$

for any $\alpha \in (0, 1)$. We assume that under the primary null hypothesis H_0, α_{obs} is uniformly distributed, so that the actual significance level can be obtained by

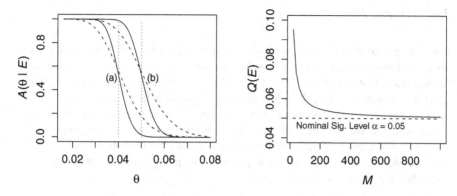

Fig. 1 Plot 1: Acceptance curves for test $Fixed(q, M)$ with (a) $q = 0.04$; (b) $q = 0.05$. The solid line represents $M = 2000$, while the dashed line represents $M = 500$. **Plot 2:** Actual significance level $Q(E)$ for design $Fixed(0.05, M)$ as a function of M. The nominal significance level of 0.05 in indicated by the dashed line

integrating $A(\theta \mid E)$ with respect to θ over the uniform density, which is the quantity $Q(E)$. Similarly, $K(E)$ is the expected stopping time under the primary null hypothesis. We also note that $D(E \mid \alpha)$ is the L_1 distance between $A(\theta \mid E)$ and $I\{\theta \le \alpha\}$, which is the ideal acceptance curve.

Let $Fixed(q, M)$ denote the secondary test using fixed sample size M with acceptance region $E = \{Z_M/M \le q\}$. The acceptance curve can be directly calculated from the binomial distribution. Examples are shown in Fig. 1 (Plot 1) for $M = 500, 2000$ and $q = 0.04, 0.05$.

As can be seen, the tests have identifiable points of symmetry, that is, the point θ_s at which $A(\theta_s \mid E) = 1/2$. The best location for this point depends on what type of test we find acceptable. Suppose we take this point to be $\theta_s = 0.05$. If the actual value of α_{obs} is $\theta = 0.05$, then we accept the test as significant with probability 1/2, and a P-value slightly larger than 0.05 will be (incorrectly) accepted as significant with probability slightly less than 1/2. Such an approach might be acceptable if (1) it is verifiable that the actual significance level $Q(E)$ equals the nominal level α; and (2) the test is understood to be a randomized test.

To investigate the first issue, we calculate the values of $Q(E)$ for test $Fixed(0.05, M)$, allowing M to vary up to $M = 1000$. Figure 1 (Plot 2) plots the resulting values. As can be seen, the actually significance level $Q(E)$ exceeds the nominal value 0.05, converging to this value as M increases. The discrepancy is quite large for $M \le 400$. Assuming the test $Fixed(\alpha, M)$ can be adjusted to achieve the correct actual significance level, we then have a randomized test, in the sense that two analyses of the same test may yield opposite conclusions, particularly when α_{obs} is close to the nominal significance level. The only reasonable alternative to such a randomized test is to render the test conservative by moving θ_s below the nominal significance level α, so that $A(\alpha \mid E)$ is close to zero. The test is still randomized, but it may at least be claimed that, with small error, any test declared significant has an observed significance level not exceeding the nominal.

2.4　Hybrid Test

It will be useful to briefly introduce a third approach to the problem, in which we obtain estimates of α_{obs} when this value is small, but stop sampling when it can be concluded that α_{obs} is too large to be of interest. While the sequential estimate described in Sect. 2.2 roughly achieves this goal, there will be an advantage, to be discussed below, to incorporating formal elements of sequential hypothesis testing.

Suppose we are given a SBP (T, Z_T) with parameter θ and, for purposes of comparison, we used a fixed sample design $T \equiv M$. We may then form estimate $\hat{\theta} = Z_M/M$ of θ. Now suppose we implement a stopping time T_M truncated at M, and define the *stopped estimate*

$$\tilde{\theta} = \hat{\theta} I\{T_M = M\} + I\{T_M < M\}. \tag{2.4}$$

The stopping time T_M is designed with somewhat different objectives than that for the fixed level test, in that we wish to stop the process when the secondary hypothesis $H_0^* : \theta \leq \alpha$ is false, but we also wish to observe a complete sample (that is, $T_M = M$) when it is true. In either case, $\tilde{\theta}$ is interpretable as a significance level for the primary hypothesis H_0, since it either estimates α_{obs}, or is set to 1 when α_{obs} is large, forcing acceptance of H_0.

3　Overview of the Sequential Probability Ratio Test

The work which follows will be based on the SPRT, so we present a brief introduction. Formally (see [17] or [15]), the SPRT tests between two simple alternatives $H_0 : \theta = \theta_0$ vs $H_1 : \theta = \theta_1$, where θ indexes a parametric family of distributions f_θ. We assume there is a sequence of *i.i.d.* observations $x_1, x_2 \ldots$ from f_θ where $\theta \in \{\theta_0, \theta_1\}$. Let $l_n(\theta)$ be the likelihood function based on $(x_1, \ldots, x_n)$, and define the likelihood ratio statistic $\lambda_n = l_n(\theta_1)/l_n(\theta_0)$. Define two constants $A < 1 < B$, and stopping time

$$T = \min\{n : \lambda_n \notin (A, B)\}. \tag{3.1}$$

It can be shown that $E_{\theta_i}[T] < \infty$, $i = 0, 1$. If $\lambda_T \leq A$ we conclude H_0, and conclude H_1 otherwise. We define errors $\alpha_0 = P_{\theta_0}(\lambda_T \geq B)$ and $\alpha_1 = P_{\theta_1}(\lambda_T \leq A)$. Errors may be approximated by:

$$\alpha_0 \approx (1 - A)/(B - A) \text{ and } \alpha_1 \approx A(B - 1)/(B - A). \tag{3.2}$$

Inverting these approximations, Wald [17] therefore recommended choosing $A = \alpha_1/(1 - \alpha_0)$ and $B = (1 - \alpha_1)/\alpha_0$. Estimates of $E_{\theta_i}[T]$ are also available:

$$E_{\theta_0}[T] = \mu_0^{-1}\{(B-1)\log A + (1-A)\log B\}/(B-A),$$

$$E_{\theta_1}[T] = \mu_1^{-1}\{A(B-1)\log A + B(1-A)\log B\}/(B-A), \text{ where}$$

$$\mu_i = E_{\theta_i}\left[\log(f_{\theta_1}(X)/f_{\theta_0}(X))\right],$$

where X is a random variable with density f_{θ_i}. For Bernoulli sequences with $E_{\theta_i}[X] = \theta_i$, we have $\mu_i = \theta_i \log(\theta_1/\theta_0) + (1-\theta_i)\log((1-\theta_1)/(1-\theta_0))$.

There are a number of factors to consider when employing a SPRT. First, hypothesis testing usually involves composite hypotheses. One method of adapting the SPRT to this case is to select surrogate simple hypotheses. For example, to test $H_0 : \theta \geq \theta'$ vs $H_1 : \theta < \theta'$ we could employ the SPRT with simple hypotheses $\theta_0 < \theta'$ and $\theta_1 \geq \theta'$. In this case, we would need to know the entire power function, which may be estimated using careful simulations.

The SPRT has an important optimal property which recommends its use in this application, proven in [18]. Let L_0 be any SPRT for deciding between two simple hypotheses H_0 and H_1, and let L_1 be any alternative test (including fixed sample as well as sequential tests). Let $\alpha_i(L_j)$, $m_i(L_j)$ be, respectively, the probability of rejecting H_i, and the expected number of observations needed to reach a decision, when H_i is true and test L_j is used. Then if $\alpha_i(L_1) \leq \alpha_i(L_0)$ it follows that $m_i(L_0) \leq m_i(L_1)$, for $i = 0, 1$.

4 Application of the SPRT to Hypothesis Tests Based on Simulated Replications of an Accept–Reject Rule

Designing a SBP (2.2) and acceptance rule for the secondary test defined in Sect. 2.3 using a SPRT is straightforward. We first assume that the sequence $U_1, U_2, \ldots$ is independent. We set the densities f_{p_i} associated with the simple hypotheses H_0, H_1, to be $Bernoulli(p_i)$ leading to likelihood sequence

$$\lambda_n = \frac{p_1^{Z_n} q_1^{n-Z_n}}{p_0^{Z_n} q_0^{n-Z_n}}, \tag{4.1}$$

where $q_0 = 1 - p_0$, $q_1 = 1 - p_1$, and Z_n is the nth cumulative sum of $i.i.d.$ Bernoulli random variables. No generality is lost by assuming $p_0 < p_1$, so we do so throughout. After a log transform (4.1) becomes a random walk

$$R_n = \sum_{i=1}^{n} V_i, \tag{4.2}$$

where $V_1, V_2, \ldots$ is an $i.i.d.$ sequence satisfying

$$P_{p_i}(V_1 = \delta^+) = p_i = 1 - P(V_1 = \delta^-), \quad i = 0, 1,$$

where

$$\delta^- = \log(q_1/q_0)$$
$$\delta^+ = \log(p_1/p_0). \tag{4.3}$$

Equivalently,

$$R_n = Z_n \delta^+ + (n - Z_n)\delta^-. \tag{4.4}$$

In our application it will be natural to modify the SPRT with truncation. Suppose a reasonable choice for a fixed sample size is M. We would then use truncated stopping time $T_M = \min\{T, M\}$ with T defined in (3.1). When $T_M = M$, we could, for example, select hypothesis H_0 if $\lambda_M \leq 1$. All such modifications could easily be incorporated into a simulated estimate of the power curve. See [15] for further discussion on the implications of truncation.

Replacing SPRT boundaries $A < 1 < B$ with $a = \log(A)$, $b = \log(B)$ we have equivalent stopping times

$$T = \inf\{n \geq 1 : R_n \notin (a, b)\},$$

$$T_M = \min\{T, M\}$$

so that H_0 is accepted if and only if $R_{T_M} \leq 0$. For our purposes, we say a *design* is specified by the quintuple $SPRT(p_0, p_1, a, b, M)$, where $a < 0 < b$.

We note that the SPRT is sometimes formulated in terms of cumulative sum Z_n. This can be obtained by applying a suitable linear transformation to R_n defined in (4.4) and to the constant boundaries, yielding an equivalent formulation in terms of Z_n paired with parallel sloping boundaries, resulting in a one-to-one reparametrization of $SPRT(p_0, p_1, a, b, M)$. Of course, the subsequent analysis may equivalently use either parameterization. We use the process defined by R_n, since it yields more natural decision rules, particularly when the truncation boundary is reached.

4.1 Single Hypothesis Test

Suppose we wish to design $SPRT(p_0, p_1, a, b, M)$ for a SBP with $\theta = \alpha_{obs}$ using secondary hypothesis $H_0^* : \theta \leq \alpha$. As discussed above we observe (T_M, Z_{T_M}), and use acceptance region $E = \{R_{T_M} \leq 0\}$, where $R_{T_M} = Z_{T_M}\delta^+ + (T_M - Z_{T_M})\delta^-$. For small values of the truncation parameter M the acceptance curve $A(\theta \mid E)$ could be calculated using sample path enumeration methods, but this would not be feasible for the larger values needed for this application, so simulations will be used.

In general, we will select M to be a number that would be suitable for a fixed sample design, leaving the problem of selecting the remaining parameters in design $SPRT(p_0, p_1, a, b, M)$. While a formal solution to the problem of optimal test design will be discussed below for a special case, a more informal approach will be useful here.

Figure 2 (Plot 1) gives $A(\theta \mid E)$ for tests $E_1 = SPRT(0.03, 0.05, -6.9, 4.6, 2000)$, $E_2 = SPRT(0.04, 0.06, -6.9, 4.6, 2000)$, and $E_3 = Fixed(0.05, 2000)$. The intent is to design an acceptance rule for a secondary hypothesis defined by $\alpha = 0.05$. We first note that the discussion regarding actual significance levels $Q(E)$ in Sect. 2.3 is equally relevant for SPRTs. As can be seen in Fig. 2 (Plot 1) the examples of acceptance curves for the fixed sample and sequential procedures have similar shapes. Thus, in either case it is important to evaluate $Q(E)$, and to otherwise

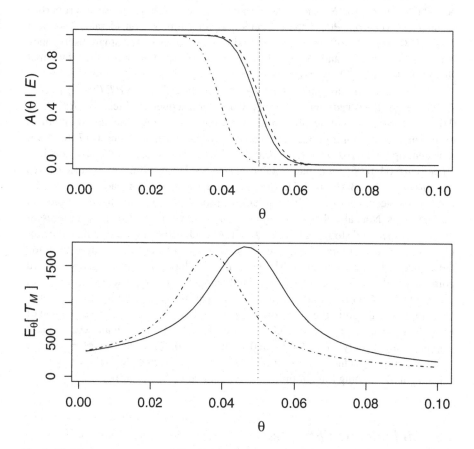

Fig. 2 Plot 1: Acceptance curve $A(\theta \mid E)$ for tests $E_1 = SPRT(0.03, 0.05, -6.9, 4.6, 2000)$ [$-\cdot-$], $E_2 = SPRT(0.04, 0.06, -6.9, 4.6, 2000)$ [$-$] and $E_3 = Fixed(0.05, 2000)$ [$- - -$]. The line $\theta = 0.05$ is indicated separately [$\cdots$]. **Plot 2:** Expected stopping times $E_\theta[T_M]$ for tests $E_1 = SPRT(0.03, 0.05, -6.9, 4.6, 2000)$ [$-\cdot-$] and $E_2 = SPRT(0.04, 0.06, -6.9, 4.6, 2000)$ [$-$]. The line $\theta = 0.05$ is indicated separately [$\cdots$]

correctly characterize the test. While it seems natural to set $p_0 < \alpha < p_1$, we may set $p_1 < \alpha$ to obtain a conservative test.

When $T_M = M$ the choice of p_0, p_1 directly defines the acceptance rule. As defined above, in this case we accept H_0^* when $R_M \leq 0$, which is equivalent to $Z_M/M \leq \rho(p_0, p_1)$, where

$$\rho(p_0, p_1) = -\delta^-/(\delta^+ - \delta^-).$$

Although $\rho(p_0, p_0)$ is not defined, it is easily shown that

$$\lim_{p_1 \to p_0} \rho(p_0, p_1) = p_0.$$

We now consider the choice of a, b. From Fig. 2 (Plot 1) we can see that the acceptance curves for E_2 and E_3 are very close (and so have the same properties of interest), yet from Plot 2 the expected stopping time is considerably less than $M = 2000$ over most of the range of θ, and never exceeds it. The small discrepancy between $A(\theta \mid E_2)$ and $A(\theta \mid E_3)$ in Fig. 2 (Plot 1) is due to the fact that $\rho(0.04, 0.06) = 0.0494 \neq 0.05$.

We then note that as $\min(-a, b) \to \infty$, the test $E_2 = SPRT(p_0, p_1, a, b, M)$ converges to $Fixed(\rho(p_0, p_1), M)$. Consider a sequence of tests $SPRT(0.04, 0.06, -t, t, M)$, and setting $M = 500, 2000$. For $-a = b = t$ the approximations of α_0 and α_1 given in (3.2) are equal. We let t range from 0.17 to 6.93, representing an approximate range for $\alpha_0 = \alpha_1$ of 0.46 to 0.001. We also consider $Fixed(\rho(0.04, 0.06), M)$, interpretable as $t = \infty$. The resulting acceptance curves are shown in Fig. 3 (Plots 1,2), superimposed separately for each value of M. Clearly, the accuracy of the SPRT quickly approaches that of the fixed sample test as t increases. Note also that stopping times increase with t, which we will represent by the expected null stopping time $K(E)$. If for each value of t we calculate the pair $(K(E), D(E, 0.05))$ (see (2.3)), we can construct a characteristic curve to guide the choice of t. These curves are shown in Fig. 3 (Plots 3,4). Table 1 gives selected points from these plots. The limiting fixed sample test is represented as $t = \infty$. The value $t = 6.93$ is the largest plotted for each M, so that the limiting case has been approached within the given rounding error. For the $M = 500$ case, a value of $t = 2.08$ gives a value of $D(E \mid 0.05)$ very close to the limit, with an expected null stopping time of only 37.14 (compared to $M = 500$). Similarly, for $M = 2000$ a value of $t = 4.16$ achieves a value of $D(E \mid 0.05)$ very close to the limit, with an expected null stopping time of only 85.8 (compared to $M = 2000$).

4.2 Multiple Hypothesis Tests

We next assume that we have K primary hypothesis tests with P-values $p_i = \alpha_{obs}^i$, $i = 1, \ldots, K$. We wish to report a global error rate, in which case the magnitude of small P-values is of importance. We will consider specifically the class of *multiple*

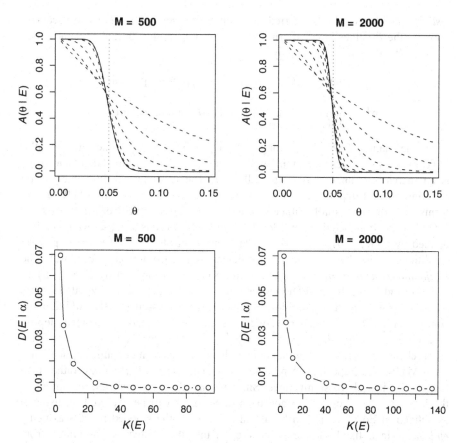

Fig. 3 Plot 1: For $M = 500$, the graph displays superimposed acceptance curves for tests $SPRT(0.04, 0.06, -t, t, M)$ with t ranging from 0.17 to 6.93 $[- - -]$, as well as for test $Fixed(\rho(0.04, 0.06), M)$ $[-]$. **Plot 2:** Same as Plot 1, but with $M = 2000$. **Plot 3:** For $M = 500$, the graph displays the curve formed by pairs $(K(E), D(E, 0.05))$ where $E = SPRT(0.04, 0.06, -t, t, M)$ with t ranging from 0.17 to 6.93. **Plot 4:** Same as Plot 3, but with $M = 2000$

Table 1 Selected points from plots in Fig. 3

t	α_i	$M = 500$		$M = 2000$	
		$K(E)$	$D(E \mid 0.05)$	$K(E)$	$D(E \mid 0.05)$
0.17	0.46	2.8	0.0696	2.81	0.0698
2.08	0.11	37.14	0.0081	85.80	0.0040
6.93	0.001	93.56	0.0077	134.76	0.0038
∞	–	500	0.0077	2000	0.0038

testing procedures (MTP) referred to as either *step-up* or *step-down* procedures. Suppose the *P*-values have ordering $v_1, \ldots, v_K$ (p_{v_i} is the *i*th ranked value), then *adjusted P*-values $p_{v_i}^a$ are given by:

$$p_{v_i}^a = \max_{j \leq i} \min(C(K, j, p_{v_j}), 1) \text{ (step-down procedure)}$$

$$p_{v_i}^a = \min_{j \geq i} \min(C(K, j, p_{v_j}), 1) \text{ (step-up procedure)},$$

where the quantity $C(K, j, p)$, along with the procedure form, defines the particular MTP. In the case of ties the ranking will not be unique. It is assumed that $C(K, j, p)$ is an increasing function of p for all K, j. The procedure is implemented by rejecting all null hypotheses for which $p_i^a \leq \alpha$. Depending on the MTP, various forms of error rate, usually either *family-wise error rate* FWER or *false discovery rate* FDR, are controlled at the α level. *Holm's procedure* is a step-down procedure defined by $C(K, j, p) = (m - j + 1)p$, and controls for FWER under general conditions (it is always at least as efficient as the Bonferroni procedure). The *Benjamini–Hochberg procedure* is a step-up procedure defined by $C(K, j, p) = j^{-1}Kp$, and controls for FDR among statistically independent hypothesis tests (some relaxation of this condition is possible, as discussed in [3]). Additionally, control for FDR under general conditions can be obtained by using a step-up procedure with $C(K, j, p) = j^{-1}Kp \sum_{i=1}^{K} i^{-1}$. See, for example [6, 7, 19] for comprehensive discussions of the topic with an emphasis on genomic applications.

In MTPs it is important to be able to estimate the small *P*-values, rather than rely on fixed levels of significance. Since we are also not interested in estimating the large *P*-values, the hybrid approach of Sect. 2.4 will be explored. In using $\tilde{p}$ as defined in Eq. (2.4), it will be useful to consider the stopped process as being *embedded* in a fixed sample procedure, in that the former is observable from the latter. If we were to evaluate a MTP using fixed sample replications, as is the usual practice, we would use estimate $\hat{p}_i$ of significance level α_{obs}^i for each test $i = 1, \ldots, K$ then substitute these estimates into the step-down or step-up procedure described in (4.5), yielding approximate adjusted *P*-values $\hat{p}_i^a$. Apart from estimation error, a global error rate can then be reported under a wide variety of conditions.

Now suppose we substitute the stopped estimates $\tilde{p}_i$ into (4.5) to obtain approximate adjusted *P*-values $\tilde{p}_i^a$. The two MTP procedures can be compared, and if we can verify a strict ordering $\tilde{p}_i^a \geq \hat{p}_i^a$, then the reportable error rates of the *fixed* MTP will also apply to the *stopped* MTP, eliminating the need for a separate analysis. We do this in the following lemma.

Lemma 4.1 *If we are given a stopped MTP embedded in a fixed MTP, then $\tilde{p}_i^a \geq \hat{p}_i^a$, $i = 1, \ldots, K$ for either a step-down or a step-up procedure in which $C(K, j, p)$ is nondecreasing in p.*

Proof Directly from (2.4) we may conclude

$$\tilde{p}_i \leq \hat{p}_i, \quad i = 1, \ldots, K$$

$$\hat{p}_i \leq \hat{p}_j \text{ implies } \tilde{p}_i \leq \tilde{p}_j, \; \forall i, j. \tag{4.5}$$

Let $v_1, \ldots, v_K$ be any ordering of $\hat{p}_1, \ldots, \hat{p}_K$. From (4.5) we conclude that $v_1, \ldots, v_K$ is also an ordering of $\tilde{p}_1, \ldots, \tilde{p}_K$. Hence, for a step-down procedure

$$\hat{p}_{v_i}^a = \max_{j \leq i} \min(C(K, j, \hat{p}_{v_j}), 1) \leq \max_{j \leq i} \min(C(K, j, \tilde{p}_{v_j}), 1) = \tilde{p}_{v_i}^a,$$

using (4.5), and a similar argument verifies the lemma for the step-up procedure.

□

By Lemma 4.1 the stopped procedure truncated at M is more conservative than the fixed MTP in which it is embedded, no matter which stopping time is used, therefore the same error rate may be reported. The remaining issue is the selection of T_M^i for the ith test which will equal M for small enough values of p_i, but will also have $E[T_M^i] \ll M$ for larger values of p_i. A simple way to achieve this is to use a *one-sided* SPRT (without lower bound), that is, $SPRT(p_0, p_1, a, b, M)$ with $a = -\infty$. In the next section we consider the problem of selecting the remain parameters in the SPRT design.

5 Optimal Design of Stopping Times Based on SPRTs

We now consider the problem of designing stopping times for the use in SBPs. Formally, we optimize among parameters for $SPRT(p_0, p_1, a, b, M)$. The objective will be to reduce the expected stopping time while maintaining a given error rate. The number of parameters in the design can be reduced with the following two lemmas.

Lemma 5.1 *Equations* (4.3) *define a one-to-one mapping from* $\mathcal{X} = \{(p_0, p_1) : 0 < p_0 < p_1 < 1\}$ *to* $\mathcal{Y} = \{(\delta^-, \delta^+) \in (0, \infty) \times (-\infty, 0)\}$.

Proof Consider the equivalent transformation

$$\delta_0^- = q_1/q_0$$
$$\delta_0^+ = p_1/p_0. \tag{5.1}$$

Let $\mathcal{Y}_0 = (0, 1) \times (1, \infty)$. It is easily verified that (5.1) maps $\mathcal{X}$ into $\mathcal{Y}_0$. For any fixed $(\delta_0^-, \delta_0^+) \in \mathcal{Y}_0$ a unique solution for $(p_0, p_1) \in \mathcal{X}$ is given by

$$p_0 = \frac{1 - \delta_0^-}{\delta_0^+ - \delta_0^-}$$

$$p_1 = \frac{\delta_0^+ - \delta_0^- \delta_0^+}{\delta_0^+ - \delta_0^-}. \tag{5.2}$$

This suffices to prove the lemma, letting $\delta^- = \log \delta_0^-, \delta^+ = \log \delta_0^+$. □

By Lemma 5.1 we may replace parameters p_0, p_1 with δ_0^-, δ_0^+. The following lemma is easily verified, and is given without proof.

Lemma 5.2 *For any design* $SPRT(p_0, p_1, a, b, M)$ *for an SBP with any parameter* θ, *the class of designs obtained by the set of transformations* $(\delta^-, \delta^+, a, b) \rightarrow (t\delta^-, t\delta^+, ta, tb)$ $\forall t \in (0, \infty)$ *forms an equivalence class with respect to the distribution of* $(T_M, t^{-1}R_{T_M})$.

Since interest in R_{T_M} will generally be in its relationship to constants a, b, Lemma 5.2 describes a true equivalence class. Thus, without loss of generality we can reduce design $SPRT(p_0, p_1, a, b, M)$ by one parameter. We will do this by setting $\delta^+ = \delta$, constraining $\delta^+ - \delta^- = 2$, and denote the resulting design $SPRT'(\delta, a, b, M)$.

5.1 Constrained Optimization

We now consider the problem of devising criteria for the selection of optimal design $SPRT'$. Unconstrained minimization of $E_\theta[T_M]$ is trivially achieved with $a = b = 0$, so this criterion must be carefully considered.

We will consider the one-sided stopping rule, setting $a = -\infty$, so we denote the design $SPRT''(\delta, b, M) = SPRT'(\delta, -\infty, b, M)$. The objective is to identify and estimate small values of θ, so we wish to select a design for which $T_M = M$ with high probability for small enough θ, and for which $E_\theta[T_M]$ is small for larger θ. Letting $D'' = (\delta, b, M)$ denote the design parameters, define *power function* and *expected stopping time function*

$$\phi(\theta \mid D'') = P_\theta(T_M = M),$$

$$\mu(\theta \mid D'') = E_\theta[T_M], \quad \theta \in (0, 1),$$

where D'' defines T_M. Let constraint $C(\theta_c, \beta)$ denote the set of designs

$$C(\theta_c, \beta) = \{D'' : \phi(\theta_c \mid D'') \geq \beta\}.$$

It will be useful to consider a hierarchical model in which θ is sampled from a two component mixture, one of which (representing the primary null hypotheses) is a uniform distribution, the other a distribution with mode at or near zero. Interest is in minimizing in some aggregate sense $\mu(\theta \mid D'')$ for tests drawn from the null component. The mean stopping time for this component for a given design is

$$\lambda(D'') = \int_0^1 \mu(\theta \mid D'')d\theta$$

so we specify the optimization problem with instance (θ_c, β):

P1(θ_c, β): Find design D'' with minimizes $\lambda(D'')$ under constraint $C(\theta_c, \beta)$.

5.2   Solution Method

Problem $P1(\theta_c, \beta)$ can be converted to an optimization problem over a single variable. First note that $\phi(\theta \mid D'')$, $\mu(\theta \mid D'')$ and $\lambda(D'')$ all increase with b. For each fixed $\delta \in (0, 2)$ we define

$$b(\delta \mid \theta_c, \beta, M) = \min\{b : D'' = (\delta, b, M) \text{ satisfies } C(\theta_c, \beta, M)\}. \quad (5.3)$$

Then minimize

$$\lambda^*(\delta \mid \theta_c, \beta, M) = \lambda(\delta, b(\delta \mid \theta_c, \beta, M), M)$$

over $\delta \in (0, 2)$. We may use the Robbins–Monro algorithm (RMA) (first proposed in [14]) to estimate $b(\delta \mid \theta_c, \beta, M)$ on a grid $\delta \in G^g$, then perform a grid search to minimize $\lambda^*(\delta \mid \theta_c, \beta, M)$.

We now briefly describe the RMA. We wish to find the solution t_0 to $g(t) = 0$, for an increasing real valued function $g(t)$. Suppose for any t we can simulate a noisy evaluation of $g(t)$:

$$G_t = g(t) + \epsilon_t, \quad E[\epsilon_t] = 0.$$

Let $a_n \to 0$ be a sequence of positive constants. The RMA is:

$$Y_{n+1} = Y_n - a_n Z_n, \quad n \geq 1, \quad (5.4)$$

where $Z_n \sim G_{Y_n}$, and the simulation is independent of process history conditional on Y_n. Then $Y_n \to t_0$ if

$$\sum_{n \geq 1} a_n = \infty \text{ and } \sum_{n \geq 1} a_n^2 < \infty. \quad (5.5)$$

Convergence was originally verified in the quadratic mean in [14], with strong convergence verified in [10].

Algorithm (5.4) can be directly applied to the calculation of $b(\delta \mid \theta_c, \beta, M)$. Fix $\delta \in G^g$, and set

$$g(b) = \phi(\theta_c \mid \delta, b, M) - \beta,$$
$$G_b = I\{T_M^* = M\} - \beta, \quad (5.6)$$

where T_M^* is a stopping time from a simulated observation of $SPRT''(\delta, b, M)$. Note that $g(b)$ may not be continuous because of the discrete structure of the process, so the algorithm may not be exact, although any overshoot error can be examined directly. Then if (5.5) holds, algorithm (5.4) converges to $b(\delta \mid \theta, \beta, M)$. Once this quantity is estimated, define grid Θ^g on $[0, 1]$. For each $\theta' \in \Theta^g$ estimate

$$\mu(\theta' \mid \delta, b(\delta \mid \theta_c, \beta, M), M)$$

by averaging stopping times T_M^* from repeated simulations of design

$$SPRT''(\delta, b(\delta \mid \theta_c, \beta, M), M)$$

for parameter θ'. This permits an estimate of $\lambda^*(\delta \mid \theta_c, \beta, M)$ by numerical integration on Θ^g.

5.3 Numerical Example

We present a numerical example, setting $M = 5000$ and using constraint $C(0.05, 0.95)$. Algorithm (5.4) was applied to the calculation of $b(\delta \mid \theta_c, \beta, M)$, for values of δ on grid $G^g = (0.02, 0.04, \ldots, 1.98)$. Figure 4 shows the resulting curve, shown in two vertical scales. There is a sharp change in slope around the point $\delta = 1.72$.

Given our estimate of $b(\delta \mid 0.05, 0.95, 5000)$, $\lambda^*(\delta \mid \theta_c, \beta, M)$ was then estimated as described in Sect. 5.2 using a grid $\Theta^g = (0.01, 0.02, \ldots, 0.99)$. The resulting estimated curve is shown in Fig. 5 (Plot 1). The minimum of this curve is approximately 527.8, and is located around $\delta = 1.72$, the same point as the feature referred to in Fig. 4. Finally, the expected stopping time as a function of θ is given for the optimal value $\delta = 1.72$ and two neighboring points in Fig. 5 (Plot 2), indicating a clear superiority of the optimal choice.

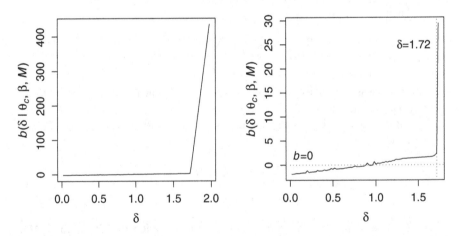

Fig. 4 Plot 1: Estimate of $b(\delta \mid \theta_c, \beta, M)$ using RMA, with $\theta_c = 0.05$, $\beta = 0.95$, $M = 5000$. **Plot 2:** Same as Plot 1, but with different vertical scales. A sharp change in slope in indicated at approximately $\delta = 1.72$

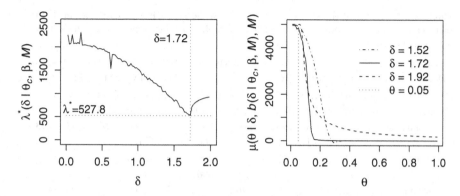

Fig. 5 Plot 1: Estimate of $\lambda^*(\delta \mid \theta_c, \beta, M)$ using RMA, with $\theta_c = 0.05$, $\beta = 0.95$, $M = 5000$. **Plot 2:** Expected stopping times $\mu(\theta \mid \delta, b(\delta \mid \theta_c, \beta, M), M)$ for selected values of δ, including the optimal value $\delta = 1.72$ derived from Plot 1

6 Examples

We will describe two applications of SPRT methodology to Monte Carlo testing problems.

6.1 Gene Set Analysis

In [2] the stopped MTP procedure was applied to the problem of *gene set analysis*, in which high dimensional arrays of gene expression data are screened for *differential expression* (DE) by comparing gene sets defined by known functional relationships. This follows the paradigm originally proposed in *gene set enrichment analysis* (GSEA) [13, 16]. Here, this takes the form of a series of two-sample multivariate tests, one for each gene set from some list of interest. FDR is reported using the Benjamini–Hochberg procedure. The test is based on a likelihood function for a Bayesian network, and significance levels are evaluated using permutation replications. Evaluation of the likelihood, although simplified, is still computationally expensive, providing a natural application for the stopped MTP.

The design $SPRT''(0.336, 4.6, 5000)$ was used and applied to eight separate gene set analyses using data and gene set lists described in [16]. Results are given in Table 2. The stopped MTP requires considerably less computation time, while giving nearly identical results to the fixed MTP. One exception, with respect to computation time, is the *Leukemia* data set. In this case nearly all null hypotheses are rejected, so the stopping time was $T_M = M$ in most cases.

Table 2 For stopped (St) and fixed (Fx) procedures the table gives computation times; mean number of replications; % gene sets completely sampled; number of pathways with P-values $\leq$ 0.01; number of such pathways in agreement

Gene set			Time (h)		Mean rep.		% Comp.		P-value ≤ 0.01		
Data	List	N	St	Fx	St	Fx	St	Fx	St	Fx	Both
Diabetes	C1	308	1.2	50.7	324.4	5000	4.5	100	2	2	2
Gender	C1	308	1.0	54.3	237.9	5000	2.9	100	5	5	5
Leukemia	C1	307	50.8	54.6	4890.0	5000	97.7	100	293	293	293
p53	C1	307	3.1	51.2	519.9	5000	8.5	100	4	3	3
Diabetes	C2	522	3.7	35.8	341.0	5000	5.4	100	6	6	6
Gender	C2	522	0.7	38.4	155.8	5000	2.3	100	3	2	2
Leukemia	C2	522	29.7	29.7	4876.2	5000	97.3	100	493	493	492
p53	C2	522	2.1	30.0	612.3	5000	10.5	100	18	19	18

6.2 Confidence Sets in Statistical Genetics

This type of method was also used in an application described in [1] involving population inference based on genetic marker frequencies. Reference data exists for populations $1, \ldots, K$. An additional sample is collected from one of these (unknown) populations, the objective being to identify the population. It will not generally be possible to identify the correct population with small enough error, so a level $1 - \alpha$ confidence set, in the form of a subset of $\{1, \ldots, K\}$, can be constructed by including all populations which do not reject a suitable two-sample test for homogeneity at a fixed significance level of α. Thus, an estimate of α_{obs} is not strictly needed, and a two sided sequential procedure would be appropriate. As in the gene set analysis application of Sect. 6.1 a considerable savings of computation time is reported, with no apparent loss of accuracy.

7 Conclusion

The *sequential probability ratio test (SPRT)* may be adapted to Monte Carlo testing applications, resulting in considerable savings in computation time with little sacrifice in accuracy or power. In particular, it proves to be straightforward to incorporate these techniques into multiple testing procedures in such a way as to preserve the reportable global error rate. Although a number of applications exist, this type of method has not been widely used in genomic applications. A considerable savings in computation time may therefore result from further adoption of these methods.

Acknowledgement This work was supported by NIH grant R21HG004648.

References

1. Almudevar, A. (2000). Exact confidence regions for species assignment based on DNA markers. *The Canadian Journal of Statistics, 28*, 81–95.
2. Almudevar, A. (2010). A hypothesis test for equality of Bayesian network models. *EURASIP Journal on Bioinformatics and Systems Biology, 2010*, 10.
3. Benjamini, Y., & Yekutieli, D. (2001). The control of the false discovery rate in multiple testing under dependency. *The Annals of Statistics, 29*, 1165–1188.
4. Besag, J., & Clifford, P. (1989). Generalized Monte Carlo significance tests. *Biometrika, 76*, 633–642.
5. Besag, J., & Clifford, P. (1991). Sequential Monte Carlo p-values. *Biometrika, 78*, 301–304.
6. Dudoit, S., Shaffer, J. P., & Boldrick, J. C. (2003). Multiple hypothesis testing in microarray experiments. *Statistical Science, 18*, 71–103.
7. Dudoit, S., & van der Laan, M. J. (2008). *Multiple testing procedures with applications to genomics.* New York: Springer.
8. Fay, M. P., & Follmann, D. A. (2002). Designing Monte Carlo implementations of permutation or bootstrap hypothesis tests. *The American Statistician, 56*, 63–70.
9. Hall, W. J. (1983). Some sequential tests for matched pairs: A sequential permutation test. In P. K. Sen (ed.), *Contributions to statistics: essays in honour of Norman L. Johnson,* (pp. 211–228). Amsterdam: North-Holland.
10. Ljung, L. (2007). Strong convergence of a stochastic approximation algorithm. *The American Statistician, 6*, 680–696.
11. Lock, R. H. (1991). A sequential approximation to a permutation test. *Communications in Statistics: Simulation and Computation, 20*, 341–363.
12. Medland, S., Schmitt, J., Webb, B., Kuo, P.-H., & Neale, M. (2009). Efficient calculation of empirical P-values for genome-wide linkage analysis through weighted permutation. *Behavior Genetics, 39*, 91–100.
13. Mootha, V. K., Lindgren, C. M., Eriksson, K. F., Subramanian, A., Sihag, S., Lehar, J., et al. (2003). PGC-1 α-responsive genes involved in oxidative phosphorylation are coordinately downregulated in human diabetes. *Nature Genetics, 100*, 605–610.
14. Robbins, H., & Monro, S. (1951). A stochastic approximation method. *The Annals of Mathematical Statistics, 22*, 400–407.
15. Siegmund, D. (1985). *Sequential analysis: tests and confidence intervals.* New York: Springer-Verlag.
16. Subramanian, A., Tamayo, P., Mootha, V. K., Mukherjee, S., Ebert, B. L., Gillette, M. A., et al. (2005). Gene set enrichment analysis: A knowledge-based approach for interpreting genome-wide expression profiles. *Proceedings of the National Academy of Sciences of the United States of America, 102*, 15545–15550.
17. Wald, A. (1947). *Sequential analysis.* New York: John Wiley and Sons.
18. Wald, A. (1948). Optimum character of the sequential probability ratio test. *The Annals of Mathematical Statistics, 19*, 326–339.
19. Yang, H. Y., & Speed, T. (2003). Design and analysis of comparative microarray experiments. In T. Speed (ed.) *Statistical analysis of gene expression microarray data* (pp. 35–92). Boca Raton, FL: Chapman and Hall.

Multistage Carcinogenesis: A Unified Framework for Cancer Data Analysis

Suresh Moolgavkar and Georg Luebeck

Abstract Traditional approaches to the analysis of epidemiologic data are focused on estimation of the relative risk and are based on the proportional hazards model. Proportionality of hazards in epidemiologic data is a strong assumption that is often violated but seldom checked. Risk often depends on detailed patterns of exposure to environmental agents, but detailed exposure histories are difficult to incorporate in the traditional approaches to analyses of epidemiologic data. For epidemiologic data on cancer, an alternative approach to analysis can be based on ideas of multistage carcinogenesis. The process of carcinogenesis is characterized by mutation accumulation and clonal expansion of partially altered cells on the pathway to cancer. Although this paradigm is now firmly established, most epidemiologic studies of cancer incorporate ideas of multistage carcinogenesis neither in their design nor in their analyses. In this paper we will briefly discuss stochastic multistage models of carcinogenesis and the construction of the appropriate likelihoods for analyses of epidemiologic data using these models. Statistical analyses based on multistage models can quite explicitly incorporate detailed exposure histories in the construction of the likelihood. We will give examples to show that using ideas of multistage carcinogenesis can help reconcile seemingly contradictory findings, and yield insights into epidemiologic studies of cancer that would be difficult or impossible to get from conventional methods. Finally, multistage cancer models provide a unified framework for analyses of data from diverse sources.

Keywords Multistage carcinogenesis · Stochastic models · Clonal expansion model · Proportional hazards model · Colon cancer · Screening · Folate supplementation

S. Moolgavkar (✉)
Fred Hutchinson Cancer Research Center, Seattle, WA, USA

Exponent, Inc., Bellevue, WA, USA

G. Luebeck
Fred Hutchinson Cancer Research Center, Seattle, WA, USA
e-mail: gluebeck@fredhutch.org

© Springer Nature Switzerland AG 2020
A. Almudevar et al. (eds.), *Statistical Modeling for Biological Systems*,
https://doi.org/10.1007/978-3-030-34675-1_7

117

1   Introduction

Originally proposed by Cox [10] for the analysis of data from clinical trials, the proportional hazards model was soon adopted by epidemiologists and today provides the conceptual framework for analyses of both cohort and case-control studies. In epidemiologic studies, age is often used as the principal time axis. The main assumption of the original proportional hazards model was that the hazard ratio remained constant with time. With age as the time axis, this assumption translates into the assumption of no effect modification by age. Although the assumption of constant hazard ratios has been relaxed and methods have been devised to test this assumption in epidemiologic studies, the proportional hazards model continues to be used in its original form for analyses of epidemiologic studies with little regard for whether constancy of hazard ratios is valid in any particular data set. Just how poor this assumption can be is seen in Fig. 1, which is a plot of a set of rate ratios from the first *Cancer Prevention Study* (CPS I) conducted by the American Cancer Society (ACS). This example shows also that tests designed to detect monotonic departures from the constant hazards ratio assumption are unlikely to detect the pattern seen in Fig. 1.

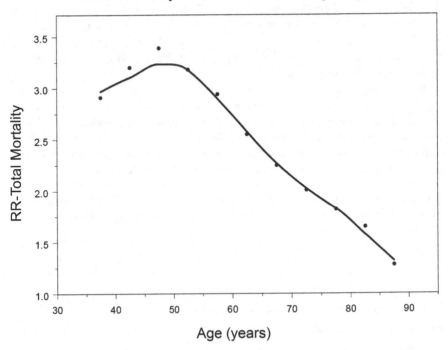

Fig. 1 Rate ratios for all-cause mortality among smokers with a two-pack a day habit as a function of age. Rate ratios are from the Cancer Prevention Study I (CPS I) cohort [8]. The line is a smoother through the observed rate ratios

For the analysis of data in cancer epidemiology multistage models provide an alternative to using the traditional statistical methods based on the proportional hazards model. Multistage models for carcinogenesis were introduced in the 1950s to explain the regular behavior of the age-specific mortality curves of many adult carcinomas. These models showed that the age-specific mortality curves for many adult carcinomas are consistent with carcinogenesis being the end result of a process of mutation accumulation. Recently, [72, 74] have also shown that cancer incidence rates in tissues are correlated with the number of stem cell divisions in those tissues. Because most mutations occur during cell division, this observation supports the idea that accumulation of randomly occurring mutations at critical gene loci is important in carcinogenesis. The best known of the multistage models is the Armitage–Doll model [1], which continues to be invoked to this day as a framework for understanding spontaneous carcinogenesis and the temporal evolution of risk with carcinogenic exposures of varying intensity (e.g., [7, 13, 65, 77]).

The two-stage clonal expansion (2SCE) model, also known as the Moolgavkar–Venzon–Knudson (MVK) model is based on three influential ideas in cancer biology [50]. In chronological order, these are (1) the concept of initiation-promotion-progression [2], which arose out of experiments in chemical carcinogenesis in the late 1940s; (2) the observation of [1] that the age-specific mortality curves of many adult carcinomas could be explained by a multistage model; and (3) Knudson's two-hit hypothesis [24, 33] for embryonal tumors, such as retinoblastoma. The 2SCE model and its generalizations allow for clonal expansion of intermediate cells on the pathway to cancer via a linear birth–death process. For a recent review of the biology as it relates to multistage models of carcinogenesis, we refer the reader to the paper by [36]. Some authors have suggested that only a few critical "driver" mutations (approximately 3 to 4) are required for carcinogenesis [32, 46, 73, 76].

In this paper, we would like to propose that the idea of multistage carcinogenesis and the multistage models derived directly from this idea provide an alternative to the traditional statistical approaches based on the proportional hazards model for analyses of data in cancer epidemiology. It is our thesis that methods of analyses based on ideas of multistage carcinogenesis can provide mechanistic insights into the data that would be difficult or impossible to obtain from traditional approaches alone. Such analyses also greatly facilitate the incorporation of temporal factors, such as age at which exposure starts and stops and, in fact, the entire exposure history, into the analyses. Such temporal factors are difficult to consider explicitly in the traditional statistical approaches. Moreover, multistage models provide a unified framework for analyses of data of different types. Thus, not only can hazard functions derived from the model be used for analyses of epidemiologic data on cancer, but also expressions for the number and size distribution of intermediate lesions on the pathway to cancer can be used for analyses of data on cancer screening and for design of optimal screening and prevention strategies.

As always, when fitting biologically based mathematical models to data, balance needs to be maintained between biological realism and simplicity for parameter identifiability and estimation. In this paper we will briefly discuss the 2SCE model and its extension, the multistage clonal expansion (MSCE) model. Parameter identifiability for these models is well understood [6, 22, 37] and rigorous statistical techniques have been developed for fitting these models to both epidemiologic (see below) and experimental data [38, 39, 42, 53, 54].

2   Brief Review of Mathematical Issues

For the mathematical development of the 2SCE model, let $X(t)$ represent the (non-random) number of susceptible cells in the tissue of interest at age t. Then we assume that the number of initiated cells arising from normal susceptible cells is a non-homogeneous Poisson process with intensity function $v(t)X(t)$, where v can be interpreted as the rate of the first "mutation." Suppose that initiated cells divide, die, and become malignant, via a linear birth-death-mutation process (with birth rate α, death rate β, and second mutation rate μ). Let $Y(t)$ and $Z(t)$ be random variables that represent, respectively, the number of initiated and malignant cells at time (age) t. Let $\Psi(y, z; t)$ be the probability generating function for the number of initiated and malignant cells at time t. Then the hazard function $h(t) = -\Psi'(1, 0; t)/\Psi(1, 0; t)$. Suppose now that $\Phi(y, z; s, t)$ is the probability generating function for the number of initiated and malignant cells at time t starting with a single initiated cell at time s, i.e., $\Phi(y, z; s, s) = 1$. Then, the process of malignant transformation is a filtered Poisson process [51, 57] and

$$\Psi(y, z; t) = \exp\left(\int_0^t vX(s)\,[\Phi(y, z; s, t) - 1]\,ds\right)$$

and thus $h(t) = -\int_0^t vX(s)[\Phi_t(1, 0; s, t)]ds$, where Φ_t is the derivative of Φ with respect to t.

Now, for any t_1 such that $0 < t_1 < t$, we have

$$h(t) = -\int_0^{t_1} vX(s)[\Phi_t(1, 0; s, t)]ds - \int_{t_1}^t vX(s)[\Phi_t(1, 0; s, t)]ds. \qquad (2.1)$$

It can be easily shown [57] that the first term of this expression $\to 0$ as $t \to \infty$ and therefore the asymptotic behavior of the hazard function depends only on the second term. This implies, in particular, that if exposure to an environmental agent modifies some or all of the parameters of the model (thus affecting the hazard function) and if these parameters revert to background levels[1] after exposure to the agent stops,

---

[1]Parameters would be expected to return to background levels with exposure to agents, such as benzene, that are rapidly cleared from the body. Other agents, such as amphibole asbestos, accumulate in tissues, and after exposure to such agents stops, the parameters of the model could

then the hazard function must approach the background hazard (i.e., the hazard in those not exposed to the environmental agent) asymptotically.

Although the 2SCE model has 4 biological parameters $(\nu, \alpha, \beta, \mu)$, [21] and [18] showed for the case of constant parameters that only 3 parameters could be identified from the hazard function, i.e., the hazard was a function of only 3 independent parameters. This result was later extended to piecewise constant parameters (the parameters need to be time-dependent to model time-dependent exposures) by Heidenreich et al. [22]. Heidenreich et al. also provided a simple recursive algorithm for expressing hazard functions in terms of time-dependent parameters. This algorithm was based on solving the forward Kolmogorov equations associated with the 2SCE model. It turns out, however, that the computations are much simpler if one uses the backward equations [12, 35, 36]. We now routinely use the algorithm provided by Crump et al. [12]. Little [35, 36] has shown that the method can be used for fairly general MSCE models.

Epidemiologic studies have shown that the hazard function for lung cancer among ex-smokers asymptotically approaches the hazard function among never-smokers. A similar pattern is seen in the hazard function for leukemia among those occupationally exposed to benzene. After exposure stops, the hazard function is back to the hazard function among the unexposed in a matter of 15–20 years [65, 75]. A second consequence of the expression (2.1) for the hazard function is that the memory of a carcinogenic exposure (to an agent that is cleared from tissues) during a specific period is gradually lost with time even if exposure continues. Thus, for example, analyses using exposure-time windows have shown that exposure to benzene more than 15–20 years earlier does not increase the risk of leukemia. Some epidemiologists explain this behavior of the hazard function by invoking repair of damage done to tissues. It must be remembered, however, that the data do not address individuals, but populations: the hazard function after exposure stops asymptotically approaches the hazard function among the unexposed. Repair of specific mutations is a highly improbable event. The observed behavior of the hazard function is a mathematical consequence of multistage carcinogenesis that does not need repair to be invoked.

The formulas given above can be easily generalized for a large class of multistage models [28, 35, 36, 38, 45, 46, 52]. The properties of the hazard function of the 2SCE model briefly discussed above hold for these more general models as well. We note here, however, that these properties hold only for the exact hazard functions derived from these models. In the past, easily computed approximations to the hazard function have been used [1, 7, 13, 49, 77] These properties do not extend to the approximate solutions. Thus, the use of these approximate solutions may yield misleading results.

---

remain altered for a long period of time and would be expected to return to background levels only slowly as the agent is excreted from tissues.

3 Construction of Likelihoods

Registry Data Multistage models have been used extensively for analyses of registry data [28, 40, 41, 45, 55]. The use of these models to replace the non-specific age effects in the conventional age-period-cohort (APC) models gets around the "arbitrary linear trends" identifiability problem of the conventional APC models (e.g., [25]). Cancer registries typically ascertain incident cases in defined geographic areas cross-tabulated by age and calendar year (i.e., period). For example, the Surveillance, Epidemiology, and End Results 9 (SEER9) registry provides cancer incidence data in 9 representative areas of the USA by 5-yr age groups (ages 0–4, 5–9,..., 80–84, 85+), by calendar year (in single years from 1973 onward), separately by sex and race [68]. In addition, population sizes (i.e., number of individuals at risk of developing cancer) for each stratum are available through the US census. Because the risk of cancer at any given time (period) is typically small, it is assumed that the number of incident cases in each cell of the contingency table follows a Poisson distribution with mean $\Lambda_{a,j} = PY_{a,j} \times h_{ij}(a)$, where $PY_{a,j}$ represents the person years at risk for individuals of age a in calendar year j, and $h_{ij}(a)$ is the model hazard function that incorporates the effect of age, birth cohort, and calendar year.

One basic approach to incorporating such cohort and period trends in the data analysis is via the AGE-PERIOD-COHORT (APC) model [26], which assumes that the hazard function at age a in birth cohort i and period j is of the form

$$h_{ij}(a) = b_i c_j \, h_{MSCE}(a), \tag{3.1}$$

where $h_{MSCE}(a)$ is the hazard function derived from a multistage model, b_i a coefficient that adjusts for birth cohort i ($i = j - a$), and the coefficient c_j adjusts for calendar year j. Because age, period, and cohort are perfectly correlated, at least one coefficient each for cohort and period must be anchored (fixed) to enforce identifiability of all other coefficients. This model isolates the impact of age on the incidence of cancer from the temporal trends described by birth cohort and period effects.

The overall likelihood, $\mathcal{L}$, for the observed incidence in an age–calendar year group is then given by

$$\mathcal{L} = \prod_{a,j} \frac{\Lambda_{a,j}^{o_{a,j}} e^{\Lambda_{a,j}}}{o_{a,j}!}, \tag{3.2}$$

where $o_{a,j}$ is the number of cases in the age group with midpoint a during calendar year j, and $\Lambda_{a,j} = PY_{a,j} \times h_{ij}(a)$ is the expected number of cases in the stratum.

The "Anatomy" of the Incidence Curve in the MSCE Model (See Fig. 2) A straightforward extension of the 2SCE model is to assume that initiation of premalignant lesions requires two rate-limiting mutational events. This extension

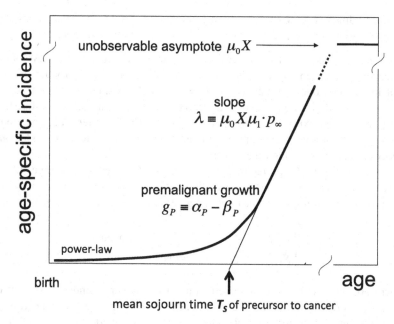

Fig. 2 Characteristic phases of incidence curves for the 3-stage model. See text for details

was first introduced and applied to colorectal cancer by [41]. According to this model, initiation of a premalignant clone occurs when a normal tissue stem cell acquires two rate-limiting mutational events with rates μ_0 and μ_1, respectively. Premalignant lesions are modeled to undergo a stochastic clonal expansion with cell division rate α_P and cell death or differentiation rate β_P. As shown in [46], the hazard function of this model has 4 basic phases, only two of which are identifiable in practice: an exponential phase reflecting the contribution of premalignant growth to the cancer process, followed by a linear phase that reflects the initiation process. The slope parameter of this linear phase, λ, is given by the product of the initiating mutations μ_0, μ_1, and the number of normal tissue stem cells, X, multiplied by the probability of clonal non-extinction, p_∞. Importantly, the intercept of the linear approximation with the time axis (indicated in Fig. 2 by an arrow) can be shown to correspond to the overall mean sojourn time T_S of a premalignant clone to cancer, conditioned on non-extinction of the clone. The asymptotic phase for this model, as shown in [46], is essentially unobservable because it occurs well beyond the life span of humans but for completeness is also shown in Fig. 2.

Mean sojourn times of premalignant clones to cancer have been estimated for colorectal and pancreatic cancer from the SEER incidence data [40]. Estimates range from 50 to 60 years, significantly longer than clinical sojourn times for most detectable precursor lesions to cancer. However, these longer sojourn times can be expected when time is measured from the (unobservable) occurrence of the premalignant founder cell to cancer. Independent efforts to estimate sojourn times have been undertaken for colorectal cancer by [4, 31] using cancer genome sequence data. The findings of these studies, although highly uncertain, are broadly consistent with long sojourn times in excess of 3 decades from initiation to cancer.

Cohort Data The likelihood is easily constructed for cohort data. Assuming independence between individuals, the cohort likelihood is the product of individual likelihoods over all subjects j, $L = \Pi L_j$. Individual likelihoods $L_j = L_j(s_j, t_j, (\cdots))$ depend on time of entry into the study s_j, censoring or failure time t_j, and on detailed exposure histories in conjunction with general dose–response models for the biological parameters in the two-stage or extended model, and possibly on a lag time or lag time distribution.

Let $P(t)$ represent the probability of occurrence of cancer by time (age) t, with survival $S(t) = 1 - P(t)$, and density $P'(t)$. The individual likelihoods for cases and survivors, including left truncation, are given by

$$L_j(t_j, s_j) = \begin{cases} P'(t_j)/S(s_j) & \text{if diagnosed with cancer,} \\ S(t_j)/S(s_j) & \text{otherwise.} \end{cases} \tag{3.3}$$

Furthermore, let $h_m(u)$ represent the individual two-stage or extended model hazard and $S_m(u)$ represent the two-stage or extended model survival at some time u. For a fixed lag time from first malignant cell to cancer incidence, t_{lag}, individual likelihoods in Eq. (3.3) are calculated using $S(t_j) = S_m(t_j - t_{\text{lag}})$; $S(s_j) = S_m(s_j - t_{\text{lag}})$; $P'(t_j) = h_m(t_j - t_{\text{lag}})S_m(t_j - t_{\text{lag}})$. For a lag time distribution, the density $P'(t_j)$ in Eq. (3.3) is given by the convolution of the 2SCE or extended model density, $P'_m(u) = h_m(u)S_m(u)$ with a lag time distribution $f(t_j - u)$,

$$P'(t_j) = \int_0^{t_j} h_m(u)S_m(u)f(t_j - u)du. \tag{3.4}$$

The survival $S(t_j)$ is calculated by convolving the two-stage or extended model probability, $(1 - S_m(u))$, with the lag time distribution up to the time of censoring,

$$S(t_j) - 1 - \int_0^{t_j} (1 - S_m(u))f(t_j - u)du. \tag{3.5}$$

Left truncation requires calculation of the survival, $S(s_j)$ at entrance into the study.

A gamma distribution, $f(X, a, b) = (b^a \Lambda(a))^{-1} x^{a-1} \exp(-x/b)$, with mean $\mu_{\text{lag}} = ab$, variance $\sigma_{\text{lag}}^2 = ab^2$, and lag time argument $x = t_j - u$ has been used for the lag time distribution.

The 2SCE model has been used for analyses of cohort data, including occupational cohort data, on lung cancer using likelihood based techniques [19, 45]. Figure 3 shows one of the results from an analysis of the CPS I data [19]. The figure shows the rate ratios for lung cancer among smokers who smoked 40 or more cigarettes per day along with the risk predicted by the fit of the 2SCE model. The pattern of risk seen in this figure is similar to that in Fig. 1. Unlike the Cox model,

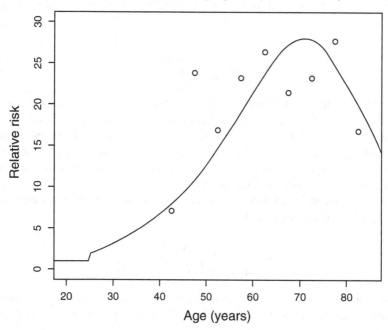

Fig. 3 Rate ratios for lung cancer among smokers with a two-pack a day smoking history in the CPS I cohort [8]. The line represents the predicted relative risks from a fit of the 2SCE model to the data [19]

the relative risk is not the target of estimation with the 2SCE model. Rather, the estimated biological parameters are used to derive the relative risks for any specific patterns of exposure to cigarette smoke. Nonetheless, the 2SCE model describes the age-dependent pattern of relative risks very well as can be seen in Fig. 3.

Case-Control Studies Multistage models have also been used for analyses of case-control data. Likelihood construction is described in [47] and [23]. In comparison to the likelihoods required for the analysis of cohort studies, where each case, survivor, or censored individual makes a simple contribution to the overall likelihood, the likelihoods for case-control studies are typically more involved [10, 63]. Cases and controls can be viewed as being drawn from a single cohort which is followed forward in time. As individuals fail, age-matched controls are chosen randomly from among those in the risk set at that time. This view of a matched case-control study as "nested" within a cohort study leads to the following likelihood contribution from a matched set. Suppose that there are M matched sets, $m = 1, \ldots, M$, with N_i controls associated with case i. Let us denote the hazard function of the ith individual by $h(t_i | x_i)$, where t_i is the age at the time of failure, and x_i a vector of covariates one component of which may represent the dose (exposure rate) the individual has been exposed to. Then the conditional likelihood contribution made by matched set i is given by

$$\mathcal{L}_i = \frac{h(t_i|x_o)}{\sum_{j=0}^{N_i} h(t_i|x_j)}. \tag{3.6}$$

The sum in the denominator is taken over the case and all controls in the risk set i. Note that in this formulation the hazard function for the controls is computed at time t_i which is the time (age) of failure of the case. Obviously, in case-control studies such precise matching is impossible, as some controls are younger and other controls are older than the case. This led to modifications of the likelihood which are discussed in the statistical literature [10].

The extension of this likelihood to a stratified set with m cases and n controls is given by Prentice and Breslow [62]

$$\mathcal{L}_s = \frac{\prod_{i=1}^m h(t_i, x_i)}{\sum_{l \in R(m,n)} \left(\prod_{i=1}^m h(t_{l_i}, x_{l_i}) \right)} \tag{3.7}$$

where $R(m, n)$ is the set of all subsets of size m from $\{1, \ldots, m + n\}$, and the subscript s denotes the stratum in question. Thus, the total likelihood is given by $\mathcal{L} = \prod_s \mathcal{L}_s$.

Two points are worth noting: (1) If the strata are large, then the likelihood given above may present a computational problem because a large number of terms ($\binom{m+n}{m}$ of them) need to be added up in the denominator of $\mathcal{L}_s$. However, a computational scheme suggested by Howard [27] and simple approximations suggested by Breslow [5], Efron and Morris [17], Peto [60], can be used to solve the computational problem. (2) Stratification is frequently carried out when it is difficult to incorporate exposures information such as age at first exposure, duration, time since last exposure, and exposure rates (as opposed to total exposures). For the MSCE model, however, such covariate information can be incorporated in a straightforward manner by means of time-dependent functions that reflect the direct impact of the covariates on the biological parameters. For example, we may assume that the rate of net cell proliferation (cell birth rate minus cell death rate) is a function of the exposure rate at time t, or a function of cumulative exposure up to time t. The 2SCE-based analysis of the Chinese Tin Miners Cohort (CTM) cohort is a case in point. Although first analyzed as a cohort [20], we subsequently analyzed it as a case-control study nested within the CTM cohort and were able to demonstrate consistency with the results obtained by analyzing the full cohort. Incorporation of the various patterns of exposure to radon, cigarette (pipe) smoke, and arsenic was easily achieved within the 2SCE model.

4 Number and Size Distribution of Intermediate (Premalignant) Lesions

Because the 2SCE and MSCE models incorporate the growth of intermediate lesions via a linear birth–death process, these models can also be used to analyze data on intermediate lesions on the pathway to cancer. For example, the rodent liver is a commonly used experimental system in chemical carcinogenesis. Data are available on the number and size distribution of enzyme-altered hepatic foci, which are considered precursor lesions in rodent liver. Another example are adenomatous polyps which are considered to be premalignant lesions in colon. In order to analyze data on the number and size distribution of intermediate lesions on the pathway to malignancy, appropriate mathematical expressions have first been developed for the 2SCE model [15] and applied for analyses of enzyme-altered foci in hepatocarcinogenesis experiments [38, 42, 53]. More recently, analogous (but more complex) expressions have been developed to analyze colorectal polyps within the context of the MSCE model [14] which was first developed to fit the age-specific incidence of colon cancer [40, 41] observed in the SEER registry. Without going into the mathematical details, which can be found in the original papers, we describe two applications of the MSCE model to human colon cancer.

5 Model for Colon Cancer

Luebeck and Moolgavkar [41] developed a four-stage model for colon cancer based on colon cancer incidence data in the SEER registry. The incidence of colon cancer in SEER was found to be most consistent with a model that posits two rare events followed by a high-frequency event in the conversion of a normal stem cell into an initiated cell, which expands clonally to give rise to an adenomatous polyp. Only one more rare event was found necessary for an initiated cell to transform into a fully malignant cell. We interpret the two rare events involved in initiation as representing homozygous loss of function of the *adenomatous polyposis coli* (APC) gene. The high-frequency event could be interpreted as a positional event, i.e., an event that moves the pre-initiated cell out of the stem cell niche into a position higher up the crypt where it is no longer under normal growth constraints. The time to the last rare event represents the waiting time distribution for an initiated cell to develop into a clinically recognizable malignant tumor. In subsequent analyses by Luebeck et al. [40], Meza et al. [46], it was shown that the mean time from the initiation of a premalignant cell to malignant transformation and cancer (conditional on non-extinction of the premalignant clone) could be directly estimated from incidence curves. For colorectal adenomas this time was found to be >50 years, roughly two to three decades longer than estimates of clinical adenoma-to-carcinoma sojourn times (see the discussion on the anatomy of incidence curves above).

5.1 Applications of the Model

5.1.1 Screening for Colon Cancer

Using a combination of analytic and simulation techniques, [30] derived expressions for the number and size distribution of adenomatous polyps based on the model for colon cancer [41] described above. These expressions were used to devise optimal screening strategies for a single screen and two distinct screens using colonoscopy, which is considered the gold standard for colon cancer screening. Assumptions had to be made regarding the smallest polyp (in terms of number of cells) that is detectable with this technique. Once a polyp was detected, it was assumed to be removed with specific efficiency, i.e., a certain fraction of cells were assumed removed by polypectomy performed through the scope. The objective was to develop screening strategies, with a single screen and two sequential screens, that would minimize the lifetime (defined as 80 years) probability of colon cancer. The results of the analyses are shown in Fig. 4 and in Table 1. The results are insensitive to a fairly wide range of assumed values for the size threshold of detectable polyps and efficiency of polyp removal.

Figure 4 shows the cancer probabilities by age 80 for a population that undergoes no screening (7% probability of colon cancer by age 80), a single screen, and two screens. For the two screens case, shown in the curve with vertical bars, for each first screening age, we optimize the second screening age so as to minimize the probability of colon cancer by age 80. The curve with vertical bars shows these optimized probabilities. The age of the screen at which the probability of cancer is minimized is shown in Table 1. If only a single screen is to be performed, then it should be performed at about age 55. A single screen at age 55 reduces the probability of colon cancer by age 80 from about 7% to about 1.5%. A two-screen strategy provides significant further benefit only if the first screen is performed between the ages of 30 and 50 and the second screening age is chosen as shown in Table 1. Such a screening strategy reduces the risk to about 0.65%. The star in Fig. 4 shows the probability of colon cancer if the current colon cancer screening guidelines are adopted. These guidelines suggest three screens 10 years apart starting at age 50. The colon cancer probability by age 80 is virtually identical to the optimal two-screen protocol shown in Fig. 4 and Table 1. Therefore, the current three-screen strategy offers little benefit over the optimal two-screen strategy described here.

5.1.2 Impact of Folate Fortification on Colon Cancer Risk

Folate is essential for the synthesis of nucleotides, which are the building blocks for DNA. Low folate status has been associated with increased risk of cancer (reviewed in [43]) suggesting a chemopreventive role for folate. Folate fortification of grain products was made mandatory in 1998 in the USA as a highly effective way to reduce neural tube defects in new born infants. However, not all studies of folate supplementation point to beneficial effects. Experimental studies show that folate

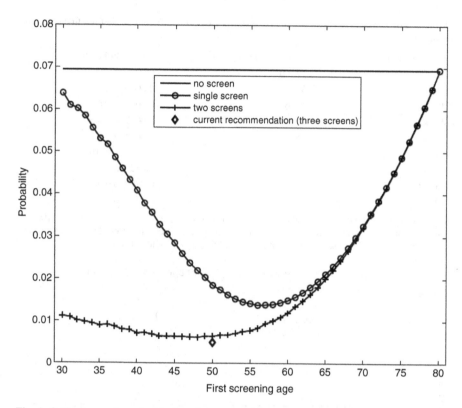

Fig. 4 Optimal screening schedule for colon cancer with one and two screens. The solid horizontal line shows the probability of colon cancer by age 80 in the unscreened population. The line with open circles shows the same probability for a single screen as a function of screening age. The line with vertical bars shows the probability for two screens. This is the optimized (minimum) probability for the first screen at the age shown on the x-axis and the second screen at a later age. For various ages at the first screen, Table 1 shows the age at second screen that minimizes the probability of colon cancer [30]. Current recommendation is 3 screens at ages 50, 60, and 70

supplementation before establishment of premalignant lesions may have a protective effect, whereas folate administration after precancerous lesions have developed may actually increase cancer risk (reviewed in [43]). Likewise, some epidemiologic evidence suggests that high levels of folate may increase cancer risk. In a controlled polyp prevention trial [9] administration of folic acid was followed by an increase in the number of adenomatous polyps. Mason et al. [44] also showed that colorectal cancer incidence in both the USA and Canada increased shortly after the initiation of fortification.

Thus, the role of folate in colon cancer remains uncertain. Because of its role in nucleotide synthesis, folate appears to decrease mutation rates. At the same time, the evidence briefly discussed above suggests that folate could be a promoter, i.e., it could increase the rate of clonal expansion of precancerous lesions, such as adenomatous polyps. Luebeck et al. [43] undertook a study of the potential impact of folate supplementation on colon cancer risk using the colon cancer model described

Table 1 For a two-screen protocol, the table shows the optimal screening age for a first screen in a specific 5 year interval

First screening age	Second screening age	Cancer probability
30–34	57–60	0.0095–0.0112
35–39	59–62	0.0080–0.0092
40–44	61–65	0.0065–0.0073
45–49	**63–65**	**0.0061–0.0065**
50–54	65–67	0.0065–0.0075
55–59	68–71	0.0078–0.0110
60–64	70–73	0.0120–0.0182
65–69	72–74	0.0202–0.0297
70–74	76–77	0.0327–0.462

The last column shows the probability of clinical colon cancer by age 80 under that specific screening protocol. The optimal screening strategy is a first screen between the ages of 45 and 49, followed by a second screen between the ages of 63 and 65 (bolded row)

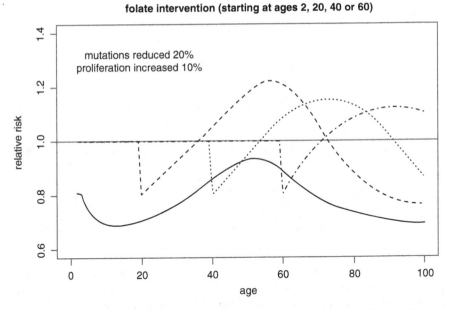

folate intervention (starting at ages 2, 20, 40 or 60)

mutations reduced 20%
proliferation increased 10%

Fig. 5 Relative risks of colon cancer for folate supplementation at various ages. These predictions are based on the four-stage model [41] and are based on the assumption that folate reduces mutation rates by 20% but increases rates of clonal expansion by 10% [43]

above. The basic assumption was that folate decreases mutation rates, but increases clonal expansion of the initiated cells. Because of this complex mode of action, the risk of colon cancer following folate administration depends in a complicated way on age at which supplementation was started. The patterns of risk for various ages at start of supplementation are shown in Fig. 5. The lesson from this figure is that unless supplementation is started early in life, folate supplementation is detrimental, leading to an increase in colon cancer risk. Although this is a hypothetical example with a number of assumptions, it provides an explanation as to why epidemiologic studies of the topic could yield inconsistent, even contradictory, results.

6 Discussion

Multistage models for carcinogenesis provide a unified framework for analyses of data from multiple sources in cancer epidemiology. As the example of colon cancer demonstrates, parameters of the model derived from age-specific incidence data from population-based cancer registries can be used to optimize screening schedules for premalignant lesions, such as colorectal polyps. Of course, the fact that optimal screening schedules can be derived from a theoretical model does not necessarily mean that these schedules will turn out to be optimal in the real world. However, we note here that the optimal screening schedule we derive is not too far from the schedule recommended by the American Cancer Society on empirical grounds. The next step in model development is to test the predictions of the model against actual data from screening trials. For example, in a recent study by Jeon et al. [29] colorectal screening data from a large flexible sigmoidoscopy (FSG) study in the Kaiser Permanente Health System were evaluated using the colon cancer multistage model described here. Importantly, carefully modeling these data made it possible to evaluate counterfactual screening modalities (e.g., full length colonoscopy vs FSG) at different ages, screening sensitivities, and different curative efficiencies for the removal of polyps (polypectomy).

The model provides a framework for understanding complex temporal patterns of risk following exposure to agents with multiple modes of action, some of which might decrease the risk of cancer and others of which might increase it. This is precisely the situation with folate supplementation. Thus, the model helps explain why epidemiologic studies of the impact of folate on colon cancer risk have yielded inconsistent results. The model shows also that epidemiologic studies of this issue would be difficult to design.

The cancer risk associated with exposure to environmental carcinogens, such as asbestos, benzene, and radiation, or to lifestyle carcinogens, such as cigarette smoking, depends upon the entire history of exposure to the carcinogen, including age at start of exposure and the time-varying intensity of exposure [16, 19, 46, 59, 64–66]. Although the importance of temporal aspects of exposure and risk has been emphasized for some time (e.g., [58, 69–71]), the vast majority of epidemiologic studies of occupational cohorts use cumulative exposure as the measure of exposure (e.g., [3, 11, 56, 67]) even when detailed exposure information is available for each member of the cohort. This is because the explicit history of exposure is difficult to incorporate in the traditional statistical models used for analyses of epidemiologic data. A recent simulation study of the proportional hazards model [48] discusses the fundamental difficulties associated with using patterns of exposure in the proportional hazards model and shows that use of cumulative exposure instead can yield misleading results. In contrast, time-dependent parameters associated with time-varying intensities of exposure can be easily accommodated in statistical analyses based on multistage models. Heidenreich et al. [22] described an algorithm for construction of the hazard function with piecewise constant parameters based on solution of the forward Kolmogorov equation. Little [35] and Crump et al. [12] provided solutions based on the backward equation, which actually leads to an easy algorithm for constructing hazard functions with continuous parameters. Thus very general time-varying histories can be explicitly considered in analyses using multistage models.

Finally, largely as a result of the focus on case-control studies, undue emphasis has been placed on the development of relative risk regression models (e.g., [34, 61]). Even when cohort data are available, epidemiologic studies have focused on estimation of relative risk when the more appropriate targets of estimation are the hazard functions for varying levels of exposure. Once the hazard functions are estimated, various measures of risk, such as the relative risk and excess risk can be easily estimated. Methods of analyses based on multistage models provide one approach to estimating hazard functions for any general time-varying exposure history.

References

1. Armitage, P., & Doll, R. (1954). The age distribution of cancer and a multi-stage theory of carcinogenesis. *British Journal of Cancer, 8*, 1–12.
2. Berenblum, I., & Shubik, P. (1947). A new, quantitative approach to the study of the stages of chemical carcinogenesis in the mouse's skin. *British Journal of Cancer, 1*, 383–391.
3. Berman, D. W., & Crump, K. S. (2008). Update of potency factors for asbestos-related lung cancer and mesothelioma. *Critical Reviews in Toxicology, 38*(Suppl 1), 1–47.
4. Bozic, I., Antal, T., Ohtsuki, H., Carter, H., Kim, D., Chen, S., et al. (2010). Accumulation of driver and passenger mutations during tumor progression. *Proceedings of the National Academy of Sciences, 107*(43), 18545–18550.
5. Breslow, N. (1974). Covariance analysis of censored survival data. *Biometrics, 30*(1), 89–99.
6. Brouwer, A. F., Meza, R., & Eisenberg, M. C. (2017). Parameter estimation for multistage clonal expansion models from cancer incidence data: A practical identifiability analysis. *PLoS Computational Biology, 13*(3), e1005431–18.
7. Brown, C. C., & Chu, K. C. (1983). Implications of the multistage theory of carcinogenesis applied to occupational arsenic exposure. *Journal of the National Cancer Institute, 70*, 455–463.
8. Burns, D. M., Shanks, T. G., Choi, W., Thun, M. J., Heath, C. W., & Garfinkel, L. (1997). The American Cancer Society Cancer Prevention Study I: 12-year followup of 1 million men and women. In D. M. Burns, L. Garfinkel, & J. M. Samet (Eds.) *Smoking and tobacco control, monograph 8* (pp. 113–304). NIH Publ. No 97-4213.
9. Cole, B. F., Baron, J. A., Sandler, R. S., Haile, R. W., Ahnen, D. J., Bresalier, R. S., et al. (2007). Folic acid for prevention of colorectal adenomas. A randomized clinical trial. *JAMA : The Journal of the American Medical Association, 297*, 2351–2359.
10. Cox, D. R. (1972). Regression models and life tables (with discussion). *Journal of the Royal Statistical Society, Series B: Statistical Methodology, 34*, 187–220.

11. Crump, K. S., Chen, C., Fox, J. F., Van Landingham, C., & Subramaniam, R. (2008). Sensitivity analysis of biologically motivated model for formaldehyde-induced respiratory cancer in humans. *The Annals of Occupational Hygiene, 52*, 481–95.

12. Crump, K. S., Subramaniam, R. P., & Landigham, C. B. (2005). A numerical solution to the nonhomogeneous two-stage MVK model of cancer. *Risk Analysis, 25*, 921–926.

13. Day, N. E., & Brown, C. C. (1980). Multistage models and the primary prevention of cancer. *Journal of the National Cancer Institute, 64*, 977–989.

14. Dewanji, A., Jeon, J., Meza, R., & Luebeck, E. G. (2011). Number and size distribution of colorectal adenomas under the multistage clonal expansion model of cancer. *PLoS Computational Biology, 7*(10), e1002213.

15. Dewanji, A., Venzon, D. J., & Moolgavkar, S. H. (1989). A stochastic two-stage model for cancer risk assessment II: The number and size of premalignant clones. *Risk Analysis, 9*, 179–186.

16. Doll, R., & Peto, R. (1978). Cigarette smoking and bronchial carcinoma: Dose and time relationships among regular smokers and life-long non-smokers. *Journal of Epidemiology and Community Health, 32*, 303–313.

17. Efron, B., & Morris, C. (1977). Comment on A Simulation Study of Alternative to Least Squares, by H. Clark and T. Schwisow. *The American Statistician, 72*, 102–109.

18. Hanin, L. G., & Yakovlev, A. Y. (1996). A nonidentifiability aspect of the two-stage model of carcinogenesis. *Risk Analysis, 16*, 711–715.

19. Hazelton, W. D., Clements, M. S., & Moolgavkar, S. H. (2005). Multistage carcinogenesis and lung cancer mortality in three cohorts. *Cancer Epidemiology, Biomarkers & Prevention, 14*, 1171–1181.

20. Hazelton, W. D., Luebeck, E. G., Heidenreich, W. F., & Moolgavkar, S. H., (2001). Analysis of a historical cohort of Chinese tin miners with arsenic, radon, cigarette smoke, and pipe smoke exposures using the biologically based two-stage clonal expansion model. *Radiation Research, 156*(1), 78–94.

21. Heidenreich, W. (1996). On the parameters of the clonal expansion model. *Radiation and Environmental Biophysics, 35*, 127–129.

22. Heidenreich, W., Luebeck, E. G., & Moolgavkar, S. H. (1997). Some properties of the hazard function of the two-mutation clonal expansion model. *Risk Analysis, 17*, 391–399.

23. Heidenreich, W., Wellmann, J., Jacob, P., & Wichmann, H. E. (2002). Mechanistic modelling in large case-control studies of lung cancer risk from smoking. *Statistics in Medicine, 21*, 3055–3070.

24. Hethcote, H. W., & Knudson, A. G. (1978). Model for the incidence of embryonal cancer: application to retinoblastoma. *Proceedings of the National Academy of Sciences of the United States of America, 75*, 2453–2457.

25. Holford, T. R. (1991). Understanding the effects of age, period and cohort on incidence and mortality rates. *Annual Review of Public Health, 12*, 425–457.

26. Holford, T. R., Zhang, Z., McKay, L. A. (1994). Estimating age, period and cohort effects using the multistage model for cancer. *Statistics in Medicine, 13*, 23–41.

27. Howard, S. (1972). Contribution to the discussion of a paper by DR Cox: Regression models and life-tables. *Journal of the Royal Statistical Society. Series B, 34*, 210–211.

28. Jeon, J., Luebeck, E. G., & Moolgavkar, S. H. (2006). Age effects and temporal trends in adenocarcinoma of esophagus and gastric cardia. *Cancer Causes Control, 17*, 971–981.

29. Jeon, J., Meza, R., Hazelton, W. D., Renehan, A. G., & Luebeck, E. G. (2015). Incremental benefits of screening colonoscopy over sigmoidoscopy in average-risk populations: a model-driven analysis. *Cancer Causes & Control, 26*(6), 859–870.

30. Jeon, J., Meza, R., Moolgavkar, S. H., & Luebeck, E. G. (2008). The evaluation of cancer screening strategies using multistage carcinogenesis models. *Mathematical Biosciences, 213*, 56–70.

31. Jones, S., Chen, W., Parmigiani, G., Diehl, F., Beerenwinkel, N., Antal, T., et al. (2008). Comparative lesion sequencing provides insights into tumor evolution. *Proceedings of the National Academy of Sciences, 105*(11), 4283–4288.

32. Knudson, A. G. (2001). Two genetic hits (more or less) to cancer. *Nature Reviews Cancer,* *1*(2), 157.
33. Knudson, A. G., Hethcote, H. W., & Brown, B. W. (1975). Mutation and childhood cancer: A probabilistic model for the incidence of retinoblastoma. *Proceedings of the National Academy of Sciences USA, 72,* 5116–5120.
34. Langholz, B. (2010). Case-control studies = odds ratios. Blame the retrospective model. *Epidemiology, 21,* 10–12.
35. Little, M. P. (1995). Are two mutations sufficient to cause cancer? Some generalizations of the two mutation model of carcinogenesis of Moolgavkar, Venzon and Knudson, and of the multistage model of Armitage and Doll. *Biometrics, 51,* 1278–1291.
36. Little, M. P. (2010). Cancer models, genomic instability and somatic cellular Darwinian evolution. *Biology Direct, 5,* 19.
37. Little, M. P., Heidenreich, W. F., & Li, G. (2009). Parameter identifiability and redundancy in a general class of stochastic carcinogenesis models. *PLoS One, 4*(12), e8520.
38. Luebeck, E. G., Buchmann, A., Stinchcombe, S., Moolgavkar, S. H., & Schwarz, M. (2000). Effects of 2,3,7,8-tetrachlorodibenzo-p-dioxin (TCDD) on initiation and promotion of GSTP-positive foci in rat liver: A quantitative analysis of experimental data using a stochastic model. *Toxicology and Applied Pharmacology, 167,* 63–73.
39. Luebeck, E. G., Curtis, S. B., Cross, F. T., & Moolgavkar, S. H. (1996). Two-stage model of radon-induced malignant lung tumors in rats: effects of cell killing. *Radiation Research, 145*(2), 163–173.
40. Luebeck, E. G., Curtius, K., Jeon, J., & Hazelton, W. D. (2013). Impact of tumor progression on cancer incidence curves. *Cancer Research, 73*(3), 2198.
41. Luebeck, E. G., & Moolgavkar, S. H. (2002). Multistage carcinogenesis and the incidence of colorectal cancer. *Proceedings of the National Academy of Sciences of the United States of America, 99,* 15095–15100.
42. Luebeck, E. G., Moolgavkar, S. H., Buchmann, A., & Schwarz, M. (1991). Effects of poly-chlorinated biphenyls in rat liver: Quantitative analysis of enzyme-altered foci. *Toxicology and Applied Pharmacology, 111*(3), 469–484.
43. Luebeck, E. G., Moolgavkar, S. H., Liu, A., & Ulrich, N. (2008). Does folic acid supple-mentation prevent or promote colon cancer? Results from model-based predictions. *Cancer Epidemiology, Biomarkers & Prevention, 17,* 1360–1367.
44. Mason, J. B., Dickstein, A., Jacques, P. F., Haggarty, P., Selhub, J., Dallal, G. et al. (2007). A temporal association between folic acid fortification and an increase in colorectal cancer rates may be illuminating important biological principles: a hypothesis. *Cancer Epidemiology, Biomarkers & Prevention, 16,* 1325–1329.
45. Meza, R., Hazelton, W. D., Colditz, G. A., & Moolgavkar, S. H. (2008a). Analysis of lung cancer incidence in the nurses' health and the health professionals' follow-up studies using a multistage carcinogenesis model. *Cancer Causes Control, 19,* 317–328.
46. Meza, R., Jeon, J., Moolgavkar, S. H., & Luebeck, E. G. (2008b). The age-specific incidence of cancer: phases, transitions and biological implications. *Proceedings of the National Academy of Sciences of the United States of America, 105,* 16284–16289.
47. Moolgavkar, S. H. (1995). When and how to combine results from multiple epidemiological studies in risk assessment. In J. Graham (ed.) *The proper role of epidemiology in regulatory risk assessment* (pp. 77–90). New York: Elsevier.
48. Moolgavkar, S. H., Chang, E. T., Watson, H. N., & Lau, E. C. (2018). An assessment of the cox proportional hazards regression model for epidemiologic studies. *Risk Analysis, 38*(4), 777–794.
49. Moolgavkar, S. H., Day, N. E., & Stevens, R. G. (1980). Two-stage model for carcinogenesis: Epidemiology of breast cancer in females. *Journal of the National Cancer Institute, 65,* 559–569.
50. Moolgavkar, S. H., Krewski, D., & Schwarz, M. (1999). Mechanisms of carcinogenesis and biologically-based models for quantitative estimation and prediction of cancer risk. In S. H. Moolgavkar, D. Krewski, L. Zeise, E. Cardis, & H. Moller (Eds.) *Quantitative estimation and prediction of cancer risk* (pp. 179–238). Lyon: IARC Scientific Publications.

51. Moolgavkar, S. H., & Luebeck, G. (1990). Two-event model for carcinogenesis: Biological, mathematical and statistical considerations. *Risk Analysis, 10*, 323–341.
52. Moolgavkar, S. H., & Luebeck, E. G. (1992). Multistage carcinogenesis: A population-based model for colon cancer. *Journal of the National Cancer Institute, 84*, 610–618.
53. Moolgavkar, S. H., Luebeck, E. G., de Gunst, M., Port, R. E., & Schwarz, M. (1990). Quantitative analysis of enzyme altered foci in rat hepatocarcinogenesis experiments. *Carcinogenesis, 11*, 1271–1278.
54. Moolgavkar, S. H., Luebeck, E. G., Turim, J., & Brown, R. C. (2000). Lung cancer risk associated with exposure to man-made fibers. *Drug and Chemical Toxicology, 23*(1), 223–242.
55. Moolgavkar, S. H., Meza, R., & Turim, J. (2009). Pleural and peritoneal mesotheliomas in SEER: Age effects and temporal trends, 1973–2005. *Cancer Causes & Control, 20*(6), 935–944.
56. Moolgavkar, S. H., Turim, J., Alexander, D. D., Lau, E. C., & Cushing, C. A. (2010). Potency factors for risk assessment at Libby, Montana. *Risk Analysis, 30*, 1240–1248.
57. Moolgavkar, S. H., & Venzon, D. J. (1979). Two-event model for carcinogenesis: Incidence curves for childhood and adult tumors. *Mathematical Biosciences, 47*, 55–77.
58. Peto, J. (2012). That the effects of smoking should be measured in pack-years: Misconceptions. *British Journal of Cancer, 107*(3), 406–407.
59. Peto, J., Seidman, H., & Selikoff, I. J. (1982). Mesothelioma mortality in asbestos workers: Implications for models of carcinogenesis and risk assessment. *British Journal of Cancer, 45*, 124–135.
60. Peto, R. (1972). Contribution to discussion paper by D. R. Cox: Regression models and life-tables. *Journal of the Royal Statistical Society, Serial B, 34*, 205–207.
61. Poole, C. (2010). On the origin of risk relativism. *Epidemiology, 21*, 3–9.
62. Prentice, R. L., & Breslow N. E. (1978). Retrospective studies and failure time models. *Biometrika, 65*(1), 153–158.
63. Prentice, R. L., & Kalbfleisch, J. D. (1979). Hazard rate models with covariates. *Biometrics, 35*(1), 25–39.
64. Rachet, B., Siemiatycki, J., Abrahamowicz, M., & Leffondre, K. (2004). A flexible modeling approach to estimating the component effects of smoking behavior on lung cancer. *Journal of Clinical Epidemiology, 57*, 1076–1085.
65. Richardson, D. B. (2008). Temporal variation in the association between benzene and leukemia mortality. *Environmental Health Perspectives, 116*, 370–374.
66. Richardson, D. B. (2009). Multistage modeling of leukemia in benzene workers: a simple approach to fitting the 2-stage clonal expansion model. *American Journal of Epidemiology, 169*, 78–85.
67. Sullivan, P. A. (2007). Vermiculite, respiratory disease, and asbestos exposure in Libby, Montana. Update of a cohort mortality study. *Environmental Health Perspectives, 115*, 579–585.
68. Surveillance, Epidemiology, and End Results (SEER) Program Research Data. (1973–2015). National Cancer Institute, DCCPS, Surveillance Research Program, released April 2018, based on the November 2017 submission. www.seer.cancer.gov.
69. Thomas, D. C. (1983). Statistical methods for analyzing effects of temporal patterns of exposure on cancer risks. *Scandinavian Journal of Work, Environment & Health, 9*, 353–366.
70. Thomas, D. C. (1988). Models for exposure-time-response relationships with applications to cancer epidemiology. *Annual Review of Public Health, 9*, 451–482.
71. Thomas, D. C. (2014). Invited commentary: Is it time to retire the pack-years variable? Maybe not! *American Journal of Epidemiology, 179*(3), 299–302.
72. Tomasetti, C., Li, L., & Vogelstein, B. (2017). Stem cell divisions, somatic mutations, cancer etiology, and cancer prevention. *Science, 355*(6331), 1330–1334.
73. Tomasetti, C., Marchionni, L., Nowak, M. A., Parmigiani, G., & Vogelstein, B. (2015). Only three driver gene mutations are required for the development of lung and colorectal cancers. *Proceedings of the National Academy of Sciences, 112*(1), 118–123.

74. Tomasetti, C., & Vogelstein, B. (2015). Variation in cancer risk among tissues can be explained by the number of stem cell divisions. *Science, 347*(6217), 78–81.

75. Triebig, G. (2010). Implications of latency period between benzene exposure and development of leukemia – A synopsis of literature. *Chemico-Biological Interactions, 184*, 26–29.

76. Vogelstein, B., & Kinzler, K. W. (2015). The path to cancer – Three strikes and you're out. *The New England Journal of Medicine, 373*(20), 1895–1898.

77. Whittemore, A. S. (1977). The age distribution of human cancer for carcinogenic exposures of varying intensity. *American Journal of Epidemiology, 109*, 709–718.

A Machine-Learning Algorithm for Estimating and Ranking the Impact of Environmental Risk Factors in Exploratory Epidemiological Studies

Jessica G. Young, Alan E. Hubbard, Brenda Eskenazi, and Nicholas P. Jewell

Abstract Epidemiological research, such as the identification of disease risks attributable to environmental chemical exposures, is often hampered by small population effects, large measurement error, and limited a priori knowledge regarding the complex relationships between the many chemicals under study. However, even an ideal study design does not preclude the possibility of reported false positive exposure effects due to inappropriate statistical methodology. Three issues often overlooked include (1) definition of a meaningful measure of association; (2) use of model estimation strategies (such as machine-learning) that acknowledge that the true data-generating model is unknown; (3) accounting for multiple testing. In this paper, we propose an algorithm designed to address each of these limitations in turn by combining recent advances in the causal inference and multiple-testing literature along with modifications to traditional nonparametric inference methods.

Keywords Machine-learning · Epidemiology · Multiple testing · Causal inference

J. G. Young
Department of Population Medicine, Harvard Medical School & Harvard Pilgrim Health Care Institute, Harvard University, Boston, MA, USA
e-mail: jyoung@hsph.harvard.edu

A. E. Hubbard (✉)
Division of Biostatistics, University of California at Berkeley, Berkeley, CA, USA
e-mail: hubbard@berkeley.edu

B. Eskenazi
Division of Environmental Health Sciences, School of Public Health, University of California at Berkeley, Berkeley, CA, USA
e-mail: eskenazi@berkeley.edu

N. P. Jewell
Division of Biostatistics, University of California at Berkeley, Berkeley, CA, USA

London School of Hygiene & Tropical Medicine, London, UK
e-mail: jewell@berkeley.edu; nicholas.jewell@lshtm.ac.uk

© Springer Nature Switzerland AG 2020
A. Almudevar et al. (eds.), *Statistical Modeling for Biological Systems*,
https://doi.org/10.1007/978-3-030-34675-1_8

137

1   Introduction

Modern epidemiology faces many challenges in identifying real exposure effects on disease risks. This is particularly true in studies of environmental chemical exposures where population effects are likely small, measurement error is large, and a priori knowledge regarding the complex relationships between the many chemicals under study limited. Clearly, successful identification of real effects must start with proper study design, including limitation of measurement error and collection of sufficient data on confounders. However, even an ideal study design does not preclude the possibility of false positive exposure effects reported in the literature due to the approach used to analyze the data. Specifically, we consider three potential pitfalls of omission with "typical" approaches to data analysis:

1. defining a meaningful measure of association (the explicit parameter of interest, not connected to a specific parametric model);
2. using model estimation strategies (such as machine-learning) that acknowledge that the true data-generating model is unknown;
3. accounting for multiple testing.

In this paper, we propose an algorithm designed to address each of these limitations in turn by combining recent advances in the causal inference and multiple-testing literature along with modifications to traditional nonparametric inference methods. Though screening algorithms are becoming more commonly used for high-dimensional data common in bioinformatic applications, as well as a recent application to data similar to that discussed in this paper [9], we believe this is one of the first systematic attempts at an algorithmic, data-adaptive approach that can be used to screen high-dimensional data, and provide a ranking for variables (risk factors) based on the semi-parametric statistical evidence of having a public health impact by removing/reducing the exposure. This algorithm does so by employing robust estimation techniques involving machine-learning, finding relatively robust marginal inference using conditional permutation methods, and finally sharp global inference using efficient multiple-testing methods, which capitalize on the obvious dependence of the association estimates. Specifically, this algorithm does the following:

1. estimates a population intervention model, for which we use a previously introduced parameter from the causal inference literature [6] naturally suited to environmental epidemiologic questions and population-based study designs;
2. provides marginal inference for this effect estimate based on a modified version of the conditional permutation test [12] to account for the presence of high-dimensional covariates via the propensity score;
3. provides joint inference for multiple effect estimates using the *quantile transformation method* [20].

Though this algorithm, like any in this context, can be improved, we think it is a significant advance over approaches like [9] that while worthy, depend heavily on parametric assumptions, particularly when simple bivariate comparisons do not address the scientific questions of interest. Specifically, we define the parameter of interest independent of any specific parametric model for the data-generating distribution. Because this type of approach does not bind itself to a particular model, the procedure can "learn" from the data, while still providing reasonable inference for candidate explanatory variables. Two downsides to more parametric approaches are (1) if the model is not data-adaptive, then the model will be misspecified to some degree, and thus one needs an interpretation of the parameters of interest (usually coefficients) in the context of a misspecified model (see, for instance, [21]). If the model is data-adaptive, while still using model coefficients as reported parameters of interest, then these do not necessarily have an invariant interpretation in repeated experiments (that is, the algorithm itself decides the parameter of interest, an awkward scenario for inference). Though under identifiability assumptions covered below, our parameter estimates can be interpreted as so-called "causal" parameters, our emphasis is more on the fact that such parameters based solely on the data-generating distribution, are (1) interesting, (2) do not depend on the definition of the models fit to the data-generating distribution, and (3) can harness available machine-learning software to aggressively look for patterns.

We describe an application of this algorithm to data collected from the Center for the Health Assessment of Mothers and Children of Salinas (CHAMACOS) project [4]. This constitutes the first reported application of such estimation and inferential methods to an environmental epidemiological data set. CHAMACOS is a longitudinal birth cohort study aimed at assessing the effects of pesticides and other environmental exposures on health outcomes in pregnant women and their children. The CHAMACOS data set contains information on birth outcomes among Latina mothers, neurobehavioral outcomes on their infants with hundreds of covariates and over one hundred exposure measurements representing at least 40 different chemicals. CHAMACOS is one of the first comprehensive studies of the impacts of chronic low-level pesticide exposure on human health, particularly in children.

In Sect. 2 we provide more background on the problems associated with "typical" data analysis as enumerated above, and briefly describe how components of the proposed machine-learning algorithm address these issues. In Sect. 3 we define the data structure and parameters of interest. In Sect. 4 we discuss the CHAMACOS data set in more detail. In Sect. 5 we describe estimation and inference for parameters of interest according to the components of the algorithm. In Sect. 6 we illustrate an application of this algorithm to the CHAMACOS data. Further remarks are provided in Sect. 7.

2   Background

1. Failure to Define a Meaningful Measure of Effect As a simple schematic example, consider a study of the effect of some baseline level of a chemical exposure A on a continuously measured disease outcome Y. Further assume that a high-dimensional covariate vector W is observed containing potential confounders of the exposure effect. A typical approach to data analysis would likely involve regressing

Y on a function of A and W. The likely reported exposure effect in this case would be the estimated coefficient on A in the postulated regression model. Assuming the regression model is correct, this coefficient represents the association between A and Y, conditional on the covariates in W. Under additional assumptions, including that W is sufficient to control confounding by unmeasured factors, then this coefficient represents the causal effect of A on Y conditional on the covariates W.

While this measure of effect has the advantage of being easily computed, it is often not very meaningful unless the typically arbitrarily chosen model is correct. The purpose of including W in our linear regression model is to adjust for confounding, and the functional manner in which these variables are included in the model are typically chosen based on convenience (e.g., all main effect terms). There is almost never any compelling background information on the functional form of this regression, and so what we would really like to do is choose a model of $h(E(Y \mid A, W))$ of the general form $m(A, W)$ from a very large space, surely to include higher order terms, splines, multiplicative interactions, etc., where $h(\cdot)$ is a suitable link function, typically assumed known. However, these models will not return directly a single coefficient that summarizes the adjusted association of interest of A and Y unless the model by sheer good luck includes only a linear term for A and no interactions of A and W. Thus, a more generally meaningful parameter of interest that can be derived from $m(A, W)$ is likely the marginal effect of A on Y, or, possibly this effect conditional on only a small subset of W, $V \in W$. For example, one might be interested in estimating the effect of exposure on the outcome separately by gender.

If A is higher dimensional, then one might want to assume the marginal means for each level of $A = a$ are connected by some model: so-called marginal structural models (MSMs). These models, introduced by Robins [11], provide an alternative to traditional regression approaches as described above. In particular, the parameters of MSMs, which model the distributions of counterfactual or potential outcomes under hypothetical exposure interventions, represent marginal (or conditional) effects directly of interest. The MSM parameters can be estimated using estimating equation-based methodology, including inverse probability weighting (IPW) or doubly robust extensions (DR-IPW) [24] and, more recently, by doubly robust targeted maximum likelihood methods [25]. The first component of the proposed algorithm discussed here consists of estimation of a relatively new class of models based on estimating equation-based methodology for MSMs; that is, population intervention models [6].

2. Failure to Use Model Estimation Strategies (such as Machine-Learning) that Acknowledge the Model is Unknown In both traditional regression approaches and estimation of MSMs, we are required to make modeling assumptions about the aspects of the data-generating distribution beyond assumptions about the model for the parameter of interest (e.g., the MSM). In traditional regression approaches, we must make assumptions regarding the distribution of Y given A and W. In estimation of MSMs, we must make assumptions regarding either the distribution (the mean) of Y given A and W and/or the distribution of A given W; i.e., the exposure mechanism. Data-adaptive approaches are a natural choice used to estimate the regressions that provide the estimates of these distributions. In the case of traditional regression approaches, machine-learning approaches are rarely used (often due to the inconvenient form of the resulting regressions for easy interpretation), and more rarely do analysts account for the contribution of the modeling uncertainty and the model selection process on the variability of parameter estimates. That is, when ad hoc informal methods of model selection are used, the most common approach is simply to ignore the impact this has on the variability of the parameter estimates in repeated experiments. Parametric assumptions (e.g., regression with only main effect terms) are commonly used with standard procedures invoked for statistical inference. Unfortunately, the assumptions behind these procedures no longer hold when the form of the regressions that provide the parameters of interest are selected data-adaptively.

The second component of the proposed algorithm makes an attempt to address this issue through the use of a version of the conditional permutation test designed to provide more robust inference for the IPW and DR-IPW population intervention model estimators.

3. Failure to Account for Multiple Testing High-dimensional epidemiological data sets often involve testing multiple exposure effects on multiple outcomes. Despite this, most reports fail to account for multiple testing in claims of statistical significance. This is mainly the result of both the overly conservative nature of well-known multiple-testing procedures (MTPs), such as the Bonferroni method, as well as a philosophical viewpoint that tends to diminish the importance of the increase in false positives that inevitably result from inferences that do not account for the number of comparisons considered [13]. This philosophy has likely contributed to what some have called an epidemic of false positive findings in empirical fields, such as epidemiology [7]. Thus, the third component of the algorithm implements a method of multiple testing that attempts to improve the power over the Bonferroni method by utilizing the potentially strong correlation of testing statistics we anticipate for typical studies of many related exposures. The particular method employed is the quantile transformation method [20], an MTP which appropriately adjusts for multiple testing (i.e., controls the appropriate type I error rate at some desired level α) and has been shown to be more powerful in simulation studies to alternative MTPs [1].

More detail regarding each of the components of the algorithm is provided in Sect. 5.

3 Data Structure and Parameters of Interest

We assume we observe n i.i.d. copies of

$$O = \{W, A, Y\} \sim P_0,$$

where W is a p-dimensional vector of covariates (potential confounders); $A = (A_1, \ldots, A_j, \ldots A_q)$, where A_j is the jth exposure of interest; $Y = (Y_1, \ldots, Y_k, \ldots, Y_r)$ where Y_k is the kth outcome of interest. We assume the variables have been chosen so that they have a particular time-ordering: W precedes A which precedes Y. In order to define our parameter of interest, we also define, for all j, k, the possibly unobserved counterfactual outcome $Y_{a_j,k}$, $a_j \in \mathcal{A}_j$: specifically, $Y_{a_j,k}$ represents the kth outcome an individual *would have* experienced had they, possibly contrary to fact, received level a_j for the jth exposure of interest with $\mathcal{A}_j$ the set of all possible levels of this exposure.

We now define the parameter of interest, for all j, k, by:

$$\psi_{a_j,k} = E[Y_{a_j,k}] - E[Y_k]. \tag{3.1}$$

A population intervention model is a model for this parameter [6]. When $a_j = 0$ represents the level *"unexposed"*, $\psi_{0_j,k}$ may be interpreted as the effect of removing the jth exposure on the mean of the kth outcome in the target population (a measure akin to attributable risk).

Here we consider only the marginal causal effect as defined by (3.1). However, as formalized in [6], (3.1) may be extended to models conditional on a subset of covariates $V \in W$:

$$\psi(V)_{a_j,k} = E[Y_{a_j,k} \mid V] - E[Y_k \mid V].$$

The parameter $\psi(V)_{0_j,k}$ is then the effect of removing the jth exposure on the mean of the kth outcome in the target population within specific strata of V (e.g., among men or among women).

We make the following three identifying assumptions in order to link the observed data to the counterfactual parameters of interest $\psi_{a_j,k}$:

1. Consistency assumption: If $A_{j,k} = a_{j,k}$, then $Y_{a_j,k} = Y_k$. The consistency assumption implies the so-called stable unit treatment value assumption (SUTVA) [14] which includes the assumptions of no multiple versions of treatment and no "interference between units" (i.e., one individual's treatment does not affect a different individual's counterfactual outcome).
2. No unmeasured confounding (sequential randomization) assumption: $Y_{a_j,k}$ is independent of $A_j \mid W$, $a_j \in \mathcal{A}_j$; this assumption will hold by design when the exposure is physically randomized within levels of W but, otherwise (as in observational studies), is not guaranteed to hold.
3. Positivity assumption:

$$\Pr[W = w] \neq 0 \implies$$

$$\Pr[A_j = a_j \mid W = w] > 0 \text{ w.p.1}, \tag{3.2}$$

or, within each group defined by unique, possibly observed covariate values $W = w$, there is a positive probability of being assigned the target level ($A_j = a_j$).

Of these identifying assumptions, only violations of the positivity assumption may be empirically examined based on the observed data.

Given these three identifying assumptions for a given j, k, we can write the counterfactual parameter of interest $\psi_{a_j,k}$ as a function of only the observed data:

$$\psi_{a_j,k}(P_0) = E_W\{E[Y_k \mid A_j = a_j, W]\} - E[Y_k]. \tag{3.3}$$

Our algorithm aims to address the challenge of estimating the observed data parameter (3.3), which links to a meaningful causal parameter of interest under explicitly stated assumptions, in realistic settings where W is high-dimensional and $q \times r$ (the number of causal effects and, in turn, hypothesis tests) is also large. Note that, for notational simplicity, in the above statement of the identifying assumptions and target function of the observed data (3.3) we define a single W regardless of exposure j and outcome k. Generally (and as we allow in the data application described below) W will depend on the index j, k and may even include one or more $A_l \, l \neq j$.

Finally, in what follows, we only consider models for a single intervention level, (3.1) or $\psi_{a_j=0,k} = \beta_{0,k}$, where by definition, the target level will be $a_j = 0$. In general, however, (3.1) may be a function of a_j such that $\psi_{a_j,k} = m(a_j \mid \beta_{j,k})$ for some parametrization $m(a_j \mid \beta_{j,k})$; see Hubbard and van der Laan [6]. Allowing (3.1) to be a function of a_j can reduce variability of estimators when a_j has many categories or is measured continuously, but this is at the expense of bias if this parametric assumption is incorrect.

4 CHAMACOS Data Description

Study participants were recruited among pregnant women initiating prenatal care at Natividad Medical Center, a county hospital in the city of Salinas, California, or at Clínica de Salud del Valle de Salinas in the Salinas Valley, California. The recruitment sites primarily serve low-income individuals, with a large proportion working in agriculture. Eligible women were less than 20 weeks gestation, 18 years or older, Medi-Cal eligible, fluent in English and/or Spanish, and planning to deliver at Natividad Medical Center. A total of 601 women were enrolled between October 1999 and October 2000. Of these, 536 continued in the study through delivery. Chemical exposure measurements were taken at approximately 26 weeks gestation and again post-delivery (not all women had both samples) via maternal blood samples. For a more detailed description of data collection procedures, see Eskenazi et al. [4], Fenster et al. [5], and Chevrier et al. [2].

A subset of the variables in the CHAMACOS data set is used for illustration purposes in this article. These consist of four birth outcomes (Y), 30 chemical exposures (A) and 13 covariates (W). Outcomes examined are birthweight (grams), gestational age (weeks), head circumference (cm), and length (cm); we only report the results for head circumference. For all results, see Young et al. [27]. Exposures include 19 polychlorinated biphenyls (PCBs) (18, 28, 44, 49, 52, 66, 74, 99, 101, 118, 138, 146, 153, 156, 180, 183, 187, 194, 201) and 11 organochlorines (OCs). The 11 OCs are (table abbreviations in parentheses): β-hexachlorocyclohexane (BHC); Dieldrin (DIE); γ-hexachlorocyclohexane (GHC); hexachlorobenzene (HCB); Heptachlor epoxide (HPE); Mirex (MIR); o,p'-DDT (ODT); Oxychlordane (OXY); p,p'-DDE (PDE); p,p'-DDT (PDT); and $trans$-Nonachlor (TNA). The primary exposure levels used for all chemicals were those taken during pregnancy. However, if this level was missing and the post-delivery level was non-missing, the latter was used. All exposure measurements were measured in ng/g and lipid-adjusted. Values below the limit of detection (LOD) were imputed as LOD/2 (see Succop et al. [17] for general discussion of handling detection limits in environmental exposure studies). Observations were assigned $A_j = 0$ if their observed level of the jth chemical was in the bottom quartile of the empirical distribution of A_j; this is thus defined as the target level for which we will estimate the effect on the mean outcomes of reducing current exposure as distributed in the population to the lowest quartile of exposure. Under this definition of A_j and under slightly modified identifying assumptions, we can more precisely interpret our effect estimates in terms of the effect on the mean outcome comparing the hypothetical intervention "assign exposure to a participant with covariate level W as a random draw from the observed distribution of exposure among those with exposure in the lowest quartile and with covariate level W" to no intervention [16, 28].

Baseline covariates were: infant sex; pre-pregnancy BMI (underweight or normal, overweight, obese); marital status (single, married/living as married); poverty level (at or below poverty level, $\geq 200\%$ poverty level); maternal education ($\leq$6th grade, 7–12th grade, $\geq$high school graduate); parity (0, $\geq$1); number of years in the US ($\leq$1, 2–5, 6–10, 11+); country of origin (US, Mexico or Other); gestational age at first prenatal visit (weeks); and maternal age at delivery (years). Poverty level was calculated by dividing household income by the number of people supported by that income and comparing this value to federal poverty thresholds [U.S. Census 18].

In estimating associations where the exposure of interest was a PCB, three OCs (p,p'-DDE; p,p'-DDT; o,p'-DDT) were considered as potential confounders (implicit components of W) in addition to the baseline covariates. For associations where this exposure was an OC, the sum of all 19 PCBs was considered in addition to the baseline covariates. Only non-missing PCBs were included in the sum for each observation.

The final analysis was limited to observations non-missing on all of these 13 covariates. This reduced the data set from 542 to 380 observations. Our results then rely on the additional assumption that the data are missing completely at random (MCAR; [8]). However, the methods presented here can easily be augmented to include procedures to account for informative missingness, including both imputation [8] and inverse probability of censoring weighting and its double-robust extension [24].

5 Methods

5.1 Estimation: DR-IPW Estimation of the Population Intervention Model

Let $a_j = 0$ and $\psi_{a_j,k} \equiv \psi_{j,k}$. Using the estimating equation methodology of van der Laan and Robins [24] the DR-IPW estimator of $\psi_{j,k}$ for the constant population intervention model $\psi_{j,k} = \beta_{j,k}$ is defined as follows [6]:

$$\hat{\psi}_{j,k} \equiv \frac{1}{n} \sum_{i=1}^{n} D(A_{ji}, Y_{ki}, W_{jki}; \hat{g}_j, \hat{Q}_{j,k})$$

$$= \frac{1}{n} \sum_{i=1}^{n} \frac{I(A_{ji} = 0)}{\hat{g}_j(0 \mid W_{jki})} Y_{ki} - Y_{ki}$$

$$- \frac{I(A_{ji} = 0) - \hat{g}_j(0 \mid W_{jki})}{\hat{g}_j(0 \mid W_{jki})} \hat{Q}_{j,k}(0, W_{jki}), \qquad (5.1)$$

where $\hat{g}_j(0 \mid W_{jki}) \equiv \hat{Pr}(A_j = 0 \mid W_{jki})$ and $\hat{Q}_{j,k}(0, W_{jki}) \equiv \hat{E}[Y_{ik} \mid A_{ji} = 0, W_{jki}]$ are, in this case, data-adaptively selected with W_{jk} a data-adaptively selected subset of W. The estimate $\hat{\psi}_{j,k}^{DR}$ will be a consistent estimator of $\psi_{j,k}$ if *either* the data-adaptively selected form of $g_j(0 \mid W_{jki})$ or $Q_{j,k}(0, W_{jki})$ is correct and the previously stated identifying assumptions hold replacing W with its data-adaptively selected subset. Note that we estimate $E[Y_{ik} \mid A_{ji}, W_{jki}]$ using all the data, but only need the predicted value when $A_{ji} = 0$—one could also more nonparametrically estimate $E[Y_{ik} \mid A_{ji}, W_{jki}]$ only among those observations with $A_{ji} = 0$ if enough observations are available.

Data-adaptively selected estimates of these nuisance parameters were obtained using the deletion/substitution/addition (DSA) algorithm [15]. The DSA algorithm is a machine-learning routine which uses cross-validation, based on the squared error loss function, to obtain a *best* model based on a set of candidate estimators. The space of candidate estimators is limited by three variables: the maximum allowable model size, the maximum order of interactions, and the maximum sum of powers on a single model term. In the estimation of both $g_j(0 \mid W)$ and $Q_{j,k}(0, W)$, values for these limiting parameters were selected as 6, 2, and 3, respectively. These should be chosen to allow for the final model to be sufficiently flexible given the sample size available and should thus ultimately grow, as a function of n, to be (nearly) nonparametric. Recent advances in machine-learning may alternatively be used for the estimation of nuisance parameters that can include many candidate algorithms, including DSA, and can further reduce mean squared error of the estimator [23].

Estimates of $g_j(0 \mid W)$ were truncated such that values were restricted to be above 0.1 in order to reduce variability associated with practical positivity violations (3.2). The possible implications of truncation, as well as the presence of practical positivity violations are discussed in Sect. 6. This truncation no doubt introduces a bias, but the decrease in the variance outweighs the bias in this case. Though the parameter we emphasize here has some advantages over other parameters with regard to positivity problems (one only needs that all units have a positive probability of getting $A_{ji} = 0$), in the context of so many comparisons, there are bound to be some observations where $g_j(0 \mid W)$ is very small. Alternative, albeit more computationally intensive, methods that are less ad hoc than weight truncation have been proposed [19], and future work can also automate the nuisance parameter and covariate selection using these methods to improve the estimation of the specific parameter of interest.

5.2 Single Exposure Inference: A Modified Conditional Permutation Test

If one assumes that the correct forms of the nuisance parameters g_j and $Q_{j,k}$ are known a priori and estimated efficiently (maximum likelihood estimates), by, say, $\hat{g}_j^*$ and $\hat{Q}_{j,k}^*$, then a simple standard error for the estimate defined by (5.1) above can be derived. Specifically, [24] shows that, under assumptions, $\sqrt{n}(\hat{\psi}_{j,k}^* - \psi_{j,k}^0) \longrightarrow N(0, \sigma_{j,k}^2)$ (where $\psi_{j,k}^0$ is the true population value of $\psi_{j,k}$) and conservative estimates of $\sigma_{j,k}^2 \equiv var(\hat{\psi}_{j,k})$ can be obtained using:

$$\hat{\sigma}_{j,k}^2 = \frac{v\hat{a}r\left\{D\left(A_{ji}, Y_{ki}, W_{jki}; \hat{g}_j^*, \hat{Q}_{j,k}^*\right)\right\}}{n}, \tag{5.2}$$

where $D - \psi_{j,k}$ is the influence curve of the estimator (5.1) treating $Q_{j,k}$ and g_j as known. These variances are estimated by the sample variances of the (plug-in) estimating function.

However, the forms of g_j and $Q_{j,k}$ are rarely known a priori and may be based on the estimator (5.1) where these forms are data-adaptively selected. Here we propose a pseudo-exact test of the null hypothesis that A_j and Y_k are independent, based on (5.1), using a modification of the conditional permutation test.

First, we define how the conditional permutation test of this null hypothesis is implemented when W is of small dimension and consists of only covariates with few categories:

- Using the original data, calculate the test statistic $T_{j,k} = \hat{\psi}_{j,k}/\sqrt{\hat{\sigma}_{j,k}^2}$, for which the components are defined by (5.1) and (5.2).
- Randomly permute the values of A_j within each stratum of W and recalculate the test statistic based on this permutation.
- Repeat the previous step B times to obtain a B-length vector of test statistics $\vec{T}_{j,k}^B = (T_{j,k,1}, \ldots, T_{j,k,B})$.

- To obtain a p-value (p), calculate the proportion of elements in $\vec{T}_{j,k}^B$ for which the absolute value exceeds the absolute value of the original test statistic, $T_{j,k}$, calculated from the observed data.

$\vec{T}_{ij}^B$ defines (via its empirical distribution) the conditional permutation distribution of T_{ij}. The conditional permutation test rejects the null hypothesis of independence between A_j and Y_k given W if $p < \alpha$ for some pre-specified level α (e.g., $\alpha = 0.05$).

Given that this is an exact test, the testing procedure retains correct type I error control at level α even when the test statistic is defined using poor estimates of $\sigma_{j,k}^2$ (which might be the case when estimators of the exposure effect are based on data-adaptively selected estimates of nuisance parameters). Note, if the null is not true, then the power of the test will suffer if, for instance, $\sigma_{j,k}^2$ is estimated poorly.

In practice, however, testing using the conditional permutation test may be infeasible. Specifically, permutation within W becomes difficult when W is high-dimensional or even has one element with continuous values. Here we present a modification of the conditional permutation test for more general W which uses $\hat{g}_j$, the estimated treatment mechanism or propensity score, when g_j is assumed to follow a logistic model. This modified approach is implemented as the second step of the proposed machine-learning algorithm. Specifically,

- Obtain an estimate of the propensity score, $\hat{g}_j(0 \mid W)$, for each observation.
- Order the data by these estimated probabilities, $\hat{g}_j(0 \mid W)$.
- Group the ordered observations so that, within each group, the minimum number with $A_j = 0$ or $A_j = 1$ is M. This grouping constitutes a new categorical variable W^*.
- Follow the steps above for performing the conditional permutation test, permuting within strata of W^* in place of W.

Use of this version of the conditional permutation test assumes that each category of W^* will contain individuals with comparable values of the covariates W. The method of grouping we propose is one of a variety of ad hoc procedures that have been suggested for obtaining categories based on the propensity score, including one originally proposed by Rosenbaum [12] based on a backtrack algorithm. Though one could certainly find exceptions, simulation studies (not shown) have suggested that defining W^* using our approach successfully controls the type I error rate at the desired level α. In the application to CHAMACOS, the conditional permutation distribution for each test was based on $B = 5000$ permutations, with categories for W^* defined by $M = 2$. In the application of these methods to the CHAMACOS data set, we found that computation time was prohibitive if $\hat{g}_j(0 \mid W)$ was reselected data-adaptively within each permutation iteration and thus the model forms (variables chosen and basis functions used) for the $\hat{g}_j$ selected using the DSA on the original data were reused, with the corresponding coefficients re-estimated using maximum likelihood within each permutation.

5.3 Joint Inference: Quantile Transformation Method

The previous section describes marginal inference based on the estimates of the effect of a single exposure A_j on a single outcome Y_k. Usually, we will be interested in testing multiple exposure-outcome associations. If we are testing hypotheses regarding all exposure-outcome combinations, we will have a total of $m = q \times r$ tests. As described in Sect. 4, the CHAMACOS data set consists of 30 chemical exposures, 4 outcomes and thus 120 tests.

While in environmental epidemiology the value of m is generally substantially greater than one, reported approaches to inference usually ignore the multiple-testing problem. In particular, these reports fail to correctly define the false positive or type I error rate in terms of the *total* number of false positives, V_n. Note that V_n is a random variable which can take on the values zero or one in the case where $m = 1$ and any value between zero and the total number of true null hypotheses among the m tests in the case where $m > 1$. There are various forms for the type I error rate, depending on the type and stringency of control desired by the investigator. For the purposes of this discussion and in the application below, we focus on controlling the family-wise error rate (FWER) or $P(V_n > 0)$ at α.

Failure to correctly define the type I error rate appropriately in most reported epidemiologic investigations is likely due to the fact that the well-known and easily implemented Bonferroni procedure is overly conservative. Specifically, it is derived under the assumption that all m tests are independent, and rejects the null hypothesis for $p < \alpha/m$. Intuitively, one can imagine that if all m tests are perfectly correlated, we should divide by one in place of m. Multiple-testing procedures (MTPs) which, in addition to correctly controlling the type I error rate below α, further maximize power through the use of information on the joint dependence structure of the test statistics, are preferable.

The third step of the algorithm implements one such approach, referred to as the quantile transformation method [20]. The quantile transformation method is a resampling-based method which essentially incorporates the desirable characteristics of currently available MTPs, including the use of information on the dependence structure of the test statistics. This approach is an extension of a resampling-based method, originally proposed by Pollard and van der Laan [10], and further developed by Dudoit et al. [3], which creates an appropriate joint null distribution using the bootstrap. The observed m test statistics are then compared to this estimated joint null to obtain p-values. P-values obtained via some MTP are generally referred to as *adjusted* and those obtained otherwise as *raw* or *marginal*.

To implement the quantile transformation method:

- Sample with replacement (or bootstrap) the observed data (W, A, Y) among the n observations.
- Calculate the m-length vector of test statistics based on this new sample.
- Repeat these first two steps B_2 times to obtain a $m \times B_2$ matrix of test statistics, $T^{\#}$, where $T^{\#}_{l,b}$ represents the test statistic obtained from the bth bootstrap sample for the lth test, $l = 1, \ldots, m, b = 1, \ldots, B_2$.

- Using the lth row of $T^{\#}$, calculate the empirical bootstrap distribution Q_{nl} for the lth test statistic T_l, where $Q_{nl}(t) = P_n(T_l < t)$. This results in a $m \times B_2$ matrix of estimated probabilities, Q_n.
- Apply the quantile-function (or inverse probability function) $Q_{0l}^{-1}(x)$ to the lth row of Q_n, defining Q_{0l}^{-1} in terms of the assumed null distribution for the lth test statistic. This maps Q_n to a new $m \times B_2$ matrix, Q_0^{-1}, representing a joint null distribution for the observed m test statistics. For example, for $x = 0.5$, $Q_{0l}^{-1}(x)$ is the median of the null distribution for the lth observed test statistic. Note that Q_{0l}^{-1} can be the inverse probability function for any desired marginal null distribution for the lth test statistic. In our case, we define the marginal null distribution in terms of the modified conditional permutation distribution described in Sect. 5.2.

Once the joint null distribution, Q_0^{-1} for the m test statistics is obtained, various MTPs can be applied to obtain an adjusted p-value. In our application to CHAMACOS we apply the single-step minP approach, which converts the matrix Q_0^{-1} to p-values based on the distribution of each row. The adjusted p-value is then obtained by comparing the lth raw p-value to the distribution of the minimum from each column. The FWER is controlled at α by rejecting the null hypothesis for a given test when the adjusted p-value is less than α.

In our data application, the joint null distribution of the test statistics was estimated using $B_2 = 5000$ bootstrap samples. As in the case of the modified conditional permutation tests, we found unreasonable computing times when the forms of $\hat{g}_j$ and $\hat{Q}_{j,k}$ were reselected using the DSA within each bootstrap iteration. Again, here, the forms of $\hat{g}_j$ and $\hat{Q}_{j,k}$ chosen by the DSA using the original sample data were re-used within each bootstrap iteration with model coefficients re-estimated using maximum likelihood based on the bootstrap sample.

6 Results

Table 1 presents the results based on estimates of (3.1) unadjusted for W. The unadjusted estimator is simply the difference between the mean of the outcome among the baseline exposure group or $A = 0$ and the overall mean. All p-values (both raw and adjusted) in Table 1 are obtained from the simple permutation distribution; that is, only the A_j's are permuted as there are no variables in W considered. Table 2 presents the results based on the DR-IPW estimates of (3.1). For comparison purposes, we also present p-values based on both the conditional permutation distribution and the standard normal. Adjusted p-values are presented based on both the Bonferroni and quantile transformation methods. Though we only show in this paper the results from one outcome (head circumference) adjusted p-values take into account all 120 tests (4 outcomes $\times$ 30 chemicals). The set chosen

Table 1 Unadjusted estimates for associations between head circumference and each exposure (exp), number of observations non-missing on exposure and head circumference (n), standard errors (SE), test statistics (T), raw, Bonferroni (Bon) adjusted, and quantile transformation method (QTM) adjusted p-values based upon the simple permutation distribution

Exp	n	Estimate	SE	T	Raw	Bon	QTM
18	360	0.119	0.149	0.797	0.381	1	1
28	370	0.157	0.154	1.017	0.242	1	1
44	306	0.169	0.169	0.998	0.25	1	1
49	323	0.106	0.16	0.665	0.461	1	1
52	336	0.104	0.159	0.649	0.473	1	1
66	359	−0.014	0.164	−0.084	0.921	1	1
74	351	0.045	0.168	0.267	0.754	1	1
99	340	−0.159	0.17	−0.936	0.256	1	1
101	311	0.089	0.177	0.505	0.551	1	1
118	347	−0.112	0.174	−0.646	0.42	1	1
138	340	−0.138	0.156	−0.888	0.324	1	1
146	324	−0.077	0.157	−0.489	0.597	1	1
153	348	−0.094	0.145	−0.644	0.497	1	1
156	353	−0.2	0.143	−1.392	0.124	1	1
180	290	0.109	0.159	0.683	0.465	1	1
183	339	−0.253	0.148	−1.706	0.064	1	0.995
187	292	−0.187	0.152	−1.228	0.19	1	1
194	343	−0.056	0.15	−0.376	0.683	1	1
201	354	−0.166	0.143	−1.159	0.213	1	1
BHC	368	−0.152	0.159	−0.957	0.251	1	1
DIE	350	−0.254	0.152	−1.665	0.056	1	0.991
GHC	364	−0.122	0.148	−0.824	0.351	1	1
HCB	370	−0.197	0.16	−1.230	0.136	1	1
HPE	353	−0.483	0.146	−3.315	<0.001	<0.001	<0.001
MIR	369	−0.038	0.149	−0.256	0.768	1	1
ODT	370	−0.236	0.163	−1.448	0.075	1	0.997
OXY	349	−0.259	0.167	−1.554	0.054	1	0.989
PDE	370	−0.021	0.154	−0.135	0.877	1	1
PDT	370	−0.149	0.162	−0.923	0.257	1	1
TNA	370	−0.157	0.154	−1.018	0.226	1	1

for adjustment of multiple testing is a philosophical decision (not a statistical one), and in our case we have chosen to adjust for all tests (for all results, see [27]).

Out of the 30 tests of association shown, the algorithm found only one significant association after adjustment for multiple testing (at FWER type I error of 0.1), namely between HPE ($j = 24$) and head circumference ($k = 3$) with the DR-IPW estimator $\hat{\psi}_{24,3} = -0.507$ and an adjusted p-value of 0.079 obtained using the quantile transformation method with marginal distribution defined by the modified conditional permutation distribution. This suggests, assuming our

Table 2 DR-IPW estimates for associations between head circumference and each exposure (exp), standard errors (SE), test statistics (T), raw p-values from conditional permutation (CPD) and standard normal (Norm) distributions, Bonferroni (Bon) adjusted and quantile transformation method (QTM) adjusted p-values based upon both distributions

Exp	Estimate[a]	SE	T	Raw CPD	Raw Norm	Bon CPD	Bon Norm	QTM CPD	QTM Norm
18	0.127	0.136	0.935	0.086	0.350	1	1	1	1
28	0.127	0.140	0.908	0.249	0.364	1	1	1	1
44	0.139	0.157	0.888	0.257	0.374	1	1	1	1
49	0.094	0.142	0.661	0.311	0.509	1	1	1	1
52	0.075	0.145	0.515	0.386	0.607	1	1	1	1
66	−0.057	0.158	−0.364	0.608	0.716	1	1	1	1
74[b]	0.079	0.163	0.488	0.546	0.626	1	1	1	1
99[b]	−0.084	0.163	−0.514	0.662	0.607	1	1	1	1
101	0.070	0.157	0.445	0.590	0.656	1	1	1	1
118[b]	−0.095	0.197	−0.484	0.669	0.629	1	1	1	1
138[b]	−0.173	0.195	−0.888	0.336	0.375	1	1	1	1
146[b]	−0.240	0.185	−1.300	0.124	0.194	1	1	1	1
153[b]	−0.216	0.193	−1.116	0.240	0.265	1	1	1	1
156[b]	−0.311	0.169	−1.841	0.040	0.066	1	1	0.977	0.998
180[b]	0.140	0.218	0.642	0.480	0.521	1	1	1	1
183[b]	−0.262	0.182	−1.438	0.107	0.150	1	1	1	1
187[b]	−0.250	0.190	−1.316	0.131	0.188	1	1	1	1
194[b]	−0.153	0.188	−0.814	0.378	0.416	1	1	1	1
201[b]	−0.332	0.157	−2.112	0.010	0.035	1	1	0.651	0.962
BHC[b]	−0.317	0.150	−2.114	0.078	0.035	1	1	1	0.961
DIE[b]	−0.107	0.176	−0.607	0.496	0.544	1	1	1	1
GHC	−0.101	0.157	−0.645	0.577	0.519	1	1	1	1
HCB[b]	−0.339	0.178	−1.904	0.033	0.057	1	1	0.955	0.995
HPE[b]	−0.507	0.172	−2.938	0.001	0.003	0.096	0.396	0.079	0.285
MIR[b]	−0.063	0.163	−0.386	0.652	0.700	1	1	1	1
ODT	−0.150	0.167	−0.900	0.322	0.368	1	1	1	1
OXY[b]	−0.411	0.200	−2.059	0.020	0.039	1	1	0.859	0.974
PDE[b]	−0.026	0.151	−0.173	0.851	0.863	1	1	1	1
PDT	0.025	0.171	0.148	0.908	0.882	1	1	1	1
TNA[b]	−0.250	0.176	−1.415	0.126	0.157	1	1	1	1

[a]See Table 1 for sample sizes (n)
[b]Estimate based on at least one weight $\hat{g}(0 \mid W_i)$, $i = 1, \ldots, n$ truncated to 0.1

identifying assumptions hold, that infant head circumference would decrease 0.507 cm on average were all maternal HCB levels changed to the bottom quartile of the observed distribution compared to the mean of observed maternal HCB levels.

Again, the consistency of the DR-IPW estimates of (3.1) relies on the positivity assumption. Our use of truncated weights is an attempt to reduce variability in the presence of possible practical positivity violations at a cost of bias resulting from

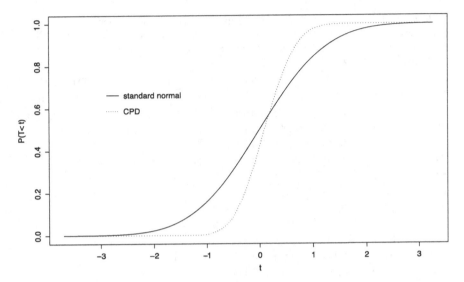

Fig. 1 Plot of the cumulative distribution function of the standard normal vs. that of the conditional permutation distribution (CPD) for the test statistic (T) based on the DR-IPW estimator of (5.1) associating head circumference and PCB 18

misspecification of the treatment mechanism (estimates in Table 2 are starred where truncation is used). There are methods to examine the bias due to the positivity assumption [26], as well as other parameters one could estimate that are less likely to suffer from such a bias [22]. For now, we take this relative simple, ad hoc approach, noting that other more detailed diagnostics and alternative parameter estimates are available. In addition, more recently developed techniques that can use model selection criteria targeted towards the parameter of interest [19, 25] can avoid this problem by selecting models for g that avoid large variances in favor of smaller biases.

To illustrate differences between the standard normal and the modified conditional permutation distribution, Fig. 1 overlays the cumulative distribution functions of these two possible distributions for the test statistic associating head circumference and PCB 18 based on the DR-IPW estimator. It is clear from this figure that the modified conditional permutation distribution is similar to that of the standard normal but with variance less than one. Table 3 presents standard errors for this same estimator based on both the influence curve defined in Eq. (5.2) and based on 5000 bootstrap samples. The standard error estimate based on the bootstrap is smaller than that based on the influence curve, suggesting the latter is overly conservative as expected given the seemingly paradoxical result that variance estimators are conservative when variability due to the estimation of nuisance parameters is ignored [24]. Similarly, we see from Table 3 that the bootstrap estimate of the variance of the test statistic (also based on 5000 samples) is smaller than one.

Table 3 Estimated standard errors (SE) of $\hat{\psi}$ for PCB 18 based on both the influence curve (IC) and the bootstrap (BS), as well as the bootstrap estimated variance (VAR) of $T = \hat{\psi}/\text{se}(\hat{\psi})$, where $\text{se}(\hat{\psi})$ is obtained using the influence curve for $\hat{\psi}$ for the association between PCB 18 and head circumference based on the DR-IPW estimator

IC SE	BS SE	BS VAR(T)
0.136	0.116	0.704

Bootstrap estimates based on 5000 bootstrap samples of (5.1)

As we can see from Table 2, several tests of association which would be classified as significant or borderline significant ($p < 0.1$) based solely on raw p-values are no longer significant after application of a MTP, regardless of type (Bonferroni or quantile transformation method). Based on the analysis, there were 5 tests with raw p-values less than 0.1. In all but one case (the association between HPE and head circumference), respective adjusted p-values were greater than 0.6. Notably, the adjusted p-value for the association between HPE and head circumference based on the DR-IPW estimator remained borderline significant whether the Bonferroni or QTM approach was used. However, for this test of association, adjusted p-values based on both MTPs were substantially larger when the standard normal was used over the conditional permutation distribution (see Table 2), indicating that inference based on the standard normal distribution tended to be more conservative than that based on the modified conditional permutation distribution. Though one cannot assert the validity of either, as both require different assumptions, it could be due to standard distributional assumptions not being met in the finite sample case when model selection procedures (machine-learning algorithms) are used to estimate the nuisance parameters; in such cases, more nonparametric inference should be preferred, i.e., the conditional permutation test.

7 Discussion

In summary, we found only one borderline significant association in the CHAMA-COS data set after the analysis with the proposed machine-learning algorithm; that between HPE and head circumference. Contrary to expectations, the direction of the estimate in this case suggests a protective effect of HPE. We consider two explanations. First, there is a violation of at least one of our identifying assumptions. It is certainly possible that our definition of W did not include all relevant confounders of the exposure effect, considering how little is known regarding the effects of organochlorines on human development and the possible relationships between different OC's and PCB's [5].

As expected, raw p-values were quite different from adjusted p-values indicating the importance of adjustment for multiple testing. In this application, conclusions were similar regardless of which multiple-testing method was used (Bonferroni or quantile transformation method). Also, as expected, results differed depending on whether a standard normal or the modified conditional permutation distribution was assumed for the test statistics, with the standard normal generally resulting in more conservative inference.

We note that results based on this analysis differ from other reported analyses of the CHAMACOS data set. In a recently published report of associations between OCs and birth outcomes in the CHAMACOS data set, Fenster et al. [5] did not find a significant relationship between HPE and head circumference. We stress that results reported by Fenster et al. [5] are not generally comparable to those reported here beyond differences in approaches to marginal inference and multiple testing. Specifically:

- our parameter is different than theirs, specifically, they use standard regression procedures, and assume a specific linear model thus deriving estimates of adjusted (for W) associations assuming the a priori model is correct, whereas ours is an estimate of the additive difference of a marginal mean (based on the mean, over W, of the predicted values of the outcome when the exposure of interest is set to its "baseline" level versus the observed marginal mean);
- we consider a slightly different set of variables in W;
- slightly different exclusionary criteria were used to obtain the final analysis sample.

Thus, the lack of correspondence of other analyses comes from both a different parameter being estimated as well as a different method used to estimate the parameter. Our method is most suitable when little existing information is available to choose models/variables a priori and thus the information for the relative contribution to variability in the outcome comes almost exclusively from the data itself. We also assert that the model, as it represents current knowledge about the data-generating distribution, is typically (nearly) unknown, or semi-parametric. In this context, one wants a procedure that produces a simple, interpretable parameter, uses flexible semi-parametric (machine-learning) methods for models when these are not known a priori (and the dimension is high) and finally returns trustworthy joint inference. Many studies pretend that knowledge of the model exists or the exploring of the data for such a model is ignored in the final inference. This generally leads to erroneous estimates (bias due to model misspecification) as well as erroneous inference (standard errors based on the assumption of a priori known models). The method described here is a potential black-box tool that can be used to screen for variables with strong evidence of adjusted associations. In addition, the algorithm is designed so that asymptotically, it converges to the "true" estimate given the identifiability assumptions (the model selection procedures based on cross-validation and a sieve approach become more nonparametric as sample size grows). Traditional ad hoc approaches have served useful purposes, but the combination of new techniques in causal inference, more powerful machine-learning tools, and fast computation that allows robust, re-sampling based inference means the practice of exploratory epidemiology should move beyond potentially misleading approaches appropriate for low dimensional problems and provide more robust results for high-dimensional studies.

References

1. Chen, J., van der Laan, M. J., Smith, M. T., & Hubbard, A. E. (2007). A comparison of methods to control type I errors in microarray studies. *Statistical Applications in Genetics and Molecular Biology, 6,* Article 28.
2. Chevrier, J., Eskenazi, B., Holland, N., Bradman, A., & Barr, D. B. (2008). Effect of exposure to polychlorinated biphenyls and organochlorine pesticides on thyroid function during pregnancy. *American Journal of Epidemiology, 68,* 298–310.
3. Dudoit, S., van der Laan, M. J., & Pollard, K. S. (2004). Multiple testing, part I. Single-step procedures for control of general type I error rates. *Statistical Applications in Genetics and Molecular Biology, 3,* Article 11.
4. Eskenazi, B., Marks, A. R., Brandman, A., Fenster, L., Johnson, C., Barr, D. B., et al. (2006). In utero exposure to dichlorodiphenyltrichloroethane (DDT) and dichlorodiphenyldichloroethylene (DDE) and neurodevelopment among young Mexican American children. *Pediatrics, 118,* 233–41.
5. Fenster, L., Eskenazi, B., Anderson, M., Bradman, A., Harley, K., Hernandez, H., Hubbard, A., Barr, D.B., (2005). Association of in utero organochlorine pesticide exposure and fetal growth and length of gestation in an agricultural population. *Environmental health perspectives, 114*(4), pp. 597–602.
6. Hubbard, A. E., & van der Laan, M. L. (2008). Population intervention models. *Biometrika, 95,* 35–47.
7. Ioannidis, J. P. A. (2005). Why most published research findings are false. *PLoS Medicine, 2,* e124.
8. Little, R. J., & Rubin, D. B. (2002). *Statistical analysis with missing data* (2nd ed.). New York: Wiley.
9. Patel, C. J., Bhattacharya, J., & Butte, A. J. (2010). An environment-wide association study (EWAS) on type 2 diabetes mellitus. *PLoS One, 5,* e10746.
10. Pollard, K. S., & van der Laan, M. J. (2003). Resampling-based multiple testing: asymptotic control of type I error and applications to gene expression data. In *Division of biostatistics,* Technical Report No. 121, University of California, Berkeley.
11. Robins, J. M. (1998). Marginal structural models. In *1997 Proceedings of the American Statistical Association, Section on Bayesian Statistical Science* (pp. 1–10). Alexandria: American Statistical Association.
12. Rosenbaum, P. R. (1984). Conditional permutation tests and the propensity score in observational studies. *Journal of the American Statistical Association, 79,* 565–574.
13. Rothman, K. J. (1990). No adjustments are needed for multiple comparisons. *Epidemiology, 1,* 43–6.
14. Rubin, D. B. (1986). Statistics and causal inference: Comment: Which 'ifs' have causal answers. *Journal of the American Statistical Association, 81,* 961–962.
15. Sinisi, S. E., & van der Laan, M. J. (2004). Loss-based cross-validated Deletion/Substitution/Addition algorithms in estimation. In *Division of biostatistics,* Technical Report No. 143, University of California, Berkeley.
16. Stitelman, O. M., Hubbard, A. E., & Jewell, N. P. (2010). *The impact of coarsening the explanatory variable of interest in making causal inferences: Implicit assumptions behind dichotomizing variables.* U.C. Berkeley Division of Biostatistics Working Paper Series. Working Paper 264.
17. Succop, P. A., Clark, S., Chen, M., & Galke, W. (2004). Imputation of data values that are less than a detection limit. *Journal of Occupational and Environmental Hygiene, 1,* 436–441.
18. United States Census Bureau (2000). *Poverty thresholds 2000, current population Survey.* www.census.gov/hhes/poverty/poverty00/pv00thrs.html
19. van der Laan, M. J., & Gruber, S. (2010). Collaborative double robust targeted maximum likelihood estimation. *The International Journal of Biostatistics, 6,* Article 17.

20. van der Laan, M. J., & Hubbard, A. E. (2006). Quantile-function based null distribution in resampling based multiple testing. *Statistical Applications in Genetics and Molecular Biology, 5*, Article 14.

21. van der Laan, M. J., Hubbard, A. E., & Jewell, N. (2010). Learning from data: semiparametric models versus faith-based inference. *Epidemiology, 21*, 479–81.

22. van der Laan, M. J., & Petersen, M. (2007). Causal effect models for realistic individualized treatment and intention to treat rules. *The International Journal of Biostatistics, 3*, Article 3.

23. van der Laan, M. J., Polley, E. C., & Hubbard, A. E. (2007). *Super Learner*. U.C. Berkeley Division of Biostatistics Working Paper Series. Working Paper 222.

24. van der Laan, M. J., & Robins, J. M. (2003). *Unified methods for censored longitudinal data and causality*. New York: Springer.

25. van der Laan, M. J., & Rubin, D. B. (2006). Targeted maximum likelihood learning. *The International Journal of Biostatistics, 2*, Article 11.

26. Wang, Y., Petersen, M. L., Bangsberg, D., & van der Laan, M. J. (2006). Diagnosing bias in the inverse probability of treatment weighted estimator resulting from violation of experimental treatment assignment. In *Division of biostatistics*, Technical Report No. 211, University of California, Berkeley.

27. Young, J., Hubbard, A. E., Eskenazi, B., & Jewell, N. P. (2009). A machine-learning algorithm for estimating and ranking the impact of environmental risk factors in exploratory epidemiological studies. In *Division of biostatistics*, Technical Report No. 250, University of California, Berkeley.

28. Young, J. G., Logan, R. W., Robins, J. M., & Hernán, M. A. (2019). Inverse probability weighted estimation of risk under representative interventions in observational studies. *Journal of the American Statistical Association, 114*, 938–947.

A Latent Time Distribution Model for the Analysis of Tumor Recurrence Data: Application to the Role of Age in Breast Cancer

Yann De Rycke and Bernard Asselain

Abstract Many popular statistical methods for survival data analysis do not allow the possibility of cure or at least of considerable lengthening of survival for breast cancer patients. Increasingly prolonged follow-up provides new data about the post-treatment outcome, revealing situations with the possibility of high cure rates for early stage breast cancer patients. Then the exclusive use of "classical" statistical models of survival analysis can lead to conclusions not completely reflecting clinical reality. It therefore appears preferable to perform additional analyses based on survival models with cured fraction to study long-term survival data, especially in oncology. This approach allows statistical methods to be adapted to the biological process of tumor growth and dissemination. After presenting the Cox model with time-dependent covariates and the Yakovlev parametric model, we study the prognostic role of age in young women (≤ 50 years) in Institut Curie breast cancer data. Age is a prognostic factor within all three models, but the interpretation is not the same. With the Cox model the younger women have a bad prognosis ($HR = 1.86$) comparing to the older one. But the HR does not verify the proportional hazard hypothesis. So the Cox model with time-dependent covariates gives a better interpretation: the age-effect decreases significantly with time. With the Yakovlev model we find that the decreasing age-effect can be viewed through the proportion of cured patients. More, there is an effect of age on survival (palliative effect) and also on the cure rates (curative effect). So the cure rate models demonstrate their utility in analyzing long-term survival data.

Keywords Cure rate · Time-dependent covariate · Survival analysis

Y. De Rycke · B. Asselain (✉)
Service de Biostatistique, Institut Curie, Paris, France
e-mail: bernard.asselain@curie.net

© Springer Nature Switzerland AG 2020
A. Almudevar et al. (eds.), *Statistical Modeling for Biological Systems*,
https://doi.org/10.1007/978-3-030-34675-1_9

1   Introduction

Survival data analysis is designed to assess the occurrence of the event studied (e.g., death, disease progression, relapse, metastasis), and the time to onset of this event. Survival data can also be analyzed in order to identify prognostic factors for epidemiological purposes: evaluation of the impact of a tumor or patient characteristic at time t_0 (e.g., age at diagnosis, stage of disease, histologic tumor grade) on the occurrence of an event of interest (metastasis, death, etc.) at time t.

Various statistical methods can be used for survival data analysis. In most cases, especially in the context of therapeutic trials, the log-rank test and semi-parametric Cox model [4] are used to compare survival between various groups and to evaluate the impact of an intervention and others covariates (histologic tumor grade at diagnosis, for example) on the survival time.

However, these methods of analysis are based on the hypothesis of proportional hazard (PHH): the ratio of the hazard of occurrence of the event (*HR*, hazard ratio) is assumed to be constant over time between the groups compared. Considerable progress in cancer therapy and especially in breast cancer now allows the possibility of cure or at least considerable lengthening of survival for patients. Increasingly prolonged follow-up provides new data about the post-treatment outcome of patients, revealing situations in which the ratio of risks of disease progression, metastasis or death is not constant over time for patients receiving different treatments or different prognostic factors [1, 11]. The PHH is therefore not valid in these situations.

High cure rates are now expected for early stage breast cancer. A proportion of patients never relapse within any reasonable period of monitoring/observation, suggesting that the survival distribution (or more precisely cause-specific survival) reaches a plateau beyond a certain time-point. However, in most of the popular parametric survival models, the survival function approaches towards 0 as t goes to infinity, but this does not correspond to the observed "plateau phase."

The exclusive use of "classical" statistical models of survival analysis can lead to conclusions not entirely reflecting clinical goals of the analysis. It therefore appears preferable to perform additional analysis based on parametric models specifically geared towards studying long-term survival data, especially in oncology [5, 8].

These parametric survival models are not necessarily based on the hypothesis of proportional hazard and are able to take into account several "explanatory covariates" that may directly influence the cure rate. This approach allows statistical methods to be adapted to the physiopathological and biological processes of the tumor. Several teams have developed and used parametric models which allow patients achieving long-term survival (or non-occurrence of the metastatic event, for example) to be considered as cured within the limits of the study duration (e.g., Gamel–Boag model [5], Yakovlev models [6, 10]). This distinction corresponds to a clinical reality, observed in patient cohorts followed after receiving treatment with potential to cure.

Nevertheless, the Cox model with constant *HR* between treatment groups is still frequently used for survival data analysis. The parametric models, even when they do not require assumption of the proportional hazards, indeed require a functional form of the survival distribution function most appropriate for the situation under consideration (e.g., exponential, gamma, log-logistic, log-normal). Maximum likelihood methods are then used to estimate the parameters corresponding to each of the covariates introduced into the model [9].

Parametric models with cured fraction therefore appear to present a real value for cancer survival data analysis. Despite being more restrictive and possibly of less flexible functional form, they allow better interpretation of the physiopathological processes. Within the limits of current knowledge in physiopathological processes, these models may therefore identify prognostic factors that are more consistent with the natural history of the cancer, to allow better information for patients and adapt post-treatment survey.

This article studies the prognostic role of age at disease detection by using several models in order to discuss their respective advantages and disadvantages. Section 2 describes a Cox model with time-dependent effect of the covariates and the Yakovlev parametric model with a log-logistic baseline distribution. Section 3 presents the results of the Cox model and the two models cited applied to Institut Curie breast cancer data and Sect. 4 comprises the discussion and conclusion.

2 Material and Methods

2.1 Cox Model

Each subject i has a particular participation time t_i, a state δ_i and a vector of k covariates Z_i. The semi-parametric Cox model assesses the hazard function for a subject i according to the unspecified basic risk function and the multiplying effect of covariates Z:

$$\lambda(t) = \lambda_0(t) \exp(\beta' Z_i).$$

Estimation of the k parameters β is obtained by maximization of the partial log-likelihood function and has all properties related to the maximum likelihood. The partial log-likelihood function is expressed by

$$LV = \sum_{i=1}^{n} \delta_i \left\{ \beta' Z_i - \log \left(\sum_{j \in R_i} \exp(\beta' Z_j) \right) \right\},$$

where R_i is all of the subjects at risk at time t_i. Only event times are taken into account. The HR between subjects i and j is expressed by

$$HR = \frac{\lambda_0(t) \exp(\beta' Z_i)}{\lambda_0(t) \exp(\beta' Z_j)} = \exp(\beta'(Z_i - Z_j)).$$

The HR is not time-dependent: we assume the proportional hazard hypothesis here.

2.2   Cox Model with Time-Dependent Covariates

When one or several covariates have a time-dependent effect, the Cox model can still be used by adding a time function. If we assume that only the covariate Z_1 has a time-dependent effect, a "new" covariate $Z_1 \times f(t)$ is added to the model, where $f(t)$ is a specified time function. The partial log-likelihood function is then expressed by

$$LV = \sum_{i=1}^{n} \left\{ \left(\beta_1 Z_{1i} + \beta_{1t} Z_{1i} \times f(t) + \beta'_{-1} Z_{-1i} \right) \right.$$
$$\left. - \log \left(\sum_{j \in R_i} \exp \left(\beta_1 Z_{1j} + \beta_{1t} Z_{1j} \times f(t) + \beta'_{-1} Z_{-1j} \right) \right) \right\},$$

where Z_{-1i} corresponds to all of the k covariates except for the first and β_{-1} corresponds to the last $k - 1$ parameters to estimate. Once again, only event times are taken into account, but the value of $f(t)$ must be known at each event time. The *HR* between subjects i and j is expressed by

$$HR = \frac{\lambda_0(t) \exp \left(\beta_1 Z_{1i} + \beta_{1t} Z_{1i} \times f(t) + \beta'_{-1} Z_{-1i} \right)}{\lambda_0(t) \exp \left(\beta_1 Z_{1j} + \beta_{1t} Z_{1j} \times f(t) + \beta'_{-1} Z_{-1j} \right)}$$
$$= \exp \left(\beta_1 (Z_{1i} - Z_{1j}) + \beta_{1t}(Z_{1i} - Z_{1j}) \times f(t) + \beta'_{-1}(Z_{-1i} - Z_{-1j}) \right).$$

As a result of this equation, *HR* is time-dependent.

The PHH can be tested for this covariate Z_1 by testing $H_0 : \beta_{1t} = 0$ versus $H_1 : \beta_{1t} \neq 0$ using one of the three usual tests. However, failing to reject H_0 does not demonstrate that the PHH is verified because the result of the test depends on the choice of the time function. Various methods can be used to specifically test the PHH or to evaluate the hazard's graphs. But even if this method resolves the problem of proportional hazard for a specified covariate, it does not necessarily estimate the effect of this covariate. However, in oncology, a high effect of a covariate is generally observed at the beginning of follow-up and this effect tends to decline with time. This kind of change over time can be estimated by introducing "simple" functions in the model such as $f(t) = t$, $f(t) = 1/t$, $f(t) = \sqrt{t}$ or $f(t) = \log(t)$.

2.3   Yakovlev Models

In the 1990s, Andreï Yakovlev et al. [12] developed a tumor growth model in breast cancer, allowing introduction of the concept of cure and prognosis studies. Suppose that treatment of the cancer left $v_i = 0, 1, 2, \ldots$ clones for subject i. These clones will continue to grow and will become important enough to be detectable at time

X_{ij}, $j = 1, \ldots, v_i$. Subject i therefore presents an event as soon as one of the clones is detected, at time $U_i = \min(X_{ij})$. X_{ij} is assumed to be independent conditionally to the subject, and we note f, the density, and F, the distribution function of X_{ij}. Moreover, we assume that v_i follows a Poisson distribution with parameter θ and that time X_{ij} and v_i are independent. The cure rate is given by the parameter of the Poisson distribution:

$$P(v_i = 0) = \exp(-\theta).$$

By convention we note that $X_{i0} = +\infty$ if $v_i = 0$.

Studying the event-free interval therefore corresponds to studying the survival function of the random variable U:

$$\overline{F_U}(u) = P(U \geq u) = \sum_{k=0}^{\infty} P(v = k) P(U \geq u) = \sum_{k=0}^{\infty} P(v = k) \prod_{j=0}^{k} P(X_j \geq u)$$

$$= \sum_{k=0}^{\infty} \frac{\theta^k}{k!} \exp(-\theta)(1 - F(u))^k$$

$$= \sum_{k=0}^{\infty} \frac{(\theta(1 - F(u)))^k}{k!} \exp(-\theta(1 - F(u))) \exp(+\theta(1 - F(u))) \exp(-\theta)$$

$$= \left(\sum_{k=0}^{\infty} \frac{(\theta(1 - F(u)))^k}{k!} \exp(-\theta(1 - F(u))) \right) \exp(-\theta F(u))$$

$$= \exp(-\theta F(u)).$$

Yakovlev initially proposed a gamma distribution for the density of X_{ij} before developing nonparametric models [10]. More recently, the works directed by Asselain et al. [2] about the risks associated with recurrences or metastases in breast cancer showed that the hazard function follows log-normal or log-logistic forms. The log-logistic law will be used here.

The k covariates Z are introduced into the model at two levels: on the cure rate part and on the survival part of the model, and either in an accelerated model or in a model under PHH:

$$\overline{F_U}(u \mid Z) = \exp\left(-\theta \exp(\alpha' Z) F(u \, \exp(\beta' Z))\right), \quad \text{or}$$

$$\overline{F_U}(u \mid Z) = \exp\left(-\theta \exp(\alpha' Z) F(u)^{\exp(\beta' Z)}\right).$$

Estimation of the $2k$ parameters α and β is obtained by log-likelihood maximization and has all of the properties related to maximum likelihood. The likelihood is expressed by

$$V = \prod_i (f_U(u \mid Z))^{\delta_i} (\overline{F_U}(u \mid Z))^{1-\delta_i} = \prod_i (\lambda_U(u \mid Z))^{\delta_i} (\overline{F_U}(u \mid Z))$$

and the log-likelihood by

$$LV = \sum_i \delta_i \log(\lambda_U(u \mid Z)) - \sum_i \Lambda_U(u \mid Z),$$

where $\lambda_U(.)$ and $\Lambda_U(.)$ are the hazard function of U and cumulative hazard function of U, respectively.

2.4 Breast Cancer Data

All patients with breast cancer, treated exclusively at the Institut Curie, have been prospectively registered since 1981. More than 25,000 patients are now registered with their demographic, clinical, and outcome characteristics. The poor prognosis of young women has been known for a long time, but only a long-term study can determine whether this effect persists or disappears with time. The prognostic role of age in young women (≤ 50 years) with no particular tumor's stage characteristics was therefore studied, by using metastasis as the event of interest.

3 Results

3.1 Dataset

A total of 4519 women with breast cancer between 1981 and 1999 were selected. The age of these women at the time of cancer diagnosis was distributed as follows (Table 1): 312 (6.9%) patients less than 33 y.o., 974 (21.6%) between 34 and 40 y.o.

Table 1 Age, T and N distribution of the patients

		Sample size (%)
Age	≤ 33	312 (6.90)
	]33–40]	974 (21.55)
	]40–50]	3233 (71.54)
T	T1	1676 (37.09)
	T2	2238 (49.52)
	T3	605 (13.39)
N	N0	2869 (63.49)
	N1	1633 (36.14)
	N2	17 (0.38)

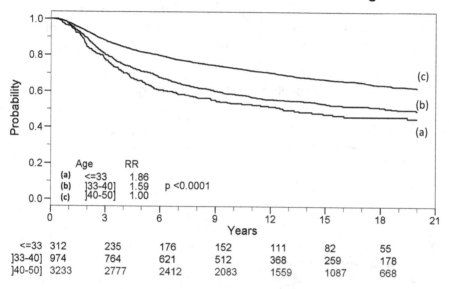

Fig. 1 Prognostic role of age on development of metastases with *RR* and *p* obtained by the log-rank test

and 3233 (71.5%) women over the age of 40 years. Almost one half of these patients presented T2 tumors, 37% presented T1 tumors, and 13% presented T3 tumors. A proportion of 2/3 did not have any clinical node involvement. The median follow-up was 15.5 years (range: 0.23–20). A total of 1474 deaths and 1623 metastases were observed in this population.

Impact of age on metastasis occurrence, assessed by the log-rank test, was statistically significant ($p < 0.0001$) between the age-groups (Fig. 1) with *RR* of 1.86 and 1.58 for the ≤33 and [33–40] age-groups, respectively, compared to the [40–50] age-group.

3.2 Cox Models

The likelihood ratio test indicated a very significant effect of age ($p < 0.0001$) (Table 2). The *HR*s were 1.86 [1.57–2.21] and 1.59 [1.42–1.78] for the ≤33 and]33–40] age-groups, respectively, compared to the]40–50] age-group. We can note that the *HR*s are close to the *RR* above. Unfortunately, although the effect is real, the estimation is incorrect because Harrell's test based on the Schoenfeld residuals was highly significant ($p < 0.01$), as clearly confirmed by the graph (Fig. 2).

Table 2 Results of the Cox model

Age	β	s_β	p	HR	IC (95%)
≤33	0.623	0.086	<0.0001	1.588	[1.575; 2.206]
]33–40]	0.463	0.057	<0.0001	1.864	[1.420; 1.776]
]40–50]	0	–	–	1	–

Schoenfeld : metastasis and age

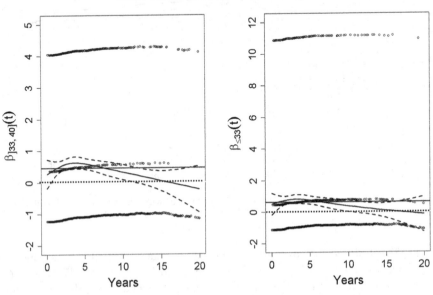

Fig. 2 Proportional Hazards on Schoenfeld's graph for the Cox model

Table 3 Results of the
time-dependent Cox model

Age	β	s_β	p
≤33	878	0.134	<0.0001
"≤33" × time	−0.0510	0.0221	0.021
]33–40]	0.631	0.091	<0.0001
"]33–40]" × time	−0.0321	0.0138	0.020
]40–50]	0	–	–

To obtain a better estimation, this proportional hazard problem must be corrected by introducing the $f(t) = t$ function into the Cox model, which still reflects this very important effect of age, but at this time PHH is verified by Harrell's test and by Schoenfeld graphs (data not shown).

The "time-dependent effect" part results in an $\exp(-0.05)$ and $\exp(-0.03)$ reduction of the effect per year for the ≤33 and]33–40] age-groups compared to the]40–50] age-group, as illustrated by the Schoenfeld graph (Table 3).

Table 4 Results of the accelerated Yakovlev log-logistic survival model

	Age	Estimation	SD	p	p^*
θ		0.556	0.025	–	
μ		2.011	0.063	–	
σ		0.657	0.020	–	
Cure	≤33	0.493	0.103	<0.0001	<0.0001
	]33–40]	0.381	0.070	<0.0001	
	]40–50]	0	–	–	
Survival	≤33	0.277	0.122	0.0226	0.0227
	]33–40]	0.176	0.084	0.0374	
	]40–50]	0	–	–	

Table 5 Estimation of cure rates using the accelerated Yakovlev log-logistic survival model

Age	Cure rate	95% CI
≤33	0.402	[0.333;0.471]
]33–40]	0.443	[0.374;0.510]
]40–50]	0.574	[0.544;0.602]

3.3 Yakovlev Model

Log-logistic survival was used in the "survival" part of the model. The existence of a cure rate in this model was first tested by comparing the Yakovlev model with a classical parametric survival model. The "cure" component must be present in the model. The contribution of age to the model was then tested by the likelihood ratio test, which demonstrated an effect of age on both cure and survival (Table 4). The Yakovlev model was also tested in the PHM, but the use of the AIC criterion showed that the accelerated model was better. The coefficients of the final model provided cure rates of 40.2%, 44.3%, and 57.4%, respectively, for the three age-groups (Table 5).

4 Discussion

Progress in the treatment of breast cancer has allowed considerable improvement of patient survival. Systematic recording of the patient's demographic and clinical data, and especially follow-up data of these patients has permitted assessment of the long-term behavior of the hazard rate of event's occurrence, such as metastasis or death. Unfortunately, these hazards are often not proportional over time. Otherwise, concerning the estimation of metastasis hazard, the use of models allowing the non-proportional hazards appears to be a reasonable solution.

The Cox model with the addition of time-dependent effect of covariates can obviously be used to take into account the non-proportional hazards. However, an appropriate time function must be selected or other nonparametric dynamic survival methods must be used, such as splines methods. These may not provide good estimation and clear interpretation of these effects. Moreover, the choice of a nonparametric hazards ratio function also increases the number of parameters to be estimated and, towards the end of follow-up when a smaller number of subjects remains, these estimates may not be very precise. Yet the goals of the analysis may include the precise interpretation of the effects specifically towards the end of follow-up. In our experience, simple functions such as that used in this study (for example, $f(t) = t$) appear to be sufficient, but when the follow-up is too long, this may produce inappropriate and meaningless interpretations towards the end of follow-up period.

In general, subjects at higher than usual risk tend to experience the events earlier than others. Consequently, in the case of long follow-up, certain covariates may no longer have any effect on the hazards of subjects still at risk after some reasonably long time period. This is particularly true when some of the subjects under certain covariate values have been cured. Beyond a certain time, the only patients remaining at risk are the non-cured patients. Many models were developed in the 1990s to take cured patients into account. Some authors [3, 7] developed semi-parametric models, while others, such as Andreï Yakovlev, developed parametric models. Most of these models can be used to estimate an effect of covariates on cure, the curative effect, and to estimate an effect of covariates on survival, the palliative effect.

One of the Andreï Yakovlev's models, used in the present study, is based on a biological approach and, from this point of view, is very relevant.

So-called "plateau" models do not resolve all of the problems. First, in the parametric case presented here, the choice of distribution must be discussed, or even compared with other distributions. Furthermore, in order to consider cure, the data must reach the cure plateau and, finally, the proportional hazard problem can persist on the palliative part of the model. Concerning the distribution, our choice of the log-logistic distribution is consistent with our knowledge of the risk of metastasis, by assuming the hypothesis that it is still true in the presence of cure. Concerning the plateau, it appears to have been reached for the ≤ 33 age-group, and has almost been reached for the other age-groups. Concerning the PH problem, an accelerated model and a model under proportional hazard assumption were tested on the palliative part of the model. Interpretation of our results is possible under these conditions. The effect of age on development of metastasis was demonstrated in both the palliative and curative parts, but the curative part was more important and was the first effect to be entered into the model so that the effect of age essentially concerns cure, although a palliative effect is also observed.

This study does not claim to provide an irreproachable and irrevocable clinical result. It is simply presented as an illustration, although the underlying question is a real and important issue. The model probably needs to be adjusted for other prognostic factors and the survival distribution used probably needs to be studied more precisely before reaching any definitive conclusions.

Acknowledgements This work is dedicated to Andreï Yakovlev, our friend, who inspired us this way of thinking, and constructing statistical models in relation with the biological knowledge.

We thank Alexia Savignoni for her careful reading and suggestions. The authors also thank the referees and the editor for their constructive comments that allowed us to improve the quality of this paper.

References

1. Appelbaum, F. R., Dahlberg, S., Thomas, E. D., Buckner, C., Cheever, M. A., Clift, R. A., et al. (1984). Bone marrow transplantation or chemotherapy after remission induction for adults with acute nonlymphoblastic leukemia. *Annals of Internal Medicine, 101*, 581.
2. Asselain, B., Rycke, Y. D., Savignon, A., & Mould, R. F. (2003). Parametric modelling to predict survival time to first recurrence for breast cancer. *Physics in Medicine and Biology, 48*, L31.
3. Broet, P., Rycke, Y., Tubert-Bitter, P., Lellouch, J., Asselain, B., & Moreau, T. (2001). A semiparametric approach for the two-sample comparison of survival times with long-term survivors. *Biometrics, 57*, 844–852.
4. Cox, D. R. (1972). Regression models and life tables (with discussion). *Journal of the Royal Statistical Society, Series B: Statistical Methodology, 34*, 187–220.
5. Frankel, P., & Longmate, J. (2002). Parametric models for accelerated and long-term survival: A comment on proportional hazards. *Statistics in Medicine, 21*, 3279–3289.
6. Hanin, L. G., Miller, A., Zorin, A., & Yakovlev, A. Y. (2006). The University of Rochester model of breast cancer detection and survival. *Journal of the National Cancer Institute Monographs, 2006*, 66–78.
7. Ibrahim, J. G., Chen, M. H., & Sinha, D. (2001). Bayesian semiparametric models for survival data with a cure fraction. *Biometrics, 57*, 383–388.
8. Nardi, A., & Schemper, M. (2003). Comparing Cox and parametric models in clinical studies. *Statistics in Medicine, 22*, 3597–3610.
9. Paoletti, X., & Asselain, B. (2010). Survival analysis in clinical trials: Old tools or new techniques. *Surgical Oncology, 19*, 55–58.
10. Tsodikov, A. D., Asselain, B., Fourque, A., Hoang, T., & Yakovlev, A. Y. (1995). Discrete strategies of cancer post-treatment surveillance. Estimation and optimization problems. *Biometrics, 51*, 437–447.
11. Wolmark, N., Yothers, G., O'Connell, M., Sharif, S., Atkins, J., Seay, T., et al. (2009). A phase III trial comparing mFOLFOX6 to mFOLFOX6 plus bevacizumab in stage II or III carcinoma of the colon: Results of NSABP Protocol C-08. *Journal of Clinical Oncology, 27*, 6s.
12. Yakovlev, A. Y., Asselain, B., Bardou, V., Fourquet, A., Hoang, T., Rochefediere, A., et al. (1993). A simple stochastic model of tumor recurrence and its application to data on premenopausal breast cancer. *Biometrie et analyse de donnees spatio-temporelles, 12*, 66–82.

Estimation of Mean Residual Life

W. J. Hall and Jon A. Wellner

Abstract Yang (Ann Stat, 6:112–116, 1978) considered an empirical estimate of the mean residual life function on a fixed finite interval. She proved it to be strongly uniformly consistent and (when appropriately standardized) weakly convergent to a Gaussian process. These results are extended to the whole half line, and the variance of the limiting process is studied. Also, nonparametric simultaneous confidence bands for the mean residual life function are obtained by transforming the limiting process to Brownian motion.

Keywords Life expectancy · Consistency · Limiting Gaussian process · Confidence bands

1 Introduction and Summary

This is an updated version of a technical report, [23], that has been referenced repeatedly in the literature— e.g., [3, 16, 17, 27, 32, 41]—although not having been published. We take this opportunity to honor the many achievements of Andrei Yakovlev and his devotion to modeling life processes throughout his career by making this work broadly available.

Let $X_1, \ldots, X_n$ be a random sample from a continuous d.f. F on $\mathbb{R}^+ = [0, \infty)$ with finite mean $\mu = E(X)$, variance $\sigma^2 \leq \infty$, and density $f(x) > 0$. Let $\overline{F} = 1 - F$ denote the survival function, let $\mathbb{F}_n$ and $\overline{\mathbb{F}}_n$ denote the empirical distribution function and empirical survival function, respectively, and let

W. J. Hall (deceased)
Department of Biostatistics and Computational Biology, University of Rochester, Rochester, NY, USA

J. A. Wellner (✉)
University of Washington, Seattle, WA, USA
e-mail: jaw@stat.washington.edu

$$e(x) \equiv e_F(x) \equiv E(X - x \mid X > x) = \int_x^\infty \overline{F} dI / \overline{F}(x), \quad 0 \le x < \infty$$

denote the *mean residual life function* or *life expectancy function* at age x. We use a subscript F or $\overline{F}$ on e interchangeably, and I denotes the identity function and Lebesgue measure on $\mathbb{R}^+$.

A natural nonparametric or life table estimate of e is the random function $\hat{e}_n$ defined by

$$\hat{e}_n(x) = \left\{ \int_x^\infty \overline{\mathbb{F}}_n dI / \overline{\mathbb{F}}_n(x) \right\} 1_{[0, X_{nn})}(x)$$

where $X_{nn} \equiv \max_{1 \le i \le n} X_i$; that is, the average, less x, of the observations exceeding x. Yang [50] studied $\hat{e}_n$ on a fixed finite interval $0 \le x \le T < \infty$. She proved that $\hat{e}_n$ is a strongly uniformly consistent estimator of e on $[0, T]$, and that, when properly centered and normalized, it converges weakly to a certain limiting Gaussian process on $[0, T]$.

We first extend Yang's results to all of $\mathbb{R}^+$ by introducing suitable metrics [50]. Her consistency result is extended in Theorem 2.1 by using the techniques of [46, 47]; then her weak convergence result is extended in Theorem 2.2 using [44] and [47].

It is intuitively clear that the variance of $\hat{e}_n(x)$ is approximately $\sigma^2(x)/n(x)$ where $\sigma^2(x) = Var[X - x \mid X > x]$ is the residual variance and $n(x)$ is the number of observations exceeding x; the formula would be justified if these $n(x)$ observations were a random sample of fixed sized $n(x)$ from the conditional distribution $P(\cdot \mid X > x)$. Noting that $\overline{\mathbb{F}}_n(x) = n(x)/n \to \overline{F}(x)$ a.s., we would then have

$$n Var[\hat{e}_n(x)] = n\sigma^2(x)/n(x) \to \sigma^2(x)/\overline{F}(x).$$

Proposition 2.1 and Theorem 2.2 validate this (see (2.4) below): the variance of the limiting distribution of $n^{1/2}(\hat{e}_n(x) - e(x))$ is precisely $\sigma^2(x)/\overline{F}(x)$.

In Sect. 3 simpler sufficient conditions for Theorems 2.1 and 2.2 are given and the growth rate of the variance of the limiting process for large x is considered; these results are related to those of [1]. Exponential, Weibull, and Pareto examples are considered in Sect. 4.

In Sect. 5, by transforming (and reversing) the time scale and rescaling the state space, we convert the limit process to standard Brownian motion on the unit interval (Theorem 5.1); this enables construction of nonparametric simultaneous confidence bands for the function e_F (Corollary 5.2). Application to survival data of guinea pigs subject to infection with tubercle bacilli as given by Bjerkedal [5] appears in Sect. 6.

We conclude this section with a brief review of other previous work. Estimation of the function e, and especially the discretized life-table version, has been considered by Chiang; see pages 630–633 of [13] and page 214 of [14]. (Also see [14, page 189], for some early history of the subject.) The basis for *marginal* inference (i.e., at a specific age x) is that the estimate $\hat{e}_n(x)$ is approximately normal with estimated standard error $S_k/\sqrt{k}$, where $k = n\overline{\mathbb{F}}_n(x)$ is the observed number of observations beyond x and S_k is the sample standard deviation of those observations. A partial justification of this is in [13, page 630], (and is made precise in Proposition 5.2 below). Chiang [14, page 214], gives the analogous marginal result for grouped data in more detail, but again without proofs; note the column $S_{\hat{e}_i}$ in his Table 8, page 213, which is based on a modification and correction of a variance formula due to [48]. We know of no earlier work on simultaneous inference (confidence bands) for mean residual life.

A plot of (a continuous version of) the estimated mean residual life function of 43 patients suffering from chronic granulocytic leukemia is given by Bryson and Siddiqui [6]. Gross and Clark [20] briefly mention the estimation of e in a life—table setting, but do not discuss the variability of the estimates (or estimates thereof). Tests for exponentiality against decreasing mean residual life alternatives have been considered by Hollander and Proschan [25].

2 Convergence on $\mathbb{R}^+$: Covariance Function of the Limiting Process

Let $\{a_n\}_{n\geq1}$ be a sequence of nonnegative numbers with $a_n \to 0$ as $n \to \infty$. For any such sequence and a d.f. F as above, set $b_n = F^{-1}(1 - a_n) \to \infty$ as $n \to \infty$. Then, for any function f on $\mathbb{R}^+$, define f^* equal to f for $x \leq b_n$ and 0 for $x > b_n$: $f^*(x) = f(x)1_{[0,b_n]}(x)$. Let $\|f\|_a^b \equiv \sup_{a\leq x\leq b} |f(x)|$ and write $\|f\|$ if $a = 0$ and $b = \infty$.

Let $\mathcal{H}(\downarrow)$ denote the set of all nonnegative, decreasing functions h on $[0, 1]$ for which $\int_0^1 (1/h)dI < \infty$.

Condition 1a There exists $h \in \mathcal{H}(\downarrow)$ such that

$$M_1 \equiv M_1(h, F) \equiv \sup_x \frac{\int_x^\infty h(F)dI/h(F(x))}{e(x)} < \infty.$$

Since $0 < h(0) < \infty$ and $e(0) = E(X) < \infty$, Condition 1a implies that $\int_0^\infty h(F)dI < \infty$. Also note that $h(F)/h(0)$ is a survival function on $\mathbb{R}^+$ and that the numerator in Condition 1a is simply $e_{h(F)/h(0)}$; hence Condition 1a may be rephrased as: there exists $h \in \mathcal{H}(\downarrow)$ such that $M_1 \equiv \|e_{h(F)/h(0)}/e_F\| < \infty$.

Condition 1b There exists $h \in \mathcal{H}(\downarrow)$ for which $\int_0^\infty h(F)dI < \infty$ and $\|eh(F)\| < \infty$.

Bounded e_F and existence of a moment of order greater than 1 is more than sufficient for Condition 1b (see Sect. 3).

Theorem 2.1 *Let* $a_n = \alpha n^{-1} \log \log n$ *with* $\alpha > 1$. *If Condition 1a holds for a particular* $h \in \mathcal{H}(\downarrow)$, *then*

$$\rho_{h(F)e/\overline{F}}(\hat{e}_n^*, e^*)$$

$$\equiv \sup\left\{ \frac{|\hat{e}_n(x) - e(x)|\overline{F}(x)}{h(F(x))e(x)} : x \le b_n \right\} \to_{a.s.} 0 \text{ as } n \to \infty. \quad (2.1)$$

If Condition 1b holds, then

$$\rho_{1/\overline{F}}(\hat{e}_n^*, e^*)$$

$$\equiv \sup\{|\hat{e}_n(x) - e(x)|\overline{F}(x) : x \le b_n\} \to_{a.s.} 0 \text{ as } n \to \infty. \quad (2.2)$$

The metric in (2.2) turns out to be a natural one (see Sect. 5); that in (2.1) is typically stronger.

Proof First note that for $x < X_{nn}$

$$\hat{e}_n(x) - e(x) = \frac{\overline{F}(x)}{\overline{\mathbb{F}}_n(x)} \left\{ \frac{-\int_x^\infty (\mathbb{F}_n - F) dI}{\overline{F}(x)} + \frac{e(x)}{\overline{F}(x)}(\mathbb{F}_n(x) - F(x)) \right\}.$$

Hence

$$\rho_{h(F)e/\overline{F}}(\hat{e}_n^*, e^*) \le \left\| \frac{\overline{F}}{\overline{\mathbb{F}}_n} \right\|_0^{b_n} \left\{ \sup_x \frac{|\int_x^\infty (\mathbb{F}_n - F) dI|}{h(F(x))e(x)} + \sup_x \frac{|\mathbb{F}_n(x) - F(x)|}{h(F(x))} \right\}$$

$$\le \left\| \frac{\overline{F}}{\overline{\mathbb{F}}_n} \right\|_0^{b_n} \cdot \rho_{h(F)}(\mathbb{F}_n, F)(M_1 + 1)$$

$$\to_{a.s.} 0$$

using Condition 1a, Theorem 1 of [46] to show $\rho_{h(F)}(\mathbb{F}_n, F) \to_{a.s.} 0$ a.s., and Theorem 2 of [47] to show that $\lim \sup_n \|\overline{F}/\overline{\mathbb{F}}_n\|_0^{b_n} < \infty$ a.s.

Similarly, using Condition 1b,

$$\rho_{1/\overline{F}}(\hat{e}_n^*, e^*) \le \left\| \frac{\overline{F}}{\overline{\mathbb{F}}_n} \right\|_0^{b_n} \left\{ \sup_x |\int_x^\infty (\mathbb{F}_n - F) dI| + \sup_x e(x)|\mathbb{F}_n(x) - F(x)| \right\}$$

$$\le \left\| \frac{\overline{F}}{\overline{\mathbb{F}}_n} \right\|_0^{b_n} \cdot \rho_{h(F)}(\mathbb{F}_n, F) \left(\int_0^\infty h(F) dI + \|eh(F)\| \right)$$

$$\to_{a.s.} 0.$$

$\square$

To extend Yang's weak convergence results, we will use the special uniform empirical processes $\mathbb{U}_n$ of the appendix of [44] or [45] which converge to a special Brownian bridge process $\mathbb{U}$ in the strong sense that

$$\rho_q(\mathbb{U}_n, \mathbb{U}) \to_p 0 \quad \text{as } n \to \infty$$

for $q \in \mathcal{Q}(\downarrow)$, the set of all continuous functions on $[0, 1]$ which are monotone decreasing on $[0, 1]$ and $\int_0^1 q^{-2} dI < \infty$. Thus $\mathbb{U}_n = n^{1/2}(\Gamma_n - I)$ on $[0, 1]$ where Γ_n is the empirical d.f. of special uniform $(0, 1)$ random variables $\xi_1, \ldots, \xi_n$.

Define the *mean residual life process* on $\mathbb{R}^+$ by

$$n^{1/2}(\hat{e}_n(x) - e(x))$$

$$= \frac{1}{\overline{\mathbb{F}}_n(x)} \left\{ -\int_x^\infty n^{1/2}(\mathbb{F}_n - F) dI + e(x) n^{1/2}(\mathbb{F}_n(x) - F(x)) \right\}$$

$$\stackrel{d}{=} \frac{1}{\overline{\Gamma}_n(F(x))} \left\{ -\int_x^\infty \mathbb{U}_n(F) dI + e(x) \mathbb{U}_n(F(x)) \right\}$$

$$\equiv \mathbb{Z}_n(x), \quad 0 \le x < F^{-1}(\xi_{nn})$$

where $\xi_{nn} = \max_{1 \le i \le n} \xi_i$, and $\mathbb{Z}_n(x) \equiv -n^{1/2} e(x)$ for $x \ge F^{-1}(\xi_{nn})$. Thus $\mathbb{Z}_n$ has the same law as $n^{1/2}(\hat{e}_n - e)$ and is a function of the special process $\mathbb{U}_n$. Define the corresponding limiting process $\mathbb{Z}$ by

$$\mathbb{Z}(x) = \frac{1}{\overline{F}(x)} \left\{ -\int_x^\infty \mathbb{U}(F) dI + e(x) \mathbb{U}(F(x)) \right\}, \quad 0 \le x < \infty. \quad (2.3)$$

If $\sigma^2 = Var(X) < \infty$ (and hence under either Condition 2a or 2b below), $\mathbb{Z}$ is a mean zero Gaussian process on $\mathbb{R}^+$ with covariance function described as follows:

Proposition 2.1 *Suppose that* $\sigma^2 = Var(X) < \infty$. *For* $0 \le x \le y < \infty$

$$Cov[\mathbb{Z}(x), \mathbb{Z}(y)] = \frac{\overline{F}(y)}{\overline{F}(x)} Var[\mathbb{Z}(y)] = \frac{\sigma^2(y)}{\overline{F}(y)} \quad (2.4)$$

where

$$\sigma^2(t) \equiv Var[X - t \mid X > t] = \frac{\int_t^\infty (x - t)^2 F(x)}{\overline{F}(t)} - e^2(t)$$

is the residual variance function; also

$$Cov[\mathbb{Z}(x)\overline{F}(x), \mathbb{Z}(y)\overline{F}(y)] = Var[\mathbb{Z}(y)\overline{F}(y)] = \overline{F}(y)\sigma^2(y). \quad (2.5)$$

Proof It suffices to prove (2.5). Let $\mathbb{Z}' \equiv \mathbb{Z}\overline{F}$; from (2.3) we find

$$Cov[\mathbb{Z}'(x), \mathbb{Z}'(y)] = e(x)e(y)F(x)\overline{F}(y) - e(x)\int_y^\infty F(x)\overline{F}(z)dz$$

$$- e(y)\int_x^\infty (F(y \wedge z) - F(y)F(z))dz$$

$$+ \int_x^\infty \int_y^\infty (F(z \wedge w) - F(z)F(w))dzdw.$$

Expressing integrals over (x, ∞) as the sum of integrals over (x, y) and (y, ∞), and recalling the defining formula for $e(y)$, we find that the right side reduces to

$$\int_y^\infty \int_y^\infty (F(z \wedge z) - F(z)F(w))dzdw - e^2(y)F(y)\overline{F}(y)$$

$$= \int_y^\infty (t - y)^2 dF(t) - \overline{F}(y)e^2(y)$$

$$= \overline{F}(y)\sigma^2(y)$$

which, being free of x, is also $Var[\mathbb{Z}'(y)]$.   □

As in this proposition, the process $\mathbb{Z}$ is often more easily studied through the process $\mathbb{Z}' = \mathbb{Z}\overline{F}$; such a study continues in Sect. 5. Study of the variance of $\mathbb{Z}(x)$, namely $\sigma^2(x)/\overline{F}(x)$, for large x appears in Sect. 3.

Condition 2a $\sigma^2 < \infty$ and there exists $q \in \mathcal{Q}(\downarrow)$ such that

$$M_2 \equiv M_2(q, F) \equiv \sup_x \frac{\int_x^\infty q(F)dI/q(F(x))}{e(x)} < \infty.$$

Since $0 < q(0) < \infty$ and $e(0) = E(X) < \infty$, Condition 2a implies that $\int_0^\infty q(F)dI < \infty$; Condition 2a may be rephrased as: $M_2 \equiv \|e_{q(F)/q(0)}/e_F\| < \infty$ where $e_{q(F)/q(0)}$ denotes the mean residual life function for the survival function $q(F)/q(0)$.

Condition 2b $\sigma^2 < \infty$ and there exists $q \in \mathcal{Q}(\downarrow)$ such that $\int_0^\infty q(F)dI < \infty$.

Bounded e_F and existence of a moment of order greater than 2 is more than sufficient for Condition 2b (see Sect. 3).

Theorem 2.2 (Process Convergence) *Let* $a_n \to 0$, $na_n \to \infty$. *If Condition 2a holds for a particular* $q \in \mathcal{Q}(\downarrow)$, *then*

$$\rho_{q(F)e/\overline{F}}(\mathbb{Z}_n^*, \mathbb{Z}^*)$$

$$\equiv \sup\left\{\frac{|\mathbb{Z}_n(x) - \mathbb{Z}(x)|\overline{F}(x)}{q(F(x))e(x)} : x \le b_n\right\} \to_p 0 \quad as \ n \to \infty. \quad (2.6)$$

If Condition 2b holds, then

$$\rho_{1/\overline{F}}(\mathbb{Z}_n^*, \mathbb{Z}^*) \equiv \sup \left\{ |\mathbb{Z}_n(x) - \mathbb{Z}(x)| \overline{F}(x) : x \leq b_n \right\} \to_p 0 \quad as \ n \to \infty. \tag{2.7}$$

Proof First write

$$\mathbb{Z}_n(x) - \mathbb{Z}(x) = \left\{ \frac{\overline{F}(x)}{\overline{\Gamma}_n(F(x))} - 1 \right\} \mathbb{Z}_n^1(x) + \left(\mathbb{Z}_n^1(x) - \mathbb{Z}(x) \right)$$

where

$$\mathbb{Z}_n^1(x) \equiv \frac{1}{\overline{F}(x)} \left\{ -\int_x^\infty \mathbb{U}_n(F) dI + e(x) \mathbb{U}_n(F(x)) \right\}, \quad 0 \leq x < \infty.$$

Then note that, using Condition 2a,

$$\rho_{q(F)e/\overline{F}}(\mathbb{Z}_n^1, 0) \leq \sup_x \frac{\left| \int_x^\infty \mathbb{U}_n(F) dI \right|}{q(F(x))e(x)} + \rho_q(\mathbb{U}_n, 0)$$

$$\leq \rho_q(\mathbb{U}_n, 0)\{M_2 + 1\} = O_p(1);$$

that $\| \overline{I}/\overline{\Gamma}_n - 1 \|_0^{1-a_n} \to_p 0$ by Theorem 0 of [47] since $na_n \to \infty$; and, again using Condition 2a, that

$$\rho_{q(F)e/\overline{F}}(\mathbb{Z}_n^1, \mathbb{Z}) \leq \sup_x \frac{\left| \int_x^\infty (\mathbb{U}_n(F) - \mathbb{U}(F)) dI \right|}{q(F(x))e(x)} + \rho_q(\mathbb{U}_n, \mathbb{U})$$

$$\leq \rho_q(\mathbb{U}_n, \mathbb{U})\{M_2 + 1\} \to_p 0.$$

Hence

$$\rho_{q(F)e/\overline{F}}(\mathbb{Z}_n^*, \mathbb{Z}^*) \leq \left\| \frac{\overline{I}}{\overline{\Gamma}_n} - 1 \right\|_0^{1-a_n} \rho_{q(F)e/\overline{F}}(\mathbb{Z}_n^1, 0) + \rho_{q(F)e/\overline{F}}(\mathbb{Z}_n^1, \mathbb{Z})$$

$$= o_p(1)O_p(1) + o_p(1) = o_p(1).$$

Similarly, using Condition 2b

$$\rho_{1/\overline{F}}(\mathbb{Z}_n^1, 0) \leq \sup_x \left| \int_x^\infty \mathbb{U}_n(F) dI \right| + \sup_x e(x) |\mathbb{U}_n(F(x))|$$

$$\leq \rho_q(\mathbb{U}_n, 0) \left\{ \int_0^\infty q(F) dI + \|eq(F)\| \right\} = O_p(1),$$

$$\rho_{1/\overline{F}}(\mathbb{Z}_n^1, \mathbb{Z}) \leq \sup_x \left| \int_x^\infty (\mathbb{U}_n(F) - \mathbb{U}(F))dI \right| + \sup_x e(x)|\mathbb{U}_n(F(x)) - \mathbb{U}(F(x))|$$

$$\leq \rho_q(\mathbb{U}_n, \mathbb{U}) \left\{ \int_0^\infty q(F)dI + \|eq(F)\| \right\} \to_p 0,$$

and hence

$$\rho_{1/\overline{F}}(\mathbb{Z}_n^*, \mathbb{Z}^*) \leq \left\| \frac{\overline{I}}{\overline{\Gamma}_n} - 1 \right\|_0^{1-a_n} \rho_{1/\overline{F}}(\mathbb{Z}_n^1, 0) + \rho_{1/\overline{F}}(\mathbb{Z}_n^1, \mathbb{Z})$$

$$= o_p(1)O_p(1) + o_p(1) = o_p(1).$$

$\square$

3 Alternative Sufficient Conditions: $Var[\mathbb{Z}(x)]$ as $x \to \infty$

Our goal here is to provide easily checked conditions which will imply the somewhat cumbersome Conditions 2a and 2b; similar conditions also appear in the work of [1], and we use their results to extend their formula for the residual coefficient of variation for large x (Eq. (3.1) below). This provides a simple description of the behavior of $Var[\mathbb{Z}(x)]$, the asymptotic variance of $n^{1/2}(\hat{e}_n(x) - e(x))$ as $x \to \infty$.

Condition 3 $E(X^r) < \infty$ for some $r > 2$.

Condition 4a Condition 3 and $\lim_{x \to \infty} d(1/\lambda(x))/dx = c < \infty$ where $\lambda = f/\overline{F}$, the hazard function.

Condition 4b Condition 3 and $\lim \sup_{x \to \infty} \{\overline{F}(x)^{1+\gamma}/f(x)\} < \infty$ for some $r^{-1} < \gamma < 1/2$.

Proposition 3.1 *If Condition 4a holds, then $0 \leq c \leq r^{-1}$, Condition 2a holds, and the squared residual coefficient of variation tends to $1/(1 - 2c)$:*

$$\lim_{x \to \infty} \frac{\sigma^2(x)}{e^2(x)} = \frac{1}{1 - 2c}. \tag{3.1}$$

If Condition 4b holds, then Condition 2b holds.

Corollary 3.1 *Condition 4a implies*

$$Var[\mathbb{Z}(x)] \sim \frac{e^2(x)}{\overline{F}(x)}(1 - 2c)^{-1} \quad as \ x \to \infty.$$

Proof Assume Condition 4a holds. Choose γ between r^{-1} and $1/2$; define a d.f. G on $\mathbb{R}^+$ by $\overline{G} = \overline{F}^\gamma$ and note that $g/\overline{G} = \gamma f/\overline{F} = \gamma\lambda$. By Condition 3 $x^r\overline{F}(x) \to 0$

as $x \to \infty$ and hence $x^{\gamma r} \overline{G}(x) \to 0$ as $x \to \infty$. Since $\gamma r > 1$, G has a finite mean and therefore $e_G(x) = \int_x^\infty \overline{G} dI / \overline{G}(x)$ is well-defined.

Set $\eta = 1/\lambda = \overline{F}/f$, and note that $\eta(x)\overline{G}(x) \to 0$ as $x \to \infty$. If $\limsup \eta(x) < \infty$, then it holds trivially; otherwise, $\eta(x) \to \infty$ (because of Condition 4a) and $\lim \eta(x)\overline{G}(x) = \lim (\eta(x)/x)(x\overline{G}(x)) = \lim \eta''(x)x\overline{G}(x) = 0$ by Condition 4a and L'Hopital's rule. Thus by L'Hopital's rule

$$0 \le \lim \frac{\eta(x)}{e_G(x)} = \lim \frac{\eta(x)\overline{G}(x)}{\int_x^\infty \overline{G} dI}$$

$$= \lim \frac{\eta(x)g(x) - \overline{G}(x)\eta'(x)}{\overline{G}(x)}$$

$$= \gamma - \lim \eta'(x) = \gamma - c \quad \text{by Condition 4a.}$$

Thus $c \le \gamma$ for any $\gamma > r^{-1}$ and it follows that $c \le r^{-1}$. It is elementary that $c \ge 0$ since $\eta = 1/\lambda$ is nonnegative.

Choose $q(t) = (1-t)^\gamma$. Then $\gamma - c > 0$, $q \in \mathcal{Q}(\downarrow)$, and to verify Condition 2a it now suffices to show that $\lim (\eta(x)/e_F(x)) = 1 - c < \infty$ since it then follows that

$$\lim \frac{e_G(x)}{e_F(x)} = \lim \frac{\eta(x)/e_F(x)}{\eta(x)/e_G(x)} = \frac{1-c}{\gamma - c} < \infty.$$

By continuity and $e_G(0) < \infty$, $0 < e_F(0) < \infty$, this implies Condition 2a. But $r > 2$ implies that $x\overline{F}(x) \to 0$ as $x \to \infty$ so $\eta(x)\overline{F}(x) \to 0$ and hence

$$\lim \frac{\eta(x)}{e_F(x)} = \lim \frac{\eta \overline{F}(x)}{\int_x^\infty \overline{F} dI} = \lim(1 - \eta'(x)) = 1 - c.$$

That (3.1) holds will now follow from results of [1], as follows: their corollary to Theorem 7 implies that $P(\lambda(t)(X - t) > x \mid X > t) \to \exp(-x)$ if $c = 0$ and $\to (1 + cx)^{-1/c}$ if $c > 0$. Thus, in the former case, F is in the domain of attraction of the Pareto residual life distribution and its related extreme value distribution. Then Theorem 8(a) implies convergence of the (conditional) mean and variance of $\lambda(t)(X - t)$ to the mean and variance of the limiting Pareto distribution, namely $(1-c)^{-1}$ and $(1-c)^{-2}(1-2c)^{-1}$. But the conditional mean of $\lambda(t)(X_t)$ is simply $\lambda(t)e(t)$, so that $\lambda(t) \sim (1-c)^{-1}/e(t)$ and (3.1) now follows.

If Condition 4b holds, let $q(F) = \overline{F}^\gamma$ again. Then $\int_0^\infty q(F)dI < \infty$, and it remains to show that $\limsup\{e(x)\overline{F}(x)^\gamma\} < \infty$. This follows from Condition 4b by L'Hopital's rule. □

Similarly, sufficient conditions for Conditions 1a and 1b can be given: simply replace "2" in Condition 3 and "1/2" in Condition 4b with "1," and the same proof works. Whether (3.1) holds when r in Condition 3 is exactly 2 is not known.

4 Examples

The typical situation, when $e(x)$ has a finite limit and Condition 3 holds, is as follows: $e \sim \overline{F}/f \sim f/(-f')$ as $x \to \infty$ (by L'Hopital's rule), and hence Conditions 4b, 2b, and 1b hold; also $\eta' \equiv (\overline{F}/f)' = [(F/f)(-f/f')] - 1 \to 0$ (Condition 4a with $c = 0$, and hence Conditions 2a and 1a hold), $\sigma(x) \sim e(x)$ from (3.1), and $Var[\mathbb{Z}] \sim e^2/\overline{F} \sim (\overline{F}/f)^2/\overline{F} \sim 1/(-f')$. We treat three examples, not all "typical," in more detail.

Example 4.1 (Exponential) Let $\overline{F}(x) = \exp(-x/\theta)$, $x \ge 0$, with $0 < \theta < \infty$. Then $e(x) = \theta$ for all $x \ge 0$. Conditions 4a and 4b hold (for all $r, \gamma \ge 0$) with $c = 0$, so Conditions 2a and 2b hold by Proposition 3.1 with $q(t) = (1-t)^{1/2-\delta}$, $0 < \delta < 1/2$. Conditions 1a and 1b hold with $h(t) = (1-t)^{1-\delta}$, $0 < \delta < 1$. Hence Theorems 2.1 and 2.2 hold where now

$$\mathbb{Z}(x) = \frac{\mathbb{U}(F(x))}{1 - (x)} - \frac{1}{1 - F(x)} \int_{F(x)}^{1} \frac{\mathbb{U}}{1 - I} dI \stackrel{d}{=} \theta \mathbb{B}(\exp(x/\theta)), \quad 0 \le x < \infty$$

and $\mathbb{B}$ is standard Brownian motion on $[0, \infty)$. (The process $\mathbb{B}_1(t) = \mathbb{U}(1-t) - \int_{1-t}^{1}(\mathbb{U}/(1-I))dI$, $0 \le t \le 1$, is Brownian motion on $[0, 1]$; and with $\mathbb{B}_2(x) \equiv x\mathbb{B}_1(1/x)$ for $1 \le x \le \infty$, $\mathbb{Z}(x) = \theta \mathbb{B}_2(1/\overline{F}(x)) = \theta \mathbb{B}_2(\exp(x/\theta))$.) Thus, in agreement with (2.4),

$$Cov[\mathbb{Z}(x), \mathbb{Z}(y)] = \theta^2 \exp((x \wedge y)/\theta), \quad 0 \le x, y < \infty.$$

An immediate consequence is that $\|\mathbb{Z}_n^* \overline{F}\| \to_d \|\mathbb{Z}\overline{F}\| \stackrel{d}{=} \theta \sup_{0 \le t \le 1} |\mathbb{B}_1(t)|$; generalization of this to other F's appears in Sect. 5. (Because of the "memoryless" property of exponential F, the results for this example can undoubtedly be obtained by more elementary methods.)

Example 4.2 (Weibull) Let $\overline{F}(x) = \exp(-x^\theta)$, $x \ge 0$, with $0 < \theta < \infty$. Conditions 4a and 4b hold (for all $r, \gamma > 0$) with $c = 0$, so Conditions 1 and 2 hold with h and q as in Example 4.1 by Proposition 3.1. Thus Theorems 2.1 and 2.2 hold. Also, $e(x) \sim \theta^{-1}x^{1-\theta}$ as $x \to \infty$, and hence $Var[Z(x)] \sim \theta^{-2}x^{2(1-\theta)} \exp(x^\theta)$ as $x \to \infty$.

Example 4.3 (Pareto) Let $\overline{F}(x) = (1 + cx)^{-1/c}$, $x \ge 0$, with $0 < c < 1/2$. Then $e(x) = (1-c)^{-1}(1+cx)$, and Conditions 4a and 4b hold for $r < c^{-1}$ and $\gamma \ge c$ (and c of Condition 4a is c). Thus Proposition 3.1 holds with $r > 2$ and $c > 0$ and $Var[\mathbb{Z}(x)] \sim c^{2+(1/c)}(1-c)^{-2}(1-2c)^{-1}x^{2+(1/c)}$ as $x \to \infty$. Conditions 1 and 2 hold with h and q as in Example 4.1, and Theorems 2.1 and 2.2 hold.

If instead $1/2 \le c < 1$, then $E(X) < \infty$ but $E(X^2) = \infty$, and Conditions 4a and 4b hold with $1 < r < 1/c \le 2$ and $\gamma \ge c$. Hence Condition 1 and Theorem 2.1 hold, but Condition 2 (and hence our proof of Theorem 2.2) fails. If $c \ge 1$, then $E(X) = \infty$ and $e(x) = \infty$ for all $x \ge 0$.

Not surprisingly, the limiting process $\mathbb{Z}$ has a variance which grows quite rapidly, exponentially in the exponential and Weibull cases, and as a power (> 4) of x in the Pareto case.

5 Confidence Bands for e

We first consider the process $\mathbb{Z}' \equiv \mathbb{Z}\overline{F}$ on $\mathbb{R}^+$ which appeared in (2.5) of Proposition 2.1. Its sample analog $\mathbb{Z}'_n \equiv \mathbb{Z}_n\overline{\mathbb{F}}_n$ is easily seen to be a cumulative sum (times $n^{-1/2}$) of the observations exceeding x, each centered at $x + e(x)$; as x decreases the number of terms in the sum increases. Moreover, the corresponding increments apparently act asymptotically independently so that $\mathbb{Z}'_n$, in reverse time, is behaving as a cumulative sum of zero-mean independent increments. Adjustment for the non-linear variance should lead to Brownian motion. Let us return to the limit version $\mathbb{Z}'$.

The zero-mean Gaussian process $\mathbb{Z}'$ has covariance function $Cov[\mathbb{Z}'(x), \mathbb{Z}'(y)] = Var[\mathbb{Z}'(x \vee y)]$ (see Eq. (2.5)); hence, when viewed in reverse time, it has independent increments (and hence $\mathbb{Z}'$ is a reverse martingale). Specifically, with $\mathbb{Z}''(s) \equiv \mathbb{Z}'(-\log s)$, $\mathbb{Z}''$ is a zero-mean Gaussian process on $[0, 1]$ with independent increments and $Var[\mathbb{Z}''(s)] = Var[\mathbb{Z}'(-\log s)] \equiv \tau^2(s)$. Hence τ^2 is increasing in s, and, from (2.5),

$$\tau^2(s) = \overline{F}(-\log s)\sigma^2(-\log s). \tag{5.1}$$

Now $\tau^2(1) = \sigma^2(0) = \sigma^2$, and

$$\tau^2(0) = \lim_{\epsilon \downarrow 0} \overline{F}(-\log \epsilon)\sigma^2(-\log \epsilon) = \lim_{x \to \infty} \overline{F}(x)\sigma^2(x) = 0$$

since

$$0 \leq \overline{F}(x)\sigma^2(x) \leq \overline{F}(x)E(X^2 \mid X > x) = \int_x^\infty y^2 dF(y) \to 0.$$

Since $f(x) > 0$ for all $x \geq 0$, τ^2 is strictly increasing.

Let g be the inverse of τ^2; then $\tau^2(g(t)) = t$, $g(0) = 0$, and $g(\sigma^2) = 1$. Define $\mathbb{W}$ on $[0, 1]$ by

$$\mathbb{W}(t) \equiv \sigma^{-1}\mathbb{Z}''(g(\sigma^2 t)) = \sigma^{-1}\mathbb{Z}'\big(-\log g(\sigma^2 t)\big). \tag{5.2}$$

Theorem 5.1 $\mathbb{W}$ *is standard Brownian motion on* $[0, 1]$.

Proof $\mathbb{W}$ is Gaussian with independent increments and $Var[\mathbb{W}(t)] = t$ by direct computation. $\qquad\square$

Corollary 5.1 *If (2.7) holds, then*

$$\rho\left(\mathbb{Z}_n^{'*}, \mathbb{Z}^{'*}\right) \equiv \sup_{x \le b_n} |\mathbb{Z}_n(x)\overline{\mathbb{F}}_n(x) - \mathbb{Z}(x)\overline{F}(x)| = o_p(1)$$

and hence $\|\mathbb{Z}_n\overline{\mathbb{F}}_n\|_0^{b_n} \to_d \|\mathbb{Z}\overline{F}\| = \sigma \|\mathbb{W}\|_0^1$ *as* $n \to \infty$.

Proof By Theorem 0 of [47] $\|\overline{\mathbb{F}}_n/\overline{F} - 1\|_0^{b_n} \to_p 0$ as $n \to \infty$, and this together with (2.7) implies the first part of the statement. The second part follows immediately from the first and (5.2). □

Replacement of σ^2 by a consistent estimate S_n^2 (e.g., the sample variance based on all observations), and of b_n by $\hat{b}_n = \mathbb{F}_n^{-1}(1 - a_n)$, the $(n - m)$−th order statistic with $m = [na_n]$, leads to asymptotic confidence bands for $e = e_F$:

Corollary 5.2 *Let* $0 < a < \infty$, *and set* $\hat{d}_n(x) \equiv n^{-1/2}aS_n/\overline{\mathbb{F}}_n(x)$. *If (2.7) holds,* $S_n^2 \to_p \sigma^2$, *and* $na_n/\log\log n \uparrow \infty$, *then, as* $n \to \infty$

$$P\left(\hat{e}_n(x) - \hat{d}_n(x) \le e(x) \le \hat{e}_n(x) + \hat{d}_n(x) \text{ for all } 0 \le x \le \hat{b}_n\right) \to Q(a) \tag{5.3}$$

where

$$Q(a) \equiv P\left(\|\mathbb{W}\|_0^1 < a\right) = \sum_{k=-\infty}^{\infty} (-1)^k\left\{\Phi((2k + 1)a) - \Phi((2k - 1)a)\right\}$$

$$= 1 - 4\left\{\overline{\Phi}(a) - \overline{\Phi}(3a) + \overline{\Phi}(5a) - \cdots\right\}$$

and Φ *denotes the standard normal d.f.*

Proof It follows immediately from Corollary 5.1 and $S_n \to_p \sigma > 0$ that

$$\|\mathbb{Z}_n\overline{\mathbb{F}}_n\|_0^{b_n}/S_n \to_d \|\mathbb{Z}\overline{F}\|/\sigma = \|\mathbb{W}\|_0^1.$$

Finally b_n may be replaced by $\hat{b}_n$ without harm: letting $c_n = 2\log\log/(na_n) \to 0$ and using Theorem 4S of [47], for $\tau > 1$ and all $n \ge N(\omega, \tau)$, $\hat{b}_n \equiv \mathbb{F}_n^{-1}(1 - a_n) \overset{d}{=}$ $F^{-1}(\Gamma_n^{-1}(1 - a_n)) \le F^{-1}(\{1 + \tau c_n^{1/2}\}(1 - a_n))$ w.p. 1. This proves the convergence claimed in the corollary; the expression for $Q(a)$ is well-known (e.g., see [4, page 79]). □

The approximation $1 - 4\overline{\Phi}(a)$ for $Q(a)$ gives 3-place accuracy for $a > 1.4$. See Table 1.

Thus, choosing a so that $Q(a) = \beta$, (5.3) provides a two-sided simultaneous confidence band for the function e with confidence coefficient asymptotically β. In applications we suggest taking $a_n = n^{-1/2}$ so that $\hat{b}_n$ is the $(n - m)$−th order statistic with $m = [n^{1/2}]$; we also want m large enough for an adequate central limit effect, remembering that the conditional life distribution may be quite skewed.

Table 1 $Q(a)$ for selected a

a	$Q(a)$	a	$Q(a)$
2.807	0.99	1.534	0.75
2.241	0.95	1.149	0.50
1.960	0.90	0.871	0.25

(In a similar fashion, one-sided asymptotic bands are possible, but they will be less trustworthy because of skewness.)

Instead of simultaneous bands for all real x one may seek (tighter) bands on $e(x)$ for one or two specific $x-$values. For this we can apply Theorem 2.2 and Proposition 2.1 directly. We first require a consistent estimator of the asymptotic variance of $n^{1/2}(\hat{e}_n(x) - e(x))$, namely $\sigma^2(x)/\overline{F}(x)$.

Proposition 5.1 *Let* $0 \leq x < \infty$ *be fixed and let* $S_n^2(x)$ *be the sample variance of those observations exceeding* x. *If Condition 3 holds then* $S_n^2(x)/\overline{\mathbb{F}}_n(x) \to_{a.s.} \sigma^2(x)/\overline{F}(x)$.

Proof Since $\overline{\mathbb{F}}_n(x) \to_{a.s.} \overline{F}(x) > 0$ and

$$S_n^2(x) = \frac{2 \int_x^\infty (y - x)\overline{\mathbb{F}}_n(y)dy}{\overline{\mathbb{F}}_n(x)} - \hat{e}_n^2(x),$$

it suffices to show that $\int_x^\infty y\overline{\mathbb{F}}_n(y)dy \to_{a.s.} \int_x^\infty y\overline{F}(y)dy$. Let $h(t) = (1 - t)^{\gamma+1/2}$ and $q(t) = (1 - t)^\gamma$ with $r^{-1} < \gamma < 1/2$ so that $h \in \mathcal{H}(\downarrow)$, $q \in \mathcal{Q}(\downarrow)$, and $\int_0^\infty q(F)dI < \infty$ by the proof of Proposition 3.1. Then,

$$\left| \int_x^\infty y\overline{\mathbb{F}}_n(y)dy - \int_x^\infty y\overline{F}(y)dy \right| \leq \rho_{h(F)}(\mathbb{F}_n, F) \int_0^\infty Ih(F)dI \to_{a.s.} 0$$

by Theorem 1 of Wellner (1977) since

$$\int_0^\infty Ih(F)dI = \int_0^\infty (I^2\overline{F})^{1/2}q(F)dI < \infty.$$

$\square$

By Theorem 2.2, Propositions 2.1 and 5.1, and Slutsky's theorem we have:

Proposition 5.2 *Under the conditions of Proposition 5.1,*

$$d_n(x) \equiv n^{1/2}(\hat{e}_n(x) - e(x))\overline{\mathbb{F}}_n^{1/2}(x)/S_n(x) \to_d N(0, 1) \quad as \ n \to \infty.$$

This makes feasible an asymptotic confidence interval for $e(x)$ (at this particular fixed x). Similarly, for $x < y$, using the joint asymptotic normality of $(d_n(x), d_n(y))$ with asymptotic correlation $\{\overline{F}(y)\sigma^2(y)/\overline{F}(x)\sigma^2(x)\}^{1/2}$ estimated by

$$\{\overline{\mathbb{F}}_n(y)S_n^2(y)/\overline{\mathbb{F}}_n(x)S_n^2(x)\}^{1/2},$$

an asymptotic confidence ellipse for $(e(x), e(y))$ may be obtained.

6 Illustration of the Confidence Bands

We illustrate with two data sets presented by Bjerkedal [5] and briefly mention one appearing in Barlow and Campo [2].

Bjerkedal gave various doses of *tubercle bacilli* to groups of 72 guinea pigs and recorded their survival times. We concentrate on Regimens 4.3 and 6.6 (and briefly mention 5.5, the only other complete data set in Bjerkedal's study M); see Figs. 1 and 2.

First consider the estimated mean residual life $\hat{e}_n$, the center jagged line in each figure. Figure 1 has been terminated at day 200; the plot would continue approximately horizontally, but application of asymptotic theory to this part of $\hat{e}_n$, based only on the last 23 survival times (the last at 555 days), seems unwise. Figure 2 has likewise been terminated at 200 days, omitting only nine survival times (the last at 376 days); the graph of $\hat{e}_n$ would continue downward. The dashed diagonal line is $\overline{X} - x$; if all survival times were equal, say μ, then the residual life function would be $(\mu - x)^+$, a lower bound on $e(x)$ near the origin. More

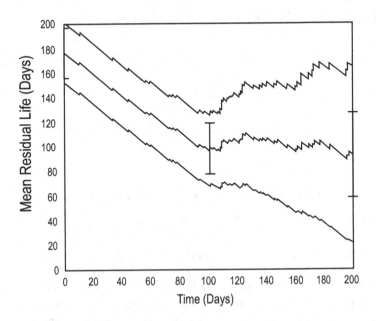

Fig. 1 90% confidence bands for mean residual life; Regimen 4.3

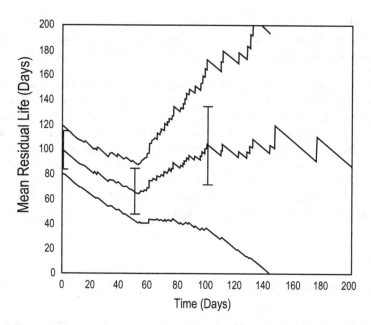

Fig. 2 90% confidence bands for mean residual life; Regimen 6.6

specifically, a Maclaurin expansion yields

$$e(x) = \mu + (\mu f_0 - 1)x + (1/2)\{(2\mu f_0 - 1)f_0 + f_0'\}x^2 + o(x^2)$$

where $f_0 = f(0)$, $f_0' = f'(0)$, if f' is continuous at 0, or

$$e(x) = \mu - x + \frac{\mu d}{r!}x^r + o(x^r)$$

if $f^{(s)}(0) = 0$ for $s < (r-1)$ (≥ 0) and $= d$ for $s = r - 1$ (if $f^{(r-1)}$ is continuous at 0). It thus seems likely from Figs. 1 and 2 that in each of these cases either $f_0 = 0$ and $f_0' > 0$ or f_0 is near 0 (and $f_0' \geq 0$).

Also, for large x, $e(x) \sim 1/\lambda(x)$, and Fig. 1 suggests that the corresponding λ and e have finite positive limits, whereas the e of Fig. 2 may eventually decrease (λ increase). We know of no parametric F that would exhibit behavior quite like these.

The upper and lower jagged lines in the figures provide 90% (asymptotic) confidence bands for the respective e's, based on (5.3). At least for Regimen 4.3, a constant e (exponential survival) can be rejected.

The vertical bars at $x = 0$, $x = 100$, and $x = 200$ in Fig. 1, and at 0, 50, and 100 in Fig. 2, are 90% (asymptotic) pointwise confidence intervals on e at the corresponding x−values (based on Proposition 5.2). Notice that these intervals are not much narrower than the simultaneous bands early in the survival data, but are substantially narrower later on.

A similar graph for Regimen 5.5 (not presented) is somewhat similar to that in Fig. 2, with the upward turn in $\hat{e}_n$ occurring at 80 days instead of at 50, and a possible downward turn at somewhere around 250 days (the final death occurring at 598 days).

A similar graph was prepared for the failure data on 107 right rear tractor brakes presented by Barlow and Campo [2, page 462]. It suggests a quadratic decreasing e for the first 1500–2000 h (with $f(0)$ at or near 0 but $f'(0)$ definitely positive), with $\overline{X} = 2024$, and with a possibly constant of slightly increasing e from 1500 or so to 6000 h. The e for a gamma distribution with $\lambda = 2$ and $\alpha = 0.001$ ($e(x) = \alpha^{-1}(\alpha x + 2)/(\alpha x + 1)$ with $\alpha = 0.001$) fits reasonably well—i.e., is within the confidence bands, even for 25% confidence. Note that this is in excellent agreement with Figures 2.1(b) and 3.1(d) of [2]. (Bryson and Siddiqui's (1969) data set was too small ($n = 43$) for these asymptotic methods, except possibly early in the data set.)

7 Further Developments

The original version of this paper, [23], ended with a one-sentence sketch of two remaining problems: "Confidence bands on the difference between two mean residual life functions, and for the case of censored data, will be presented in subsequent papers." Although we never did address these questions ourselves, others took up these further problems.

Our aim in this final section is to briefly survey some of the developments since 1979 concerning mean residual life, including related studies of median residual life and other quantiles, as well as developments for censored data, alternative inference strategies, semiparametric models involving mean or median residual life, and generalizations to higher dimensions. For a review of further work up to 1988 see [21].

7.1 Confidence Bands and Inference

Csörgő et al. [16] gave a further detailed study of the asymptotic behavior of the mean residual life process as well as other related processes including the Lorenz curve. Berger et al. [3] developed tests and confidence sets for comparing two mean residual life functions based on independent samples from the respective populations. These authors also gave a brief treatment based on comparison of *median residual life*, to be discussed in Sect. 7.3 below. Csörgő and Zitikis [17] introduced weighted metrics into the study of the asymptotic behavior of the mean residual life process, thereby avoiding the intervals $[0, x_n]$ changing with n involved in our Theorems 2.1 and 2.2, and thereby provided confidence bands and intervals for e_F in the right tail. Zhao and Qin [51] introduced empirical likelihood methods to the study of the mean residual life function. They obtained confidence intervals and confidence bands for compact sets $[0, \tau]$ with $\tau < \tau_F \equiv \inf\{x : F(x) = 1\}$.

7.2 Censored Data

Yang [49] initiated the study of estimated mean residual life under random right censorship. She used an estimator $\hat{F}_n$ which is asymptotically equivalent to the Kaplan–Meier estimator and considered, in particular, the case when X is bounded and stochastically smaller than the censoring variable C. In this case she proved that $\sqrt{n}(\hat{e}(x) - e(x))$ converges weakly (as $n \to \infty$) to a Gaussian process with mean zero. Csörgő and Zitikis [17] give a brief review of the challenges involved in this problem; see their page 1726. Qin and Zhao [41] extended their earlier study [51] of empirical likelihood methods to this case, at least for the problem of obtaining pointwise confidence intervals. The empirical likelihood methods seem to have superior coverage probability properties in comparison to the Wald type intervals which follow from our Proposition 5.2. Chaubey and Sen [9] introduced smooth estimates of mean residual life in the uncensored case. In [7] they introduce and study smooth estimators of e_F based on corresponding smooth estimators of $\overline{F} = 1 - F$ introduced by Chaubey and Sen [8].

7.3 Median and Quantile Residual Life Functions

Because mean residual life is frequently difficult, if not impossible, to estimate in the presence of right censoring, it is natural to consider surrogates for it which do not depend on the entire right tail of F. Natural replacements include median residual life and corresponding *residual life quantiles*. The study of median residual life was apparently initiated in [43]. Characterization issues and basic properties have been investigated by Gupta and Langford [22], Joe and Proschan [30], and Lillo [34]. Joe and Proschan [29] proposed comparisons of two populations based on their corresponding median (and other quantile) residual life functions. As noted by Joe and Proschan, "Some results differ notably from corresponding results for the mean residual life function." Jeong et al. [28] investigated estimation of median residual life with right-censored data for one-sample and two-sample problems. They provided an interesting illustration of their methods using a long-term follow-up study (the National Surgical Adjuvant Breast and Bowel Project, NSABP) involving breast cancer patients.

7.4 Semiparametric Models for Mean and Median Residual Life

Oakes and Dasu [39] investigated a characterization related to a *proportional mean residual life* model: $e_G = \psi e_F$ with $\psi > 0$. Maguluri and Zhang [38] studied several methods of estimation in a semiparametric regression version of the proportional mean residual life model, $e(x \mid z) = \exp(\theta^T z)e_0(x)$ where $e(x \mid z)$ denotes the conditional mean residual life function given $Z = z$. Chen et al. [12] provide a nice review of various models and study estimation in the same semiparametric proportional mean residual life regression model considered by Maguluri and Zhang [38], but in the presence of right censoring. Their proposed estimation method involves inverse probability of censoring weighted (IPCW) estimation methods [26]; [42]. Chen and Cheng [10] use counting process methods to develop alternative estimators for the proportional mean residual life model in the presence of right censoring. The methods of estimation considered by Maguluri and Zhang [38], Chen et al. [12], and Chen and Cheng [10] are apparently inefficient. Oakes and Dasu [40] consider information calculations and likelihood based estimation in a two-sample version of the proportional mean residual life model. Their calculations suggest that certain weighted ratio-type estimators may achieve asymptotic efficiency, but a definitive answer to the issue of efficient estimation apparently remains unresolved. Cheng [11] proposed an alternative additive semiparametric regression model involving mean residual life. Ma and Yin [37] considered a large family of semiparametric regression models which includes both the additive model proposed by Chen and Cheng [11] and the proportional mean residual life model considered by earlier authors, but advocated replacing mean residual life by median residual life. Gelfand and Kottas [19] also developed a median residual life regression model with additive structure and took a semiparametric Bayesian approach to inference.

7.5 Monotone and Ordered Mean Residual Life Functions

Kochar et al. [32] consider estimation of e_F subject to the shape restrictions that e_F is increasing or decreasing. The main results concern ad-hoc estimators that are simple monotizations of the basic nonparametric empirical estimators $\hat{e}_n$ studied here. These authors show that the nonparametric maximum likelihood estimator does not exist in the increasing MRL case and that although the nonparametric MLE exists in the decreasing MRL case, the estimator is difficult to compute. Ebrahimi [18] and Hu et al. [27] study estimation of two mean residual life functions e_F and e_G in one- and two-sample settings subject to the restriction $e_F(x) \leq e_G(x)$ for all x. Hu et al. [27] also develop large sample confidence bands and intervals to accompany their estimators.

7.6 Bivariate Residual Life

Jupp and Mardia [31] defined a multivariate mean residual life function and showed that it uniquely determines the joint multivariate distribution, extending the known univariate result of [15]; see [24] for a review of univariate results of this type. See [35, 36] for further multivariate characterization results. Kulkarni and Rattihalli [33] introduced a bivariate mean residual life function and propose natural estimators.

References

1. Balkema, A. A., & de Haan, L. (1974). Residual life time at great age. *Annals of Probability*, *2*, 792–804.
2. Barlow, R. E., & Campo, R. (1975). Total time on test processes and applications to failure data analysis. In *Reliability and fault tree analysis (Conference, University of California, Berkeley, 1974); Conference on Reliability and Fault Tree Analysis* (pp. 451–481). Society for Industrial and Applied Mathematics; Philadelphia.
3. Berger, R. L., Boos, D. D., & Guess, F. M. (1988). Tests and confidence sets for comparing two mean residual life functions. *Biometrics, 44*, 103–115.
4. Billingsley, P. (1968). *Convergence of probability measures*. New York: Wiley.
5. Bjerkedal, T. (1960). Acquisition of resistance in guinea pigs infected with different doses of virulent tubercle bacilli. *American Journal of Hygiene, 72*, 130–148.
6. Bryson, M. C., & Siddiqui, M. M. (1969). Some criteria for aging. *Journal of the American Statistical Association, 64*, 1472–1483.
7. Chaubey, Y. P., & Sen, A. (2008). Smooth estimation of mean residual life under random censoring. In *Beyond parametrics in interdisciplinary research: Festschrift in honor of professor Pranab K. Sen*. Institute of Mathematical Statistics Collection (Vol. 1, pp. 35–49). Beachwood: Institute of Mathematical Statistics.
8. Chaubey, Y. P., & Sen, P. K. (1998). On smooth estimation of hazard and cumulative hazard functions. In *Frontiers in probability and statistics (Calcutta, 1994/1995)* (pp.91–99). New Delhi: Narosa.
9. Chaubey, Y. P., & Sen, P. K. (1999). On smooth estimation of mean residual life. *Journal of Statistical Planning and Inference, 75*, 223–236.
10. Chen, Y. Q., & Cheng, S. (2005). Semiparametric regression analysis of mean residual life with censored survival data. *Biometrika, 92*, 19–29.
11. Chen, Y. Q., & Cheng, S. (2006). Linear life expectancy regression with censored data. *Biometrika, 93*, 303–313.
12. Chen, Y. Q., Jewell, N. P., Lei, X., & Cheng, S. C. (2005). Semiparametric estimation of proportional mean residual life model in presence of censoring. *Biometrics, 61*, 170–178.
13. Chiang, C. L. (1960). A stochastic study of the life table and its applications: I. Probability distributions of the biometric functions. *Biometrics, 16*, 618–635.
14. Chiang, C. L. (1968). *Introduction to stochastic processes in biostatistics*. New York: Wiley.
15. Cox, D. R. (1962). *Renewal theory*. London: Methuen.
16. Csörgő, M., Csörgő, S., & Horváth, L. (1986). *An asymptotic theory for empirical reliability and concentration processes*. Lecture Notes in Statistics (Vol. 33). Berlin: Springer.
17. Csörgő, M., & Zitikis, R. (1996). Mean residual life processes. *Annals of Statistics, 24*, 1717–1739.
18. Ebrahimi, N. (1993). Estimation of two ordered mean residual lifetime functions. *Biometrics, 49*, 409–417.

19. Gelfand, A. E., & Kottas, A. (2003). Bayesian semiparametric regression for median residual life. *Scandinavian Journal of Statistics, 30*, 651–665.
20. Gross, A. J., Clark, V. A. (1975). *Survival distributions: Reliability applications in the biomedical sciences*. New York: Wiley.
21. Guess, F., & Proschan, F. (1988). Mean residual life: Theory and application. In *Handbook of statistics: Quality control and reliability* (Vol. 7, pp. 215–224). Amsterdam: North-Holland.
22. Gupta, R. C., & Langford, E. S. (1984). On the determination of a distribution by its median residual life function: A functional equation. *Journal of Applied Probability, 21*, 120–128.
23. Hall, W. J., & Wellner, J. A. (1979). Estimation of mean residual life. In *Technical Report, Department of Statistics*, University of Rochester.
24. Hall, W. J., & Wellner, J. (1981). Mean residual life. In *Statistics and related topics (Ottawa, Ont., 1980)* (pp. 169–184). Amsterdam: North-Holland.
25. Hollander, M., & Proschan, F. (1975). Tests for the mean residual life. *Biometrika, 62*, 585–593.
26. Horvitz, D. G., & Thompson, D. J. (1952). A generalization of sampling without replacement from a finite universe. *Journal of the American Statistical Association, 47*, 663–685.
27. Hu, X., Kochar, S. C., Mukerjee, H., & Samaniego, F. J. (2002). Estimation of two ordered mean residual life functions. *Journal of Statistical Planning and Inference, 107*, 321–341.
28. Jeong, J.-H., Jung, S.-H., & Costantino, J. P. (2008). Nonparametric inference on median residual life function. *Biometrics, 64*, 157–163.
29. Joe, H., & Proschan, F. (1984). Comparison of two life distributions on the basis of their percentile residual life functions. *The Canadian Journal of Statistics, 12*, 91–97.
30. Joe, H., & Proschan, F. (1984). Percentile residual life functions. *Operations Research, 32*, 668–678.
31. Jupp, P. E., & Mardia, K. V. (1982). A characterization of the multivariate Pareto distribution. *Annals of Statistics, 10*, 1021–1024.
32. Kochar, S. C., Mukerjee, H., & Samaniego, F. J. (2000). Estimation of a monotone mean residual life. *Annals of Statistics, 28*, 905–921.
33. Kulkarni, H. V., & Rattihalli, R. N. (2002). Nonparametric estimation of a bivariate mean residual life function. *Journal of the American Statistical Association, 97*, 907–917.
34. Lillo, R. E. (2005). On the median residual lifetime and its aging properties: a characterization theorem and applications. *Naval Research Logistics, 52*, 370–380.
35. Ma, C. (1996). Multivariate survival functions characterized by constant product of mean remaining lives and hazard rates. *Metrika, 44*, 71–83.
36. Ma, C. (1998). Characteristic properties of multivariate survival functions in terms of residual life distributions. *Metrika, 47*, 227–240.
37. Ma, Y., & Yin, G. (2010). Semiparametric median residual life model and inference. *The Canadian Journal of Statistics, 38*, 665–679.
38. Maguluri, G., & Zhang, C.-H. (1994). Estimation in the mean residual life regression model. *Journal of the Royal Statistical Society: Series B (Methodological), 56*, 477–489.
39. Oakes, D., & Dasu, T. (1990). A note on residual life. *Biometrika, 77*, 409–410.
40. Oakes, D., & Dasu, T. (2003). Inference for the proportional mean residual life model. In *Crossing boundaries: Statistical essays in honor of Jack Hall*. IMS Lecture Notes Monograph Series (Vol. 43, 105–116). Beachwood: Institute of Mathematical Statistics.
41. Qin, G., & Zhao, Y. (2007). Empirical likelihood inference for the mean residual life under random censorship. *Statistics & Probability Letters, 77*, 549–557.
42. Robins, J. M., & Rotnitzky, A. (1992). Recovery of information and adjustment for dependent censoring using surrogate markers. In *AIDS epidemiology, methodological issues* (pp. 297–331). Boston: Birkhauser
43. Schmittlein, D. C., & Morrison, D. G. (1981). The median residual lifetime: A characterization theorem and an application. *Operations Research, 29*, 392–399.
44. Shorack, G. R. (1972). Functions of order statistics. *Annals of Mathematical Statistics, 43*, 412–427.

45. Shorack, G. R., & Wellner, J. A. (1986). *Empirical processes with applications to statistics.* Wiley Series in Probability and Mathematical Statistics: Probability and Mathematical Statistics. New York: Wiley.

46. Wellner, J. A. (1977). A Glivenko-Cantelli theorem and strong laws of large numbers for functions of order statistics. *Annals of Statistics, 5,* 473–480.

47. Wellner, J. A. (1978). Limit theorems for the ratio of the empirical distribution function to the true distribution function. *Zeitschrift für Wahrscheinlichkeitstheorie und verwandte Gebiete, 45,* 73–88.

48. Wilson, E. B. (1938). The standard deviation of sampling for life expectancy. *Journal of the American Statistical Association, 33,* 705–708.

49. Yang, G. (1977/1978). Life expectancy under random censorship. *Stochastic Processes and Their Applications, 6,* 33–39.

50. Yang, G. L. (1978). Estimation of a biometric function. *Annals of Statistics, 6,* 112–116.

51. Zhao, Y., & Qin, G. (2006). Inference for the mean residual life function via empirical likelihood. *Communications in Statistics Theory Methods, 35,* 1025–1036.

Likelihood Transformations and Artificial Mixtures

Alex Tsodikov, Lyrica Xiaohong Liu, and Carol Tseng

Abstract In this paper we consider the generalized self-consistency approach to maximum likelihood estimation (MLE). The idea is to represent a given likelihood as a marginal one based on artificial missing data. The computational advantage is sought in the likelihood simplification at the complete-data level. Semiparametric survival models and models for categorical data are used as an example. Justifications for the approach are outlined when the model at the complete-data level is not a legitimate probability model or if it does not exist at all.

1 Artificial Mixtures

Let $p(x \mid z)$ be a family of probability distributions describing a model for the random response X regressed on covariates z. The idea of an artificial mixture is to represent $p(x \mid z)$ as a marginal probability (a mixture model)

$$p(x \mid z) = E\{p(x \mid z, U) \mid z\}, \qquad (1.1)$$

where U is a mixing variable, possibly a vector, representing artificial missing data, and $p_0 = p(\cdot \mid \cdot, U)$ are some complete-data probabilities conditional on U. The expectation is taken conditional on z implying that U is generally itself a regression

A. Tsodikov (✉)
University of Michigan, School of Public Health, Department of Biostatistics,
Ann Arbor, MI, USA
e-mail: tsodikov@umich.edu

L. X. Liu
Amgen, South San Francisco, CA, USA
e-mail: lyricaliu@gmail.com

C. Tseng
H2O Clinical, LLC, Hunt Valley, MD, USA
e-mail: carol.tseng@h2oclinical.com

© Springer Nature Switzerland AG 2020
A. Almudevar et al. (eds.), *Statistical Modeling for Biological Systems*,
https://doi.org/10.1007/978-3-030-34675-1_11

191

on z. In other words, an artificial mixture model is considered such that one gets the original target model when missing data are integrated out.

As we will see later, representation (1.1) can be considered a form of an integral transform of r.v. U. This makes finding the complete-data model $p(x \mid z, U)$ a matter of inverting the transform. The utility of (1.1) that is exploited in this paper is the simplicity of the complete-data model contrasted with the complexity of the original one. Specifically, let the vector of model parameters be partitioned into the parameters of interest β (regression coefficients), and a high-dimensional nuisance parameters α. The original MLE problem is based on maximization of the (now marginal) likelihood

$$L = \prod_i p(x_i \mid z_i), \quad \ell = \log L = \sum_i \log p(x_i \mid z_i) \qquad (1.2)$$

over α and β, a challenge due to the dimensionality of α. We are looking for the representation (1.1) to resolve the curse of dimensionality through a simple form of the complete-data likelihood

$$L_0(\alpha, \beta) = \prod_i p(x_i \mid z_i, U_i), \quad \ell_0(\alpha, \beta) = \log L_0(\alpha, \beta) \qquad (1.3)$$

such that the solution for the problem $\max_\alpha \ell_0(\alpha, \beta)$ with respect to the high-dimensional parameter α at the complete-data level is simple, and the kernel of the complete-data likelihood that depends on parameters of interest is linear in U. This simplification (1.3) allows us to fit the original model iteratively by using the EM idea:

Algorithm

1. Set initial α_0, hold out fixed β.
2. **E-Step.** Perform imputation of U_i for each data point i using the model at α_0, β_0.
3. **M-Step.** Maximize $\ell_0(\alpha, \beta)$ with respect to α. This is supposed to be a simple problem resulting in α_1.
4. Set α_0 to the α_1 found at the M-Step, check convergence, and go back to the E-Step if the convergence criterion is not satisfied.

This procedure converges to some $\hat\alpha(\beta)$, and generates the profile likelihood $\ell_{pr}(\beta) = \ell(\hat\alpha(\beta), \beta)$ that can be maximized by generic methods with respect to the low-dimensional parameter β.

Alternatively, maximization with respect to β can be done concurrently with α by using the M-Step:

3. **M-Step.** Maximize $\ell_0(\alpha, \beta)$ with respect to α, β resulting in α_1, β_1, provided this is still a simple problem.

Efficient execution of the E-Step is a crucial part of the above procedure. When the model is specified using the left side of (1.1), it represents a transform of the r.v. U. Obtaining the distribution of missing data U by inverting the transform (1.1), and then taking expectations over $\{U_i\}$ conditional on observed data $\{x_i\}$ as integrals over the distribution per classical EM algorithm recipe can quickly defeat the purpose of the self-consistency idea. First, inverting the transform (finding the form of the distribution of U) can prove difficult even with simple models. Second, numerical or Monte-Carlo integration at the E-Step can slow down the procedure considerably. The solution comes through recognition that imputed U is some kind of a conditional moment. Moments of random variables can be found by differentiating a transform (a pgf or a characteristic function). The form of the transform is readily available by the nature of the problem and is represented by the left part of (1.1). This observation translates into a closed form E-Step providing computational efficiency for the above procedure. This same observation allows one to generalize the procedure to the case when ℓ_0 is not a legitimate likelihood or when the r.v. U in the sense of (1.1) does not exist. The generalization is based on the so-called quasi-expectation operator (QE).

2 The Quasi-Expectation Operator and the Quasi-EM (QEM) Algorithm

Probabilities that define a marginal mixed model (1.1) can be viewed as an integral transform of the distribution of missing data. As a motivation, suppose U is non-negative and consider the Laplace transform of U

$$\mathcal{L}(s) = \mathrm{E}\left\{e^{-Us}\right\}. \tag{2.1}$$

Derivatives of the Laplace transform can be used to find moments of the corresponding random variable as well as expectations of the form

$$\frac{\partial^k \mathcal{L}(s)}{\partial s^k} = (-1)^k \mathrm{E}\left\{U^k e^{-Us}\right\}, \qquad \mathrm{E}\left\{U^k\right\} = (-1)^k \left.\frac{\partial^k \mathcal{L}(s)}{\partial s^k}\right|_{s=0}. \tag{2.2}$$

Imputation and the QEM construction (Q stands for *quasi*) is based on the quasi-expectation operator QE. Let $e_k(u, s)$ be some basis functions, where u is an argument of the function, and s is a parameter. Motivated by (2.2) we will consider

$$e_k(u, s) = u^k e^{-us}, \quad k = 0, 1, 2, \ldots. \tag{2.3}$$

We shall follow the rule that integral operators like E act on e as a function of u, while differential operators will act on e as a function of s, i.e., derivatives will be with respect to s unless noted otherwise. This convention follows the basic idea of integral transforms that replace the problem in the space of functions of u equipped with an integral operator by the one of s equipped with a differential operator. The space of admissible functions will be a linear span of the basis function up to a certain order K. As we will see later, multinomial and univariate survival models will require an order up to $K = 2$, while the multivariate survival models will need larger K dependent on the dimension of the survival response.

Note the following important properties of the basis functions.

- For our method to work the linear span of the basis functions with appropriately chosen parameter s must include the complete-data likelihood contributions p_0 or its kernel depending on the parameters of interest, and the functions submitted as an argument to QE at the E-Step as we will see later.
- The kernel that depends on s is log-linear in u. As model parameters will enter the formulation through s, this yields a complete-data log-likelihood kernel that is linear in u. This reduces the E-Step to the imputation of u by a conditional QE.
- Next function is a consequence of the derivative of the previous

$$e_{k+1} \equiv -e'_k, \tag{2.4}$$

where $\equiv$ denotes the uniform equality with respect to s. This ensures that such functions can be cloned based on specifying the first one and generating the others recurrently by differentiation. Also, this ensures that the values of the QE on basis functions can be cloned in a similar fashion provided QE and differentiation are interchangeable.

Formally, QE is defined as a linear operator on the linear span of $\{e_k\}$ such that

$$\text{QE}\{e_0(s)\} = \mathcal{L}(s), \tag{2.5}$$

$$\frac{\partial}{\partial s}\text{QE}\{e_k(s)\} = \text{QE}\left\{\frac{\partial}{\partial s}e_k(s)\right\}. \tag{2.6}$$

Obviously, QE is a generalization of E as its axioms (2.5), (2.6) are a subset of that of E. The generalization is achieved at the cost of a restricted set of admissible functions that QE can act on. To mimic the random variable notation used with the expectation we will write $\text{QE}\{f(U)\}$ for the result of QE acting on an admissible function $f(u)$. In the particular case when QE is an E, U is interpreted as a random variable. In the operator QE notation u is an argument of an admissible function. With this notation, and using (2.4), (2.5), and (2.6) we obtain the QE values on the set of basis functions (2.3) as

$$\text{QE}\{e_k(s)\} = (-1)^k \mathcal{L}^{(k)}(s) = \text{QE}\{U^k e^{-Us}\}, \tag{2.7}$$

where $(\cdot)^{(i)}$ is a derivative of the order i with respect to s (compare with (2.2)). In (2.5) and (2.7) $\mathcal{L}$ is not necessarily a Laplace transform, but rather a k-differentiable model-generating function (mgf) satisfying some properties motivated by the

Laplace transform. There are two reasons behind the name. First, note that if QE is an expectation, then $\mathcal{L}$ is a Laplace transform, and $\mathcal{L}(-\log x)$ is a probability-generating function (pgf) of U. Second, as follows from (2.7), the whole construction is determined by specifying the function $\mathcal{L}$ and differentiating it.

For any functions f, g, and their product $f \cdot g$ in the $\{e_k\}$ linear span, conditional QE is defined as

$$QE\{f \mid g\} = \frac{QE\{f \cdot g\}}{QE\{g\}}. \tag{2.8}$$

The QEM algorithm is a procedure designed to maximize a likelihood of the form

$$L(\omega) = QE\{L_0(\omega \mid u)\}, \tag{2.9}$$

over the model parameter vector ω, where $L_0(\omega \mid u)$, as a function of u, is admissible (belongs to the linear span of the basis functions that define the QE). The following recurrent expression defines the procedure:

$$\omega^{(m+1)} = \arg\max_{\omega} QE\left\{ \log L_0(\omega \mid u) \,\middle|\, L_0(\omega^{(m)} \mid u) \right\}, \tag{2.10}$$

where all the arguments of the conditional QE are supposed to be admissible, and m counts iterations. The conditional QE in the right part of (2.10) corresponds to the conditional expectation of the complete-data log-likelihood given observed data known from the classical EM framework. This interpretation turns into the fact when QE is an E.

When $\log L_0(\omega \mid u)$ is linear in u, the imputation is reduced to taking a conditional QE (2.8) with $f = e_1(u, 0) = u$ and $g = L_0(u)$. Similar to the EM algorithm, the QEM algorithm is monotonic when QE satisfies the operator Jensen inequality, with f, g representing complete-data likelihood contributions that depend on the parameters of interest, that is

$$QE\{\log(f) \mid g\} < \log[QE\{f \mid g\}]. \tag{2.11}$$

The proof of the monotonicity essentially repeats the classical argument of the EM algorithm of [4]. It was shown in [11] that for a univariate frailty model in survival analysis a characterization of the Jensen inequality (2.11) can be obtained through a simple property of the imputation operator. In this context, (2.11) must hold when $f = e_0(u, a)$ and $g = e_k(u, b)$, $k = 0, 1$, uniformly over non-negative a, b, u. To see this, and now for arbitrary non-negative integer k, consider a general EM/QEM monotonicity argument. Suppose, a contribution to the complete-data likelihood is given by some $L_0 = e_k(U, \omega) = U^k \exp(-u\omega)$ for some parameter of interest ω. In the EM/QEM algorithm, at the ith iteration, two instances of ω will be in play, $\omega^{(i)}$ defining the model for the imputation of U at the E-Step, and ω (that will become the next iteration $\omega^{(i+1)}$) over which maximization at the M-Step occurs. Based on the algorithm (2.10), the difference between the two log-likelihoods at iteration $i+1$ and i ωs in the EM/QEM algorithm has the form

$$\ell^{(i+1)} - \ell^{(i)} = \log \text{QE} \left\{ \frac{L_0^{(i+1)}}{L_0^{(i)}} \,\Big|\, L_0^{(i)} \right\} > \text{QE} \left\{ \log \frac{L_0^{(i+1)}}{L_0^{(i)}} \,\Big|\, L_0^{(i)} \right\},$$

where the inequality is supposed to be a consequence of (2.11), and the right part is positive because $\text{QE}\{\log L_0^{(i+1)} \mid L_0^{(i)}\}$ is the maximized value of the complete-data likelihood at the M-Step, while $\text{QE}\{\log L_0^{(i)} \mid L_0^{(i)}\}$ is some non-optimal (smaller) value in the same context. Now,

$$\frac{L_0^{(i+1)}}{L_0^{(i)}} = \frac{U^k \exp(-U\omega^{(i+1)})}{U^k \exp(-U\omega^{(i)})} = \exp(-U(\omega^{(i+1)} - \omega^{(i)})) = e_0(U, \omega^{(i+1)} - \omega^{(i)}).$$

This explains why we only need the Jensen inequality to work for $f = e_0(u, a)$ and $g = e_k(u, b)$ to ensure monotonicity of the resultant algorithm.

Define the imputation operator Θ of order k as

$$\Theta(s \mid k) = \text{QE}\{U \mid e_k\} = \frac{\text{QE}\{U^{k+1} \exp(-sU)\}}{\text{QE}\{U^k \exp(-sU)\}} = -\frac{\mathcal{L}^{(k+1)}(s)}{\mathcal{L}^{(k)}(s)}. \qquad (2.12)$$

A sufficient condition for the validity of the Jensen inequality defined above, and its characterization through the imputation, is that (2.12), expressing imputations of U conditional on observed data, represents non-increasing functions of s for any k, which is a subject of the proposition below. We will consider Θ defined for all k up to some K, and will choose K according to the specific model. When QE operators are expectations, Cauchy inequality yields the stated result. When it is not known whether QE as defined by $\mathcal{L}$ is an expectation, we have the following general proposition.

Proposition 1 *Let*

$$\Theta(s \mid K) = -\frac{\mathcal{L}^{(K+1)}(s)}{\mathcal{L}^{(K)}(s)}$$

be a non-increasing positive function of s, and $\mathcal{L}^{(0)}(s) := \mathcal{L}(s) > 0$. Then

(1) $(-1)^k \mathcal{L}^{(k)} > 0$, and $(-1)^k \mathcal{L}^{(k)}$ is log-convex, $k = 0, 1, \ldots, K$;
(2) $\Theta(s \mid k) = -(d/ds) \log \left[(-1)^k \mathcal{L}^{(k)} \right] > 0$, $k = 0, 1, \ldots, K$;
(3) $\log \text{QE}(e_0 \mid e_k) > \text{QE}(\log e_0 \mid e_k)$, $k = 0, 1, \ldots, K$;
(4) $\Theta(s \mid k) > \Theta(s \mid k - 1)$, $k = 1, \ldots, K$.

The proof of items (1) and (2) is based on the ideas of [9], while item (3) is a relatively straightforward extension of [11]. Item (4) follows from the following recurrent relationship

$$\frac{d}{ds} \log \left[\Theta(s \mid k) \right] = \Theta(s \mid k) - \Theta(s \mid k + 1). \tag{2.13}$$

The conditions in the above proposition are analogous to the so-called *multivariate totally positive of order 2 condition* (MTP2) imposed on Archimedean copulas to induce positive dependence [9].

Note that when QE is an E, the functions Θ given by (2.12) represent conditional expectations of random variables U given observed data. The role of conditioning on observed data is played by e_k in the condition representing complete-data likelihood contribution of the observed event. In the QE form this statement retains the meaning of quasi-imputation of U. In the sequel we will give interpretations of this statement in the context of different models.

3 Multinomial Regression

The multinomial regression example is based on [12]. The *multinomial-Poisson* (MP) transformation has been used to simplify maximum likelihood estimation (Baker [2]). The approach works by substituting a Poisson likelihood for the multinomial likelihood at the cost of augmenting the model parameters by auxilary ones. The disadvantage is that this only works when covariates are discrete. The artificial mixture approach we proposed works with arbitrary covariate structures.

Multinomial probabilities $p_i(z)$, $i = 1, \ldots, M$ are modeled using log-linear predictors $\theta_i(z)$ specific to categories i and conditional on a vector of covariates z. Multinomial logit model is constructed by normalization

$$p_i(z) = \frac{\theta_i(z)}{\sum_{k=1}^{M} \theta_k(z)}, \tag{3.1}$$

where without loss of generality θ_1 is restricted to 1 for identifiability.

Each θ_i can be parameterized using a vector of regression coefficients β_i so that

$$\theta_i(z) = \exp\left\{ \beta_i^{\mathrm{T}} z \right\}. \tag{3.2}$$

Data can be represented as a set $\{y_{ij}, z_j\}$, $i = 1, \ldots, M$, $j = 1, \ldots, N$, where $y_{ij} = 1$ if observation j with covariates z_j falls into category i, and $y_{ij} = 0$, otherwise.

The log-likelihood kernel corresponding to the above formulation can be written as

$$\ell_M = \sum_{ij} y_{ij} \left\{ \log\left(\theta_{ij}\right) - \log\left[\sum_{k=1}^{M} \theta_{kj}\right] \right\}, \tag{3.3}$$

where $\theta_{ij} = \theta_i(z_j)$.

With categorical covariates z summarized into groups, y_{ij} becomes a count of observations in response category i belonging to group j. The parameters in the model form a potentially large matrix $B = [\beta_{ij}]$, which may create computational problems.

Using the self-consistency idea we factorize the complete-data likelihood into $M - 1$ independent Poisson regressions each with only a subset β_i of parameters.

In the multinomial model we are looking for an artificial mixture formulation such that p on the left of (1.1) corresponds to the multinomial distribution, while the p on the right of (1.1) yields Poisson complete-data likelihood. Once the transformation is defined, we can construct an EM algorithm with E-Step solving the problem of imputation of U, while M-Step deals with maximizing a log-likelihood obtained from the complete-data model $p(x \mid z, U)$.

The construction is motivated by the Laplace transform of an exponentially distributed random variable $U \propto \mathrm{Exp}(1)$ with expectation of 1

$$\mathcal{L}(s) = \mathrm{E}\{\exp(-Us)\} = \frac{1}{1+s}. \tag{3.4}$$

Therefore, noting that $\theta_1 := 1$, the multinomial probabilities (3.1) can be written in the artificial mixture form

$$p_i(z) = \theta_i(z)\mathrm{E}\left\{\exp\left[-U\sum_{k=2}^{M}\theta_k(z)\right]\right\}, \tag{3.5}$$

with $p(\cdot \mid \cdot, U)$ obtained by dropping the "E" symbol from the above expression

$$p_i(z \mid U) = \theta_i(z)\exp\left(-U\sum_{k=2}^{M}\theta_k(z)\right) = \theta_i(z)e_1(s), \tag{3.6}$$

where

$$s = \sum_{k=2}^{M}\theta_k(z). \tag{3.7}$$

Thus, the multinomial model likelihood contribution is described as a $p = \mathrm{QE}\{ce_1(s)\}$ on the linear span of e_1 with appropriately chosen constant $c = \theta_i(z)$ and argument s given by (3.7). Formally, (3.6) gives rise to a sum of Poisson complete-data log-likelihoods parameterized in a disjoint fashion

$$\ell_0(B) = \sum_{i=1}^{M} \sum_{j=1}^{N} \log \left\{ p_i(z_j \mid U_j) \right\} = \sum_{i=2}^{M} \ell_{0,i}(\beta_i), \tag{3.8}$$

where i goes over categories and j goes over subjects, and

$$\ell_{0,i} = \sum_{j=1}^{N} \left\{ y_{ij} \log \left(\theta_{ij} \right) - y_{*j} U_j \theta_{ij} \right\} \tag{3.9}$$

is a Poisson likelihood specific to the ith category parameterized by β_i, the vector represented by the transposed ith row of the parameter matrix B. Note that we used the identifiability restriction $\theta_1 := 1$ in the derivation of (3.8) that resulted in the summation over i starting at $i = 2$. The term $y_{*j} = \sum_i y_{ij}$ is equal to one if j has the meaning of subjects. In the discrete formulation where j refers to a group of subjects with a common covariate vector, y_{*j} is the count of subjects in group j.

The imputation of U is readily obtained from (2.12) and the mgf $\mathcal{L}$ (3.4) as

$$\hat{U} = -\frac{\mathcal{L}'(s)}{\mathcal{L}(s)} = \frac{1}{1+s}. \tag{3.10}$$

Hence, the QEM algorithm proceeds by computing $\hat{U}_j$ for each subject with the subject-specific s given by (3.7) at the E-Step, maximizing $M - 1$ likelihoods (3.9) at the M-Step, iteratively until convergence.

Note that the complete-data "probabilities" (3.6) are not really probabilities and do not correspond to any probabilistic model on the complete-data level (it suffices to note that $p_i(\cdot \mid \cdot, U)$ have the range of $(0, \infty)$). The QEM theory provides a justification for the algorithm in this case.

4   Nonlinear Transformation Models (NTM)

Consider univariate semiparametric survival models induced by the $\mathcal{L}$–transformation as

$$G(t \mid z) = \mathcal{L}(H \mid \beta, z) = \mathrm{QE}(\exp(-UH) \mid \beta, z), \tag{4.1}$$

where G is the model family survival function conditional on covariates z, H is unspecified baseline cumulative hazard, and $\mathcal{L}$ is a non-increasing mgf parameterized by regression coefficients β and dependent on z. A detailed study of such models and the QEM approach was given in [11].

The complete-data likelihood contributions are

$$p_0 = \exp(-UH)(Uh)^{\delta}, \tag{4.2}$$

where δ is an indicator of failure (=1 if failure, =0 if censoring), and h is the jump of H at the point of failure. Note that p_0 is a survival function ($\delta = 1$) or pdf ($\delta = 1$) based on the proportional hazards (PH) model with hazard ratio given by U, and also that p_0 is in the span of e_0 for a censoring, and in the span of e_1 for a failure. Again, the imputation of U is readily obtained from (2.12) and the mgf defined by (4.1) as

$$\hat{U} = \Theta(H(t) \mid \delta, \beta, z) = -\frac{\mathcal{L}^{(\delta+1)}(H \mid \beta, z)}{\mathcal{L}^{(\delta)}(H \mid \beta, z)}, \tag{4.3}$$

and this defines the E-Step. For algorithm stability reasons, use of profile likelihood has been advocated. With this approach β is held out, and QEM iterations invoke the M-Step specific to H represented by the Nelson–Aalen–Breslow estimator

$$H^{(m+1)}(t) = \int_0^t \frac{dN(x)}{\sum_i Y_i(x)\Theta(H^{(m)}(x) \mid \delta_i, \beta, z_i)}, \tag{4.4}$$

where $N(t)$ is the counting process of failures, and $Y_i(t)$ is the ith subject-specific at risk process (an indicator that subject i is at risk at $t-$), and m is the QEM iteration count. Once the procedure converges at some solution $H(\beta)$, the profile likelihood is determined as $\ell(\beta, H(\beta))$, and maximized by a general numerical method.

5   Copula Models

Now, consider a multivariate (K survival times) survival model

$$G(t_1, \ldots, t_K \mid \beta, z) = C(M_1, \ldots, M_K),$$

where G is the multivariate survival function, C is a survival copula [10], and M_k are marginal survival functions. Motivated by C induced by a shared frailty model, we consider Archimedean C in which case we can write

$$G(t_1, \ldots, t_K \mid \beta, z) = \mathcal{L}(\mathcal{L}^{-1}(M_1) + \cdots + \mathcal{L}^{-1}(M_K)) = \mathcal{L}(H_1 + \cdots + H_K \mid \beta, z), \tag{5.1}$$

where $\mathcal{L}^{-1}$ is the inverse function, and $H_k = H(t_k)$ is the baseline cumulative hazard. The marginal survival functions are NTM models

$$M_k = \mathcal{L}(H_k).$$

This framework is thus a direct multivariate generalization of the QEM for univariate NTM models (4.1).

Generally the dimension of survival response may be variable as in the case of clusters of survival data of variable size. The motivating shared frailty model assumes that all observations in the same cluster c share a common random variable U, and that given U survival times in the cluster are independent with conditional survival functions

$$G(t_k \mid U) = \exp(-U H_k), \quad k = 1, \ldots, K_c,$$

where K_c is the size of the cluster. The indicators of failure δ_i^c will now be specific to the cluster c and to the observation i within the cluster, $i = 1, \ldots, K_c$.

The contribution of cluster c to the likelihood is a joint survival/pdf function derived from (5.1) by applying a negative partial derivative with respect to t_k for every failure in the cluster ($k : \delta_k^c = 1$) resulting in

$$p = \prod_{k=1}^{K_c} \left[-\frac{\partial}{\partial t_k} \right]^{\delta_k^c} G(t_1, \ldots, t_{K_c} \mid \beta, z) = (-1)^{\delta_*^c} \mathcal{L}^{(\delta_*^c)}(H_*^c) \prod_{k=1}^{K_c} [h_k]^{\delta_k^c}, \qquad (5.2)$$

where $\delta_*^c = \sum_{k=1}^{K_c} \delta_k^c$ is the number of failures in the cluster, $H_*^c = \sum_{k=1}^{K_c} H_k$, and $[-\partial/\partial t_k]^0$ is interpreted as no derivative with respect to t_k. This contribution (5.2) can be written as a quasi-expectation by virtue of (2.7),

$$p = \mathrm{QE} \left\{ C U^{\delta_*^c} \exp(-Us) \right\}, \quad s = H_*^c, \quad C = \prod_{k=1}^{K_c} [h_k]^{\delta_k^c},$$

implying that the imputation of the shared U based on observations in cluster c is given by

$$\hat{U}^c = \Theta(H_*^c \mid \delta_*^c, \beta, z) = -\frac{\mathcal{L}^{(\delta_*^c + 1)}(H_*^c \mid \beta, z)}{\mathcal{L}^{(\delta_*^c)}(H_*^c \mid \beta, z)}, \qquad (5.3)$$

according to (2.12), and this defines the E-Step. Note the resemblance with the univariate imputation operator (4.3).

Maximizing the complete-data likelihood based on contributions

$$p_0 = \prod_{k=1}^{K_c} [h_k]^{\delta_k^c} U^{\delta_*^c} \exp(-U H_*^c)$$

over the baseline hazard $H^{(m+1)}$ in the $(m+1)$th iteration version, and substituting $\hat{U}$ computed using the mth iteration version of H, we arrive at the M-Step's Nelson–Aalen–Breslow iterative estimator

$$H^{(m+1)}(t) = \int_0^t \frac{dN(x)}{\sum_c Y^c(x) \Theta\left(H_*^{c(m)} \mid \delta_*^c, \beta, z\right)}, \qquad (5.4)$$

where $Y^c(x)$ is the count of subjects at risk at time x in cluster c. Note that this estimator is similar to (4.4). The difference is that because survival times in the same cluster share the same U, the subjects $\{i\}$ of cluster c in the denominator of (4.4) will contribute the same $\hat{U}^c = \Theta$, and the cluster-specific $\hat{U}^c = \Theta$ is simply weighted by Y_i^c as in $\sum_{i \in c} Y_i \Theta = Y^c \Theta$.

By the QEM theory, each iteration (5.4) will improve the likelihood.

If marginal distributions are dependent on covariates z through the predictors $\theta = e^{\beta z}$, estimating equations are similar with H_*^c replaced by $\sum_{i \in c} H_i^c \theta_i^c$, and h_i by $\theta_i^c h_i$, where the quantities are specific to the subjects i in the cluster c. The estimator (5.4) will become

$$H^{(m+1)}(t) = \int_0^t \frac{dN(x)}{\sum_c \left[\sum_{i=1}^{K_c} Y_i^c(x) \theta_i^c \right] \Theta \left(\sum_{i=1}^{K_c} \theta_i^c H_i^{c(m)} \mid \delta_*^c, \beta, z \right)}. \tag{5.5}$$

Note the connection of QEM monotonicity and the characterization of positive dependence by the MTP2 assumption [9]. Namely, when the EM algorithm for a shared frailty model is generalized into the QEM algorithm for an Archimedean copula, monotonic convergence is not guaranteed unless the Archimedean copula models positive dependence, the latter is of course the case for the frailty model. This is a multivariate equivalent to the characterization of monotonic convergence by a decreasing $\Theta(H)$ for a univariate NTM model. The meaning of the univariate characterization is that imputed risk U for the subject decreases if the subject survives longer, a natural property of the univariate frailty model. We have shown that this property becomes positive dependence in the multivariate case.

Essentially the QEM generalization applied to the multivariate survival case is about characterizing the difference between Archimedean copulas induced by shared frailty models and the general ones. We have shown that the EM framework based on the shared frailty model is generalizable as long as the Archimedean copula preserves positive dependence. In this respect it is interesting to note that the generalization will vanish as the maximal cluster size $\max_c K_c$ increases. Indeed with K_c unlimited, Proposition 1 will have to be enforced for $K \to \infty$. As this happens, the K-alternating property represented by the first statement in item (1) of the Proposition 1 will turn into complete monotonicity [5], which corresponds to $K = \infty$. Then by the Bernstein theorem [5] $\mathcal{L}$ is a Laplace transform of some random variable U. In other words, a positively dependent Archimedean copula that serves clusters of any size without limit must be induced by a shared frailty model.

In the context of the QE operator this means that if QE is defined on the basis functions e_k of unlimited order $k = 0, 1, 2, \ldots, \infty$, and is assumed to satisfy the condition of the Proposition 1 (ensuring the Jensen inequality), then such operator is a good old mathematical expectation. The generality of QE as compared to E is thus inversely related to the complexity of the space of admissible functions on which it is supposed to work. The difference between the QEM and EM algorithms is the largest in problems where the basis of the space of admissible functions is the smallest, such as the multinomial logistic example and the univariate survival example. And there is no difference between them at all in the multivariate survival case with unlimited cluster sizes.

6 Example

6.1 A Composition of Gamma and Positive Stable Shared Frailty Models

As we discussed in Sect. 5, QEM construction for a clustered survival data problem starts with the choice of the model-generating function $\mathcal{L}$ that in case of a mixture model represents a Laplace transform of a shared frailty variable. Two most popular choices in the literature are the gamma shared frailty model [3, 10] with the Laplace transform

$$\mathcal{L}_\Gamma(s) = \left(\frac{1}{1 + \frac{s}{\eta}}\right)^\eta, \tag{6.1}$$

where $\eta = 1/\mathrm{Var}(U) \geq 0$ is the dependence parameter, and the positive stable frailty model [6]

$$\mathcal{L}_\Gamma(s) = \exp(-s^\mu), \tag{6.2}$$

where $0 \leq \mu \leq 1$ is the dependence parameter. The dependence parameters η and μ can include covariates shared by observations in the same cluster. Independence is a special case of $\eta = \infty$ and $\mu = 1$, respectively. The gamma frailty model is an example of early dependence, while the positive stable model is associated with late dependence [7]. Both models are being included as special cases in the *power variance family* (PVF) that has a three-parameter Laplace transform [7]. However, if one just wants to combine the distinct features of gamma and positive stable families, a new parsimonious model is a more natural choice, parameterized by η, μ such that it would turn into a gamma frailty if $\mu = 1$, and into a positive stable frailty model if $\eta = \infty$. Such a model can conveniently be built using the composition device. Indeed, one can show that compound random variable

$$U = \sum_{i=1}^{V} \xi_i, \tag{6.3}$$

where ξ_i are independent identically distributed random variables with Laplace transform $\mathcal{L}_2$, and V is an independent discrete variable with Laplace transform $\mathcal{L}_1$, has the Laplace transform given by the composition [1]

$$\mathcal{L}(s) = \mathcal{L}_1 \left(-\log[\mathcal{L}_2(s)] \right). \tag{6.4}$$

The composition device is more general than the form of the U in (6.3) suggests as $\mathcal{L}_1$ may not correspond to a discrete random variable, while (6.4) remains formally valid. In this case it will actually be unknown whether the resultant $\mathcal{L}(s)$ is a Laplace transform at all, i.e., if a random variable U exists, and even if it does, the form of the distribution of U will be unknown as (6.3) would no longer be a valid representation. At this point, the conventional EM construction will be stuck as it relies on the existence of U, and the availability of the explicit form of its distribution to be used in the integration at the E-Step of the algorithm. In other words the situation represents a good example for the QEM construction, where we do not have to rely on the existence of U and can instead treat it as an operator argument to the QE. The procedure is as follows.

First, we define the desired shared frailty model combining features of gamma and positive stable frailties by specifying the model-generating function as

$$\mathcal{L}(s) = \mathcal{L}_S \left(-\log[\mathcal{L}_\Gamma(s)] \right) = \exp\left\{ -\left[\eta \log\left(1 + \frac{s}{\eta} \right) \right]^\mu \right\}. \tag{6.5}$$

Second, using the recurrent relationship (2.13), we obtain the imputation operators Θ by differentiation

$$A(s) = \log(\eta + s) - \log(\eta)$$

$$B(s) = \eta^\mu \mu A^\mu(s) - \mu - 1 + A(s)$$

$$\Theta(s|0) = \frac{\eta^\mu \mu A^{\mu-1}(s)}{\eta + s} \tag{6.6}$$

$$\Theta(s|1) = \frac{B(s)}{(\eta + s)A(s)}$$

$$\Theta(s|2) = \frac{B(s)}{(\eta + s)A(s)} - \frac{\eta^\mu \mu^2 A^\mu(s) + A(s) - A(s)B(s) - B(s)}{(\eta + s)A(s)B(s)}.$$

Third we need to check that the Θ's are non-decreasing in s. Note that according to Proposition 1, it suffices to check the last one $\Theta(s|2)$, and we have computed them up to order $K = 2$ that will serve the bivariate survival data example that follows. This can be done analytically or numerically. Alternatively, we can prove that U exists as a random variable, in which case all desired properties will follow automatically. Composition of Laplace transforms is again a Laplace transform if the first derivative

$$\varphi(s) = \frac{d}{ds} \{ -\log[\mathcal{L}_2(s)] \} = \frac{d}{ds} \{ -\log[\mathcal{L}_\Gamma(s)] \} = \frac{1}{1 + \frac{s}{\eta}}$$

is a completely monotonic function, i.e., satisfies $\varphi(s) > 0$, $(-1)^k \varphi^{(k)}(s) > 0$, $k = 1\ldots, \infty$ [5]. This condition is clearly satisfied here.

Finally, it follows that QEM iterations (5.5) are fully specified for the composition model and will improve the likelihood at each step.

6.2 Retinopathy Application

We use a subset of the data on time to blindness in diabetic retinopathy patients [8] as available in the R package **timereg**. The subset includes 83 adult patients with 166 observations of time to blindness in the two eyes per patient. Laser treatment (yes/no) served as a covariate delaying the onset of blindness. Patient serves as a cluster of size 2 represented by data on the two eyes. The follow-up times are measured in months divided by 100. This dataset serves as a real data example as well as a basis for the numerical and simulation studies.

The composition model used in the example has the survival function given by (6.5) with s replaced by $\theta_1 H(t_1) + \theta_2 H(t_2)$, where the index points at the eye within the patient, t_i is the time to blindness specific to the eye, and θ is a predictor for the marginal distribution. Marginal predictors include treatment as an indicator covariate $\theta_i = \exp(\beta z_i^c)$, where c is patient (cluster), i is the eye within the patient, and z_i^c is an indicator of laser treatment. The two dependence parameters η, μ did not include covariates.

The results of model fitting are presented in Table 1. Confidence intervals and standard errors are based on the curvature of the profile likelihood surface evaluated numerically using an iterative interpolation algorithm with error control. The profile likelihood emerges when the QEM iterations (5.5) converge.

Based on the 95% confidence intervals we conclude that the treatment effect is highly significant and that the gamma frailty model is a better fit to this data. The stable frailty dependence parameter is rather close to 1 (the value when the composition model turns gamma frailty), and the confidence interval does not exclude this value.

Figure 1 shows Kaplan–Meier curves (dashed) representing marginal time to blindness distributions overlaid with their expected counterparts based on the fitted composition model.

Table 1 The results of the composition model fit

Parameter	Point estimate	Standard error	Confidence interval (profile likelihood)
Treatment effect β	−1.537	0.319	(−2.162, −0.912)
Dependence (gamma), η	2.559	4.495	(0, 11.369)
Dependence (positive-stable) μ	0.897	0.147	(0.61, 1)

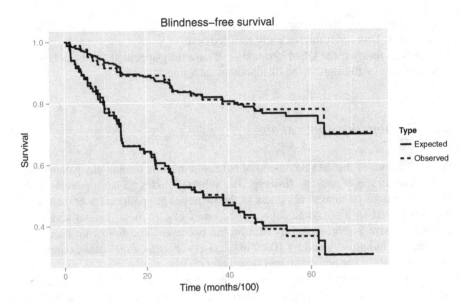

Fig. 1 Observed and expected survival curves based on the composition model fitted to diabetic retinopathy bivariate survival data

6.3 Numerical Experiments and Simulations

In order to study the performance of the proposed algorithms, we fitted a parametric version of the composition model to the diabetic retinopathy data. The baseline cumulative hazard function H was specified according to a two-parameter Weibull distribution. We then used the parametric model to generate samples of size 100, 300, 500, 700, and 1000 patients. With each of the samples likelihood maximization was performed with varying levels of tolerance 0.1, 10^{-2}, 10^{-3}, ..., 10^{-9}, 10^{-10}.

Measuring methods precision was done as follows. In view of the fact that different convergence criteria and different structures of the likelihood maximization algorithms lead to different stopping points, we did not use tolerance as a measure of precision. Instead, estimates obtained with 10^{-10} tolerance were treated as virtual true values. Then errors were defined as differences vs. these true values. Error measures for the log-likelihood and the model parameters represented such log transformed absolute differences.

Method complexity was evaluated as a count of operations needed to achieve a certain precision. This approach is taken to avoid dependence of time-based measures on the specific equipment and technical issues of programming languages, coding implementations, compilers, etc. In the study we defined a computation of the function $\mathcal{L}$, an imputation operator Θ, or of the predictor $\theta = \exp(\beta z)$ as one operation, while all other quantities needed in the procedure are combinations of these.

Two algorithms were compared. One is a QEM algorithm used to compute the profile likelihood as described above. Profile likelihood was maximized by a generic conjugate gradients method. The second algorithm expanded the set of parameters by the jumps of the cumulative hazard function ΔH_k assessed at the points of failure, and maximized the full likelihood over $(\beta, \eta, \mu, \{\Delta H_k\})$ using the same conjugate gradients procedure.

Shown in Fig. 2 is the complexity by error curves resulting from the above experiment. Based on the figure we conclude that the QEM algorithm is faster in the sense that it allows to achieve better precision with the same number of operations or takes less operations to get the result with a specified precision. This follows from the fact that QEM solid curves are lower than the dashed curves corresponding to the full likelihood maximization. We also see the difference between methods increases with sample size while it is small and then stabilizes. This is perhaps a consequence of the adverse effect of dimension (sample size) on the full maximization and increasing complexity of the QEM as well, and the favorable effect of sharpening of the likelihood peak that makes it easier to reach at the same time. Overall we found both methods to be stable in operation converging 100% of the time in our experiments. We also observe that the dashed curves start at a later point on the error scale that indicates that QEM is more precise a solution even when the user decides to achieve the best precision at any cost.

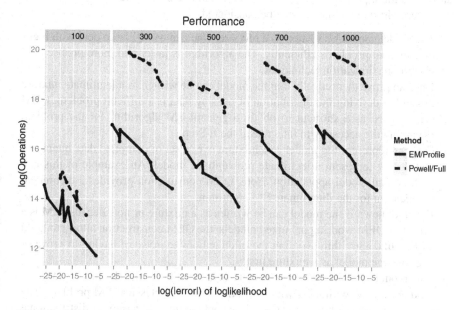

Fig. 2 Performance of the EM-type algorithm vs. a generic fit of the full model by conjugate gradients (Powell method)

7 Discussion

We have presented an approach that allows to construct and justify EM-like algorithms that do not require integration at the E-Step or the specific formulation of the mixture model down to the known distribution of missing data.

The approach works best if the model is given by a model formula through some link function, and the model formulation does not involve reference to missing data. The advantages of the proposed approach are both numerical and analytical, and their realization depends on the specific modeling situation.

Generally, from the computational point of view, the QEM algorithm works as an EM, and there is no computational advantage vs. the EM when the latter is applicable. However, there may be a theoretical advantage over the EM as QEM construction gives the general ready-to-use E-Step that under an EM framework could be a considerable theoretical problem. An example is a general QEM for an arbitrary shared frailty model written out in Sect. 5 vs. the case-by-case EM constructions, each for a specific frailty distribution. In problems where there is no missing data, and EM is not applicable, full model maximization by a generic algorithm is often the only choice available. Here, QEM construction (when the model satisfies appropriate conditions for monotonicity) can offer the advantage of a structured stable solution, devoid of the curse of dimensionality, computationally faster and more precise, as we showed in the numerical example.

Several different situations can be considered:

1. The model cannot be written as a mixture. One can use QEM to deal with high-dimensional problems efficiently. The algorithm is faster and more precise than generic maximization techniques.
2. One can prove theoretically that the model can be written as a legitimate mixture model but deriving the distribution of missing data is a difficult problem. QEM can be used as a shortcut to obtain a general EM algorithm for the problem bypassing the necessity to integrate at the E-Step.
3. One can represent the model as a mixture over a set of complete-data quantities that do not represent a legitimate probabilistic model. An example of this is a multinomial model application. Here, QEM construction provides a theoretical justification for the illegitimate EM algorithm.
4. It is not known if the model can be written as a mixture or not. Because EM is a subset of QEM working on mixed models, the QEM construction allows to build an EM-like algorithm avoiding the often tedious exercise of having to prove or disprove the model as a mixture first.
5. The model is built from a complete-data one by explicitly incorporating a mixing variable with a known distribution. This is a classical EM problem. The QEM construction here allows painless application of a variety of distributional assumptions about the mixing variable with little analytical overhead, provided the Laplace transforms of those distributions are known.

That said, QEM fails if one does not have a model yielding decreasing imputation operators $\Theta(s, k)$. Despite the fact that obtaining Θs is a simple exercise in elementary differentiation, resulting expressions could be cumbersome, and checking whether Θs are non-decreasing could be difficult. For problems requiring a data-driven order of Θs such as clustered survival data where order is up to the size of the cluster, Θs need to be pre-calculated when a recurrent automatic procedure is not feasible. Because calculating Θ involves differentiation, when recurrent relationships for derivatives of arbitrary order cannot be obtained but are needed, symbolic differentiation programs may be used to deal with the problem. QEM inherits the properties of an EM and they will be slow converging if the missing information fraction is high (was not the case with all models we considered).

Acknowledgement This research is supported by National Cancer Institute grant U01 CA97414 (CISNET).

References

1. Aalen, O. (1992). Modelling heterogeneity in survival analysis by the compound Poisson distribution. *The Annals of Applied Probability, 4*, 951–972.
2. Baker, S. G. (1994). The multinomial-Poisson transformation. *Statistician, 43*, 495–504.
3. Clayton, D. (1978). A model for association in bivariate life tables and its application in epidemiological studies of familial tendency in chronic disease incidence. *Biometrika, 65*, 141–151.
4. Dempster, A. P., Laird, N. M., & Rubin, D. B. (1977). Maximum likelihood from incomplete data via the EM algorithm. *Journal of the Royal Statistical Society, Series B: Statistical Methodology, 39*, 1–38.
5. Feller, W. (1971). *An introduction to probability theory and its applications*. New York: Wiley.
6. Hougaard, P. (1986). Survival models for heterogeneous populations derived from stable distributions. *Biometrika, 73*, 387–396.
7. Hougaard, P. (2000). *Analysis of multivariate survival data*. New York: Springer.
8. Huster, W., Brookmeyer, R., & Self, S. (1989). Modelling paired survival data with covariates. *Biometrics, 45*, 145–156.
9. Müller, A., & Scarsini, M. (2005). Archimedean copulae and positive dependence. *Journal of Multivariate Analysis, 93*, 434–445.
10. Oakes, D. (1989). Bivariate survival models induced by frailties. *Journal of the American Statistical Association, 84*, 487–493.
11. Tsodikov, A. (2003). Semiparametric models: A generalized self-consistency approach. *Journal of the Royal Statistical Society, Series B: Statistical Methodology, 65*, 759–774.
12. Tsodikov, A., & Chefo, S. (2008). Generalized self-consistency: Multinomial logit model and Poisson likelihood. *Journal of Statistical Planning and Inference, 138*, 2380–2397.

On the Application of Flexible Designs When Searching for the Better of Two Anticancer Treatments

Christina Kunz and Lutz Edler

Abstract In search for better treatment, biomedical researchers have defined an increasing number of new anticancer compounds attacking the tumour disease with drugs targeted to specific molecular structure and acting very differently from standard cytotoxic drugs. This has put high pressure on early clinical drug testing since drugs may need to be tested in parallel when only a limited number of patients—e.g., in rare diseases—or limited funding for a single compound is available. Furthermore, at planning stage, basic information to define an adequate design may be rudimentary. Therefore, flexibility in design and conduct of clinical studies has become one of the methodological challenges in the search for better anticancer treatments. Using the example of a comparative phase II study in patients with rare non-clear cell renal cell carcinoma and high uncertainty about effective treatment options, three flexible design options are explored for two-stage two-armed survival trials. Whereas the two considered classical group sequential approaches integrate early stopping for futility in two-sided hypothesis tests, the presented adaptive group sequential design enlarges these methods by sample size recalculation after the interim analysis if the study has not been stopped for futility. Simulation studies compare the characteristics of the different design approaches.

Keywords Clinical phase II trial · Survival data · Flexible design · Sample size recalculation

This work has been done in memoriam of Andrei Yakovlev (1944–2008) whom the second author and the Department of Biostatistics of the German Cancer Research Center is obliged for a more than 15 years collaboration and friendship, remembering with great thanks his always vivid and enthusiastic stimulations and contributions giving guidance to good statistical science and practice.

C. Kunz · L. Edler (✉)
German Cancer Research Center, Heidelberg, Germany
e-mail: Christina.Kunz@mri.bund.de; lutzedler@gmx.de

© Springer Nature Switzerland AG 2020
A. Almudevar et al. (eds.), *Statistical Modeling for Biological Systems*,
https://doi.org/10.1007/978-3-030-34675-1_12

1 Introduction

The gold standard of searching for new treatments in medicine is through clinical prospective trials and a rigid statistical concept of decision-making where the control of the false positive error ranges on top of statistical quality control measures. Recently, this paradigm has been challenged in oncology by the advent of an increasing number of new anticancer compounds from which one knows more on their molecular target than has been known for the traditional cytotoxic drugs. An intrinsic property of many targeted drugs is their specificity to disease and, in particular, to subtypes of diseases where one could expect better clinical outcome compared with traditional treatment approaches. This raises the question for designs appropriate for patient populations of limited size, such as patients with rare tumours or rare subtypes, for whom no standard treatment has yet been identified. Studies of rare entities have problems to find sufficient funding since the cost–pay back relationship is often not very favourable for pharmaceutical industries. The development of a design is then challenged not only by statistical and ethical requirements, but also by economical and practical conditions. Furthermore, in trials for rare entities, even more than in standard trials, one has to face the situation of only limited information on the potential effects of new drugs. This is an important factor when choosing between different treatment and design options for a comparative clinical trial.

Our work was motivated when designing a prospective, open label, randomized multicentre phase II study to evaluate progression-free survival (PFS) in patients with locally advanced or metastatic non-clear cell renal cell cancer (ncc-RCC) to be investigated for receiving either temsirolimus or sunitinib as chemotherapy. Some limited data on the efficacy of temsirolimus or sunitinib in ncc-RCC patients had shown interesting outcomes for both agents [29, 31]. Since no randomized clinical trial was performed yet in this specific patient population, a study was proposed to find evidence on superiority of one of the two treatments in first-line therapy of ncc-RCC. Accounting for the rarity of the disease and the limited funding an investigational randomized phase II trial not exceeding approximately $N = 100$ patients was considered as both reasonable and realistic from a practical point by the clinical investigators and the two sponsors involved for the two drugs, respectively. A two-stage procedure was therefore set up as follows:

Stage 1 (feasibility step).
Recruit about $N_1 = 30$ patients from maximum $N_1 + N_2$ patients in stage 1 and stage 2. Stop (solely) for futility after stage 1 when no sufficient difference in PFS is observed between the two arms. Otherwise, a sample size recalculation using accumulated information is performed and the new sample size $N_{2,new}$ for stage 2 is determined. If $N_{2,new}$ is such that the trial remains feasible within 3 years of patient accrual in total (stage 1 and stage 2), it continues with stage 2, otherwise it stops at stage 1.

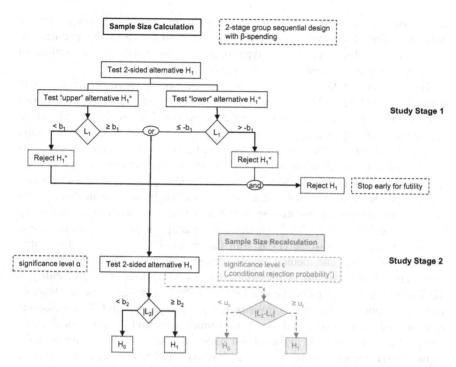

Fig. 1 Diagram of the statistical decision process during the course of the study. Group sequential and adaptive decision criteria are depicted. L_i denotes the logrank statistic in the i-th analysis, $i = 1, 2$

Stage 2 (adaptive extension step).
Continue the trial until either a maximum of $N_{2,new}$ additional patients are recruited or until in total 36 months accrual followed by 12 months follow-up.

Therefore, we investigated two-stage group sequential designs (GSD) as well as more recently developed adaptive designs (AD) for their applicability to develop a study design fitting to this set-up. In order to comply with the specific two-stage structure of the hypothesis testing (see Fig. 1) we combined elements of GSDs with those of ADs into a comprehensive flexible two-stage design for a two-sided test problem with futility testing at the first and final testing at the second stage. Furthermore, the survival type of the primary clinical endpoint had to be taken into account. In an extensive simulation approach we analysed, under the restrictions described above, the critical design characteristics (average sample size, average number of events for the second stage and both stages in total as well as the accrual rates needed to achieve (re)calculated sample sizes in stage 2). The application of this approach into a practical clinical design for the motivating example of the above-mentioned prospective randomized ncc-RCC trial has been reported previously [36]. That paper also described analytical details both of the construction of the flexible two-stage design as well as of the implementation of the simulation approach. This contribution presents comprehensive information on the wide variety of scenarios to be considered for this enterprise and provides such a general platform for identifying a flexible and practical design for the investigation of drugs for rare diseases or rare subtypes of a more common disease accounting, in particular, for time-to-event as primary endpoint.

The structure of the paper is as follows. Section 2 provides the analytical details of two-stage GSDs which consider two-sided tests for survival endpoints and include early futility stopping (Sect. 2.1). Section 2.2 introduces ADs which extend the flexibility of the group sequential approach by sample size recalculation if the trial had not been stopped for futility after the first stage. Adaptive approaches for survival data, which allow for this flexibility, considered so far only one-sided test decisions. We present in Sect. 2.2 two different approaches which suit our or a similar study question. Section 2.2.1 introduces briefly adaptive methods which are based on p-value combination rules and which can also easily be applied to the two-sided setting. In Sect. 2.2.2, we develop an adaptive group sequential design which is based on two-stage GSDs with the possibility to stop early for futility after the first stage. The method enables design adaptations under control of the type I error rate by means of preserving the conditional rejection probability in the second study stage given the test results of the first stage. Using an elaborate simulation approach, this AD is investigated and compared with GSDs presuming either a fixed accrual rate over the total study time—denoted by GSD1—or allowing for an adjusted accrual rate in the second stage—denoted by GSD2—to recruit the required study sample number, e.g., by inclusion of further study centres. The three designs are analysed by simulations under different scenarios (Sect. 3).

2 Flexible Two-Stage Designs in Survival Trials

Flexibility in the design of a clinical study and especially adapting its sample size can be achieved either by applying conventional group sequential designs with the possibility to stop early or by more flexible adaptive designs which use the information of the ongoing trial to redesign the remainder of the trial. The simplest 'reaction' on information upcoming during an ongoing study is to stop for futility if no relevant difference between the treatments is observed, or to stop for efficacy if the difference between the treatment arms becomes sufficiently large. In both cases the statistical error probabilities must be controlled.

In this work we will allow stopping for futility after the first stage in a two-stage trial. This strategy fits best to the aim of investigating the effects of two new compounds in a rare disease given a large uncertainty on the outcome of different treatments and given restricted funding in this early stage of clinical research. Stopping for futility will then prevent patients from potential harm of being treated with an ineffective drug. This can be realized under control of the type I error rate in group sequential trials (see, e.g., [17]). If the trial has not been stopped for futility and one is affirmed not to treat patients with an ineffective drug, one may use the first stage information to execute a sample size recalculation by means of a more flexible design method. Knowledge about the treatments will have accumulated during the first stage such that the recalculated sample size is expected to fit better the trial's aim than that planned at the beginning of the study. Furthermore, in the case when promising treatments are observed in the first stage the power for the second stage could be scaled up, so that the evidence for a potential treatment effect will be stronger at the final analysis.

Next, we present adaptive two-stage approaches for survival trials which allow for two-sided tests, early stopping only for futility as well as sample size recalculation if no stopping has occurred. At first, we describe adaptive designs which base the final decision on a combined p-value. Further, we extend the adaptive conditional rejection probability principle of Schäfer and Müller [28] to the two-sided setting.

2.1   Two-Stage Group Sequential Survival Trials

Progression-free survival is considered to be a valuable and relevant clinical endpoint in early oncological trials. Therefore we focus on the development of flexible designs on the endpoint survival time which covers overall survival as well as any appropriately defined time-to-event, such as progression-free survival. The Cox proportional hazards model is used to derive the respective score test statistics and to construct two-sided decision regions for early stopping in the first stage and rejection of the null hypothesis of no treatment effect in the final analysis in a two-stage trial. Study time can then be scaled either as calendar time t (starting at study start with $t = 0$ and usually counting in weeks, months or years) or as information time t^I describing the proportion of observed events at the observation time point (cf. [20]).

2.1.1   Cox Proportional Hazards Model to Describe Survival Times and Score Statistics

We assume that N patients enter the trial in a staggered way during the recruitment period of length a ($0 \leq t \leq a$), i.e., the study starts at $t = 0$ and recruitment ends at $t = a$. Denote by Z_k the treatment indicator and by $\mathbf{W}_k$ the ($p \times 1$) vector of covariates. The event rate of patient k ($k = 1, \ldots, N$) at time point x follows the proportional hazards model of Cox [6]

$$\lambda_k(x|Z_k, \mathbf{W}_k) = \lambda_0(x) \exp(\theta Z_k + \boldsymbol{\beta}^T \mathbf{W}_k), \tag{2.1}$$

where θ describes the treatment effect equalling the log hazard ratio, $\boldsymbol{\beta}$ the coefficient vector of the covariates and $\lambda_0(x)$ the baseline hazard function. We assume stochastic independence of Z and W and non-informative censoring which is independent from survival time, given Z and W [12].

The log partial likelihood $l_N(t; \theta, \boldsymbol{\beta})$ is used to derive the score processes $S_N(t; \theta, \boldsymbol{\beta}) = \frac{\partial}{\partial \theta} l_N(t; \theta, \boldsymbol{\beta})$ and $U_N(t; \theta, \boldsymbol{\beta}) = \frac{\partial}{\partial \boldsymbol{\beta}} l_N(t; \theta, \boldsymbol{\beta})$ to test the null hypothesis $H_0\colon \theta = \theta_0 = 0$, see [7].

Under H_0 is $\{S_N(t; \theta_0, \hat{\boldsymbol{\beta}}_t), t \geq 0\}$ asymptotically for $N \to \infty$ a Gaussian process with independent increments and mean zero. Under an alternative $\theta \neq 0$ the score process has asymptotically the distribution of a Brownian motion with linear drift under an adapted time scale [12, 33].

In the case of no covariates the score statistic reduces to the logrank statistic. In the following we will use the scaled score statistic $L(t) = \frac{1}{\sigma_Z} S_N(t; \theta_0, \boldsymbol{0})$, where σ_Z is the standard deviation of Z. Note, $\sigma_Z = \frac{1}{2}$ for balanced randomization in two treatment arms.

Using the information time t^I and the number of observed events, we can set the analysis times of a two-stage study at the information times $t_j^I = \frac{d_j}{d_2}$, $j = 1, 2$, where d_j is the number of observed events at calendar time t_j, $j = 1, 2$. $L_j = L(t_j)$, $j = 1, 2$, then denotes the logrank statistics values of the interim analysis time t_1 and the final analysis time t_2 with $L_1 = L(t_1) = L(t_1^I) = L(d_1)$ and $L_2 = L(t_2) = L(t_2^I) = L(d_2)$.

Using the asymptotic normality of the score processes, the joint asymptotic multivariate normal distribution of L_j, $j = 1, 2$, is characterized by

$$L_j \sim \mathcal{N}(\sigma_Z \theta d_j, d_j)$$
$$L_2 - L_1 \sim \mathcal{N}(\sigma_Z \theta (d_2 - d_1), d_2 - d_1) \tag{2.2}$$

with asymptotically independent increments under $H_0\colon \theta = 0$, i.e., $\mathrm{Cov}(L_2 - L_1, L_1) = 0$.

2.1.2 Statistical Decision in Interim and Final Analysis of Survival Trials

We consider a two-stage design where early stopping after the first stage is only possible in favour of H_0, and H_0 can exclusively be rejected in the final analysis in favour of the two-sided alternative H_1. Using the stopping and rejection boundaries b_1 and b_2 the trial will be stopped early if

$$|L_1| < b_1 \quad (b_1 > 0).$$

In the final analysis H_0 will be rejected if the trial has not been stopped early and

$$|L_2| \geq b_2 \quad (b_2 > 0).$$

H_0 can only be rejected in the final analysis; therefore, a type I error can solely occur at the end of the trial and is controlled at level α. Since early stopping for futility is an option, type II error can occur at the interim analysis and at the final analysis. To control the type II error probability of the trial we use the error spending approach of Lan and deMets [19], where a β-spending function $B(t^I)$ is defined as non-decreasing function on [0, 1] with $B(0) = 0$ and $B(1) = \beta$. $B(t^I)$ is chosen from the γ-family defined by Hwang et al. [13] as

$$B(t^I) = \beta \frac{1 - \exp(-\gamma t^I)}{1 - \exp(-\gamma)} \qquad \text{for } 0 \leq t^I \leq 1 \qquad (\gamma \neq 0). \qquad (2.3)$$

One may prefer to specify the amount β_1 of type II error which can be spent in the interim analysis at t_1^I. In terms of the spending function, the fraction of spent type II error is β_1/β with $\beta_1 = B(t_1^I)$. In our examples where $\gamma = -4$, we obtain $\beta_1/\beta = 0.043, 0.074, 0.119$ for $t_1^I = 0.3, 0.4, 0.5$, respectively.

In order to develop the GSD for survival endpoints we adapted the method of Chang et al. [5] to survival data and determined simultaneously the maximum number of events $d_{max}(= d_2)$ of the trial and the acceptance and rejection boundaries b_1 and b_2 using the β-spending function. This was done under the condition that $\theta_a > 0$ is the minimal relevant treatment effect and by making use of the asymptotic distribution properties of the logrank statistics. In general, for two-sided group sequential trials, where the null hypothesis $H_0 : \theta = 0$ is rejected if the test statistic exceeds some boundary, the continuation and rejection regions can be obtained as superposition of two one-sided tests—a one-sided test of an 'upper' alternative $H_1^> : \theta \geq \theta_a$ and a one-sided test of the 'lower' alternative $H_1^< : \theta \leq -\theta_a$ (see, e.g., [17]). In early interim analyses this could lead to empty stopping regions in favour of H_0 [27], i.e., in the considered setting there would be no opportunity to stop for futility in the first interim analysis. Therefore, non-empty stopping regions, i.e., acceptance and rejection boundaries b_1 and b_2, respectively, are deduced by the modified method of Chang et al., see Wunder et al. [36]. This means that closed formulas for b_1 and b_2 cannot be given but they are determined iteratively.

2.2 Adaptive Design Approaches for Two-Stage Survival Trials

In this section, we introduce two basic principles for adaptive designs applicable to our specific two-stage survival trials: One principle is the combination of p-values which is shortly described; the other principle is the conditional rejection probability (CRP) principle which is finally applied in the ncc-RCC trial example reported here. For a discussion on history and role of these two principles of adaptive designs, their interrelationship and background literature, see [1].

In short: The p-value combination rules base their final decision on a combination of the p-values of the first and the adapted second stage (Sect. 2.2.1). The CRP principle bases the new study plan for the second stage on the conditional type I error probability for the second stage given the observations of the first stage (Sect. 2.2.2).

The CRP principle was found most suitable for sample size recalculation in the ncc-RCC trial, since, allowing also unplanned interim analyses, it is more flexible than the p-value combination rules and is therefore best suited for the unexplored study setting of the ncc-RCC trial. The next two subsections adapt the principles to survival data.

2.2.1 P-value Combination Rules to Combine Results from Both Study Stages

The principle of p-value combination assumes that the test statistics, calculated separately from the subsamples of the two stages, are stochastically independent. The final test decision is then derived from a predefined function that combines the p-values p_1 and p_2 from the two stages into a single decision criterion, say $C(p_1, p_2)$. The combination rule can have different functional forms and is chosen such that the overall significance level is controlled (see, e.g., [4]). In a two-stage survival trial the test statistic of the second stage uses information from patients which were already under risk at the first stage. Therefore the test statistics from the two stages are no longer independent. Instead of the test statistic L_2, the increment of the logrank statistics ($L_2 - L_1$) is used to get (asymptotically) stochastically independent p-values p_1 and p_2. Two different adaptive approaches were presented by Desseaux and Porcher [9] and Wassmer [35], based on different p-value combination functions.

Desseaux and Porcher [9] use Fisher's product test as proposed by Bauer and Köhne [2] to combine test results from two different study stages. Three decisions after the first study stage are possible:

- if $p_1 < \alpha_1$, the trial is stopped with rejection of H_0 ('stopping for efficacy')
- if $p_1 \geq \alpha_0$, the trial is stopped with acceptance of H_0 ('stopping for futility')
- if $\alpha_1 \leq p_1 < \alpha_0$, the trial is proceeded to the second stage where Δd additional events are expected to be observed. This number is recalculated in the interim analysis to obtain a prespecified power in stage 2.

If, as in our case, only early stopping for futility should be possible, α_1 would be set to 0. H_0 is rejected after the second stage if the trial has not been stopped early and

$$C(p_1, p_2) = p_1 \cdot p_2 \leq c_\alpha \quad \text{with} \quad c_\alpha = \exp\left(-\frac{1}{2}\chi^2_{4,1-\alpha}\right),$$

where $\chi^2_{4,q}$ denotes the q-quantile of the χ^2-distribution with four degrees of freedom. Here, c_α/p_1 can be seen as modified significance level for the second stage, denoted as $\tilde{\alpha}(p_1)$. The number of events Δd to be additionally observed to reach conditional power $1 - \beta_{cond,2}$ in the second study stage (conditional on the outcome p_1 of the first stage) is then

$$\Delta d = \frac{4([z_{1-\tilde{\alpha}(p_1)} + z_{1-\beta_{cond,2}}]^+)^2}{\theta_a^2},$$

where z_q denotes the q-quantile of the standard normal distribution and $[\cdot]^+$ the positive part of the argument.

Wassmer [35] extended the weighted inverse normal method for adaptive designs of Lehmacher and Wassmer [21] to survival data, using the combination rule

$$C(p_1, p_2) = 1 - \Phi[w_1 \Phi^{-1}(1 - p_1) + w_2 \Phi^{-1}(1 - p_2)]$$

with weights $0 < w_i < 1$ and $w_1^2 + w_2^2 = 1$ and the standard normal distribution function Φ. The weights are chosen at the start of the trial and remain fixed throughout the trial. The control of type I error rate can thus be ensured despite data-dependent changes after the first stage. The test statistic for the second stage therefore is calculated as standardized increment of the logrank statistics

$$Z_2 = \frac{L_2 - L_1}{\sqrt{d_2 - d_1}},$$

and is approximately standard normally distributed.

Wassmer [35] proposed to recalculate the sample size to reach a conditional power $1 - \beta_{cond,2}$ in the subsequent stage. The number of additional events Δd which is necessary in the subsequent stage of a balanced trial is then given by

$$\Delta d = \frac{9}{2\theta_a^2} \left(\left(\left(\sum_{j=1}^2 w_j^2 \right)^{1/2} \frac{b_2}{\sqrt{d_2}} - w_1 z_1 \right) \middle/ w_2 + \Phi^{-1}(1 - \beta_{cond,2}) \right)^2.$$

2.2.2 The Adaptive CRP Principle to Extend Group Sequential Trials

We use for our work the conditional rejection probability (CRP) principle of Schäfer and Müller [24, 28] and extend the two-stage GSD to an adaptive group sequential design (denoted AD) which allows sample size recalculation for the second study stage under the condition that the trial has not been early stopped for futility. Therefore, we extend the method of Schäfer and Müller [28] for two-sided tests, possible early stopping for futility and the implementation of the score statistics (for more details see [36]). Next, we specify the adaptive CRP principle for two stages and two-sided tests. The CRP principle is a means to derive test procedures which control the overall type I error of the final test decision in a clinical trial irrespective of potential earlier design adaptations whether being scheduled or unscheduled. We aim at changing the sample size during the interim analysis executed at t_1^I with $t_1^I < t_2^2 = 1$ using the test statistics $L(t_1^I) = \tau$ at t_1^I. The conditional rejection probability $\epsilon(\tau, t_1^I)$ is then the probability of erroneously rejecting the null hypothesis in the final analysis given $L(t_1^I) = \tau$. The trial can be continued with an adapted second stage for a significance level $\epsilon(\tau, t_1^I)$. If the adaptations are based on $L(t_1^I)$ the overall type I error is controlled (cf. [3]). Given $L(t_1^I) = \tau$, the CRP can be calculated as

$$\epsilon(\tau, t_1^I) = P_{H_0}(|L_1| \geq b_1, |L_2| \geq b_2 \quad |L_1 = \tau)$$

$$= 1_{\{|\tau| \geq b_1\}} \cdot \left\{ 1 - \Phi\left(\frac{b_2 - \tau}{\sqrt{d_2 - d_1}}\right) + \Phi\left(\frac{-b_2 - \tau}{\sqrt{d_2 - d_1}}\right) \right\}$$

using the asymptotic normal distribution properties of the logrank statistics. Thus, the two-sided rejection boundary u_ϵ for the final analysis is calculated from

$$P_{H_0}(|L_2 - L_1| \geq u_\epsilon) = \epsilon(\tau, t_1^I)$$

which is equivalent to $u_\epsilon = z_{1-\epsilon(\tau, t_1^I)/2}\sqrt{\Delta d}$. Here, Δd denotes again the number of additional events in the second study stage and $|L_2 - L_1|$ serves as test statistic at the end of the second study stage. Δd is powered for $1 - \beta_{cond,2}$ under the alternative $H_1 : |\theta| = \theta_a$ $(\theta_a > 0)$ and given $L_1 = \tau$, it is the solution of

$$1 - \beta_{cond,2} = P_{H_1}(|L_2 - L_1| \geq u_\epsilon)$$

$$= \Phi\left(-z_{1-\epsilon(\tau,t_1^I)/2} + \frac{1}{2}\theta_a\sqrt{\Delta d}\right) + \Phi\left(-z_{1-\epsilon(\tau,t_1^I)/2} - \frac{1}{2}\theta_a\sqrt{\Delta d}\right).$$

$$(2.4)$$

It can be shown that there exists a solution of (2.4) if and only if the expression on the right in (2.4) is $\geq \epsilon(\tau, t_1^I)$. Wunder et al. [36] show that this solution is unique.

Contrarily to [36] where the conditional power $1 - \beta_{cond,2}$ for stage 2 was chosen as fixed, we here chose $1 - \beta_{cond,2}$ such that a fixed overall power $1 - \beta_{cond}$ is obtained under the relevant treatment effect θ_a, see formula (2.5). Through this, we ensure a specific overall power. In contrast, holding fixed the conditional power for stage 2 neglects the possible type II error when stopping after the first stage, and thus results in a slightly lowered overall power.

Implementing β-spending according to the γ-family (2.3) to obtain conditional overall power $1 - \beta_{cond}$, the conditional power $1 - \beta_{cond,2}$ must be chosen as

$$1 - \beta_{cond,2} = \frac{1 - \beta_{cond}}{1 - P_{\theta_a}(|L_1| < b_1)} = \frac{1 - \beta_{cond}}{1 - \beta\frac{1-\exp(-\gamma t_1^I)}{1-\exp(-\gamma)}} = \frac{1 - \beta_{cond}}{1 - \beta_1}. \quad (2.5)$$

2.2.3 Sample Size Determination

For sample size recalculation in the adaptive design we have to account for expected observed events—denoted by $E(D_{old})$—from patients which were recruited in the first study stage and are still under observation in the second stage ('overrunning patients'). To estimate this number the survival function for each overrunning patient is used as given by Tsiatis [32]. To estimate the proportion $\psi(t_o)$ ('event rate') of patients which will have suffered an event up to time t_o, the formulas of Rubinstein et al. [26] and Lachin and Foulkes [18] are implemented and the maximum sample size for the GSD was calculated to

$$N_{max} = \frac{d_{max}}{\psi(t_o)} = \frac{d_{max}}{\psi(a + f)}$$

presuming a study accrual period a and a follow-up period f.

The recalculated sample size for the second stage of the AD is then given by

$$N_{2,new} = \frac{\max\{\Delta d - E(D_{old}), 0\}}{\psi(a_2 + f_2)} \tag{2.6}$$

if an accrual time a_2 and a follow-up time f_2 are assumed for the second study stage.

Figure 1 depicts the flow of the two-stage study and the statistical decision process during the interim and final analysis—both for the conventional GSD (black lines) and the adaptive design AD (grey lines). The group sequential decision process is the same whether the accrual rate is fixed for both stages (GSD1) or allowed to change in the second stage (GSD2). All three designs share the same first stage and the interim analysis which takes place after the first d_1 events. The final analysis after stage 2 occurs in GSD1/GSD2 after the next $(d_2 - d_1)$ events and in AD after the recalculated number Δd of events in stage 2. Given that our study duration aims for 48 months (36 for accrual and 12 for follow-up), the actually observed number of events may differ from the planned one. The early acceptance/stopping boundary b_1 for the first stage of GSD1/GSD2 and the final rejection boundary b_2 for their second stage are determined iteratively together with the maximum number of events d_{max} ($= d_2$) by the modified method of Chang et al. [5]; for numerical values of b_1, b_2 and d_{max} see Appendix, Table 3.

3 Simulation Studies

In the following the conventional group sequential setting (GSD) (see Sect. 2.1) and the adaptive group sequential design (AD) based on the CRP principle (presented in Sect. 2.2.2) are compared and analysed using simulation studies (using the software R, see [25]). We choose simulation scenarios with parameter combinations which are relevant for the motivating ncc-RCC trial and we discuss the simulation results under the specific restrictions for this study. The simulation studies extend the results of [36], where a more detailed description of the design of the simulation studies can be found.

3.1 Design of Simulation Studies

In the simulation studies the interim analysis takes place at a prespecified information time t_1^I, i.e., after a prespecified number of events; t_1 denotes the corresponding calendar time. We assume that the final analysis is executed either 12 months after the last patient entered the trial or after an observation time fixed in advance for stage 2 (including accrual and follow-up phase), whatever occurs first. Since in the ncc-RCC trial no covariates were considered, survival times are simulated as exponentially distributed. The assumptions under which the simulation studies are executed are summarized in Table 1.

For sample size recalculation in AD we adapt the conditional power $1 - \beta_{cond,2}$ for the second stage such that the overall conditional power equals $1 - \beta_{cond} = 0.8$ (see Tables 4, 5, 6, and 7 in Appendix). Furthermore, we reestimate the sample size $N_{2,new}$ (see (2.6)), the event rate and the accrual rate n_2 for the second stage. In AD this accrual rate is $n_2 = N_{2,new}/(a_2 - t_1)$. As minimum requirement for the simulation studies we claim that at least one additional event has to be observed in stage 2. In practice, the observed event number in stage 2 will trivially fulfil this condition.

Keeping the accrual rate in the group sequential setting GSD1 fixed at $n = 2$ often leads to very poor power only due to the fact that too few patients are included and not enough events can be observed during the fixed study time of 48 months. Therefore, we additionally consider the slightly modified setting GSD2, which

Table 1 Characteristics of two-armed trials in simulation studies

Treatment allocation	Balanced randomization in treatment arms A and B
Median survival times	$M_A = 7$ [months], $M_B = 11$ [months]
Hazard rates	$\lambda_A = 0.099, \lambda_B = 0.063$
Hazard ratio	$11/7 = 1.57$
Total accrual period	$a = 36$ [months]
Total follow-up period	$f = 12$ [months]
Accrual period in stage 2	$a_2 = \max(6, a - t_1)$ [months]
Follow-up period in stage 2	$f_2 = 12$ [months]
Accrual rate in stage 1	$n = 2$ [patients/month]
β-Spending	According γ-family (2.3) with $\gamma = -4$
Loss to follow-up	$\sim \exp(0.005)$ ($\approx$5–10% drop-outs in total)

Table 2 Simulated design strategies and potential adaptations in stage 2

GSD1	GSD2	AD
Group sequential design	Group sequential design	Adaptive group sequential design
No adaptations at interim	New accrual rate $n_2 = N_2/(a - t_1)$	Recalculated sample size $N_{2,new}$
		New accrual rate $n_2 = N_{2,new}/(a - t_1)$
		New test: rejection if $\lvert L_2 - L_1 \rvert \geq u_\epsilon$

allows changing the accrual rate in the second stage to $n_2 = N_2/(a - t_1)$, where N_2 denotes the fixed sample size for stage 2. The three different design strategies are summarized in Table 2.

GSD1, GSD2 and AD are analysed under the null hypothesis $\theta = 0$ and under the alternative hypotheses $\theta = \log(10/7), \log(11/7), \log(12/7)$ or $\log(14/7)$. The sample size is powered for $\theta_a = \log(11/7)$ as relevant treatment effect. The error probabilities are chosen in $(\alpha, \beta) \in \{(0.05, 0.2), (0.1, 0.2), (0.2, 0.2), (0.05, 0.3), (0.1, 0.3), (0.2, 0.3), (0.3, 0.3)\}$. Additionally, GSD1/GSD2 for $(\alpha, \beta) = (0.3, 0.2)$ for comparison to the final design in the ncc-RCC trial. The interim analysis is executed for three differently chosen information times $t_1^I = 0.3, 0.4$ or 0.5 of the underlying GSD. For each scenario 1000 simulation iterations are performed and the average sample number (ASN), the proportion of early stops and the overall power are reported as main design characteristics.

3.2 Simulation Results

In this section we report the results of the simulation studies designed as described in the precedent section. Figures 2, 3, and 4 show the results of the three flexible designs GSD1, GSD2 and AD separately, whereas Figs. 5 and 6 compare the three designs for ASN and the average reestimated accrual rate $\overline{n_2}$ for the second stage, respectively. Numerical results are given in Tables 4, 5, 6, and 7 in the Appendix.

Figures 2, 3, 4, 5, and 6 depict simulation results for selected error probability combinations. The figures allow to compare 'conventional' error probability settings with (α, β) equal to $(0.05, 0.2)$ or $(0.1, 0.2)$ and settings which accommodate the investigational character of the study with (α, β) equal to $(0.2, 0.2)$ or $(0.3, 0.3)$.

Figures 2, 3, and 4 display the simulated power and type I error rates for strategies GSD1, GSD2 and AD, respectively. For specific significance levels α and power $1 - \beta$ (inserted as grey lines) the simulated power is shown under the alternatives $\theta = \log(10/7), \log(11/7), \log(12/7)$ and $\log(14/7)$ (black lines). The observed

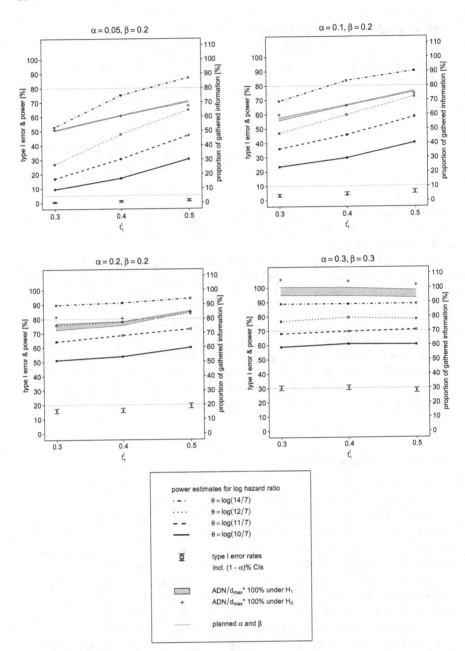

Fig. 2 Simulation results of GSD1 for selected scenarios with sample size powered under $\theta = \log(11/7)$. Depicted are power estimates under different alternatives, type I error rates including $(1 - \alpha)\%$ confidence intervals. Additionally, the proportion of gathered information $ADN/d_{max} * 100\%$ is shown as grey shaded area under all alternatives and marked with '+' under the null hypothesis. The results are obtained for 1000 simulation iterations

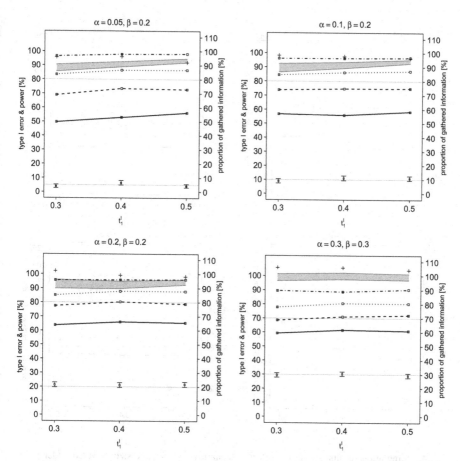

Fig. 3 Simulation results of GSD2 for selected scenarios with sample size powered under $\theta = \log(11/7)$. Depicted are power estimates under different alternatives, type I error rates including $(1 - \alpha)\%$ confidence intervals. Additionally, the proportion of gathered information $ADN/d_{max} * 100\%$ is shown as grey shaded area under all alternatives and marked with '+' under the null hypothesis. The results are obtained for 1000 simulation iterations

type I error rate and $(1 - \alpha)\%$ confidence intervals under the null hypothesis are drawn as box with whiskers. The actually observed proportion of total information is displayed as ratio of the average total number of observed events ADN and the maximum event number d_{max}, i.e., $ADN/d_{max} * 100\%$. Under the alternatives this proportion is enclosed in the grey shaded area and under the null hypothesis it is

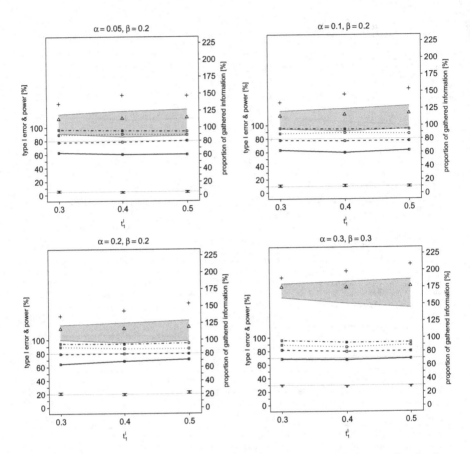

Fig. 4 Simulation results of AD for selected scenarios with sample size powered under $\theta = \log(11/7)$. Depicted are power estimates under different alternatives and type I error rates including $(1 - \alpha)\%$ confidence intervals. Additionally, the proportion of gathered information $ADN/d_{max} * 100\%$ is shown as grey shaded area under all alternatives, marked with '$\triangle$' under $\theta = \log(11/7)$ (for which the study is powered) and marked with '+' under the null hypothesis. The results are obtained for 1000 simulation iterations

assigned with '+' symbols. For AD (Fig. 4), this proportion is additionally plotted for the scenario with correctly assumed treatment effect $\theta = \log(11/7)$ and marked with '$\triangle$' symbols. The scale for type I error rates and power estimates is drawn on the left side of the plots, and the scale for the proportion of total information is adjoined on the right side.

The average sample number (ASN) curves are shown in Fig. 5: black lined for strategy AD under the null and the four alternative hypotheses and enclosed in the

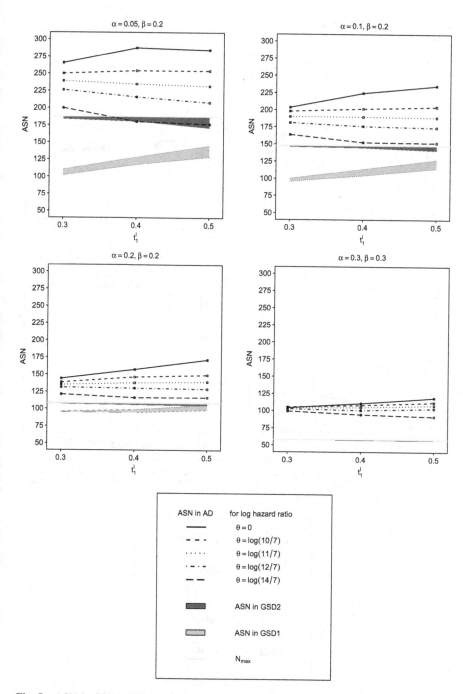

Fig. 5 *ASN* in GSD1, GSD2 and AD for selected scenarios. *ASN* in AD is depicted as black lines, *ASN* in GSD1 and GSD2 is shown as grey shaded areas and N_{max} of the group sequential setting as grey line. The results are obtained for 1000 simulation iterations

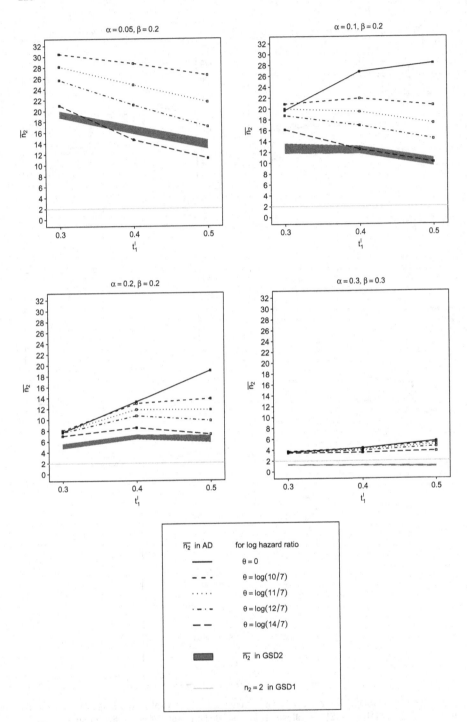

Fig. 6 (Average) accrual rates for stage 2 in GSD1, GSD2 and AD for selected scenarios. Average reestimated accrual rates $\overline{n_2}$ in AD are depicted as black lines and in GSD2 as grey shaded area. The constant accrual rate $n_2 = 2$ in GSD2 is shown as grey line. The results are obtained for 1000 simulation iterations

dark grey area for GSD2 and the light grey area for GSD1. The maximum sample size N_{max} of the two-stage group sequential setting is added for comparison.

Similarly, Fig. 6 displays the average reestimated accrual rates $\overline{n_2}$ in stage 2 for AD under the null and the four alternative hypotheses (black lines). Also drawn is the area which contains $\overline{n_2}$ for GSD2 (dark grey shaded area) and the constant accrual rate $n_2 = n = 2$ in GSD1.

A complete listing of the numerical results of the simulation is included in Appendix in Tables 4 and 6 for GSD1 and GSD2, and Tables 5 and 7 for AD. In total, strategies GSD1, GSD2 and AD reveal control of type I error (see Tables 4 and 5).

Further to note, in AD there is a scenario where the confidence interval for the type I error does not cover the corresponding prespecified significance level. However, increasing for this scenario the iteration number from 1000 to 10,000 leads to a type I error estimate and corresponding confidence interval which enclose the desired level ($\alpha = \beta = 0.2, t_1^I = 0.5$: $\alpha_{est} = 20.1\%, [CI_{\alpha,low}, CI_{\alpha,up}] = [19.6\%, 20.6\%]$).

3.3 Evaluation of Simulation Results

Next, we summarize the major findings obtained by these extensive simulation studies. If one considers GSD1 presuming 'conventional' error probabilities ($\alpha \in \{0.05, 0.1\}, \beta = 0.2$) one can observe an average total number of observed events ADN between 50% and 76% of the determined event number d_{max} for the group sequential plan. That the necessary number of events cannot be reached during the study duration is due to the fixed low accrual rate of $n = n_2 = 2$ patients per month for the prespecified accrual and total trial duration of 36 and 48 months (accrual and 12 months follow-up), respectively. Hence, the power for these design scenarios is far below the desired one of 80%. Under the aimed treatment effect $\theta = \log(11/7)$ it ranges between 16.9% and 58.3% (see Fig. 2). Even allowing for larger error probabilities ($\alpha, \beta \in \{0.2, 0.3\}$), and thus smaller necessary event numbers the proportion of actually observed events stays below 85% of d_{max} ($\alpha = \beta = 0.2$) and mostly below 100% ($\alpha = \beta = 0.3$) under the different alternatives (cf. grey shaded areas in Fig. 2). The corresponding power under $\theta = \log(11/7)$ lies between 64.1% and 72.7% for $\alpha = \beta = 0.2$ and between 68.0% and 70.7% for $\alpha = \beta = 0.3$.

GSD2 corrects for the early too low patient accrual by allowing an adjusted accrual rate n_2 in the second stage. In practice, this can be achieved, e.g., through inclusion of further study centres to increase the number of patients recruited per month. Indeed, in GSD2 we can observe for conventional error settings ($\alpha \in \{0.05, 0.1\}, \beta = 0.2$) that the proportion of gathered information $ADN/d_{max}*100\%$ can considerably be augmented to 86–98%. Under the aimed treatment effect of $\theta = \log(11/7)$ the power estimates are between 68.9% and 72.9% for $\alpha = 0.05$,

$\beta = 0.2$ and between 73.9% and 74.8% for $\alpha = 0.1$, $\beta = 0.2$ (see Fig. 3). Under the different alternatives, for more exploratory errors $\alpha = \beta = 0.2$ the proportion of gathered information reaches 90–96% and the power is between 77.5% and 78.8% under $\theta = \log(11/7)$, whereas for $\alpha = \beta = 0.3$ a power between 68.6% and 72.1% is attained for in average 100% d_{max} observed events.

In ADs with sample size recalculation under conditional overall power of $1 - \beta_{cond} = 0.8$ the power estimates for the depicted scenarios in Fig. 4 are always between 76.9% and 81.9% and thus near the focussed 80% for the aimed effect. Thus, implementing the AD approach results in all scenarios in a slight to considerably increase of power compared to GSD2 (see Fig. 4). However, this gain is paid by an increase in ADN of 113–119% of d_{max} for scenarios with initial choice of $\beta = 0.2$ and an increase of 143–177% if $\beta = 0.3$, leading similarly to increased sample sizes. Scenarios with initially presumed power $1 - \beta = 0.7$ are extended in the adaptive case to designs with overall power $1 - \beta_{cond} = 0.8$ which increases the sample size considerably. Further, the price for the flexibility of an AD with sample size recalculation is an increased sample size to achieve power comparable to the group sequential setting. This was already discussed in Denne [8] for normal data, for further discussion see Wunder et al. [36]. For low presumed error probabilities $(\alpha, \beta) = (0.05, 0.2)$ and thus high sample size in the group sequential setting, we can even observe a reduction in ASN for AD compared to GSD2 for highly underestimated treatment effects in case of the true effect $\theta = \log(14/7)$ and later interim analyses $t_1^I \in \{0.4, 0.5\}$.

3.4 Application of Simulation Results

The simulation results reported above were used to establish the study design and schedule of analyses for the ncc-RCC study accounting for the investigational character of the study and the practical restrictions. These restrictions meant that the total sample size should be limited to $N \approx 100$ and that the sample size in the first stage should be kept near 30 patients, i.e., around one-third of the number of patients achievable in total (see Sect. 1). The restriction in total sample size implies in operational terms that in AD the recalculated number of patients for the second stage should not widely cross that number of 100 patients which was thought to be the limit achievable by the study group. Therefore, neither GSD2 nor AD presuming error probabilities $\alpha = 0.05$, $\beta = 0.2$ or $\alpha = 0.1$, $\beta = 0.2$ came into consideration since they exceeded an average sample number of 100 by far (see Fig. 5). In the scenario with $\alpha = \beta = 0.2$ the average sample number of GSD2 is in fact only slightly higher than 100 ($ASN \approx 107$) but the average accrual rate $\overline{n_2}$ for stage 2 between 5 and 8 patients per month would outrun the maximum recruitable number of 3–4 patients per month.

Finally, we proposed for the trial the AD with the initial group sequential trial plan with $\alpha = \beta = 0.3$, an interim analysis after 30% of the planned events ($t_1^I = 0.3$) and the expected maximum number of events $d_{max} = 48$. This means that the interim look should be executed after 15 observed events achieved after 17 months with an expected number of 34 patients. Assuming some detention of recruitment in the first weeks of the study, in the study protocol the interim analysis was planned to be executed after 18 months. Acceptance and rejection boundaries for the first and the second stage are $b_1 = 0.090$ and $b_2 = 7.115$, respectively. The ASN of the simulation studies equals 105.1, 103.8, 102.5 and 99.2 for $\theta = \log(10/7)$, $\log(11/7)$, $\log(12/7)$ and $\log(14/7)$, respectively. For this plan, the study is expected to recruit 106 patients, 53 patients per treatment arm. The corresponding power estimates lie between 68.6% and 96.1%, with an average power of 81.9% under correctly assumed effect. The reestimated accrual rate for stage 2 assumes accrual of on average 3.5–3.7 patients per month. Assuming fixed conditional power $1 - \beta_{cond,2}$ for the second stage as in [36]—instead of fixed overall power $1 - \beta_{cond}$ as reported here—one gets slightly lower ASN and power in AD (ASN between 98.8 and 101.5 patients and power between 68.6% and 95.2% with 78.9% power under $\theta = \log(11/7)$). This design was finally applied in the study protocol of the ncc-RCC trial. We estimated the AD strategy best suited to our specific study situation since it shows the ability to deal with the unexpected—which is assessed indispensable for this unexplored study setting—e.g., in terms of treatment difference, recruitment or event rate.

4 Discussion

In studies for rare patient entities, one faces specific limitations such as limited total achievable sample, slow accrual and sparse information on the potential of possible drugs. Study plans presuming conventional error rates would require unrealistically large sample sizes and would be based on conventional study designs which are unable to adjust false planning assumptions and may therefore fail the study objective. This was demonstrated above in our investigations of the two-stage GSD1 design for too slow patient accrual. Thus, a more flexible framework is required for rare diseases, in particular, for rare cancer types, which allows tailored sample size reassessment. Using survival as endpoint, our recalculation method depends on estimates of 'nuisance' parameters such as the event rate or the expected number of events in stage 2 from 'overrunning' patients. This may cause a possible distortion of type I error since strong control of type I error is only warranted for adaptations solely conditioned on the interim test statistics. Nevertheless, our simulations show that the type I error is well controlled for the adaptive method chosen. This result is confirmed in more general simulation settings, which analysed the developed adaptive recalculation method for two-stage survival trials when covariate information is included (results not shown). Accounting for covariates reduced the expected sample size of the study if the covariate effect increases the hazard rate. In general, the inclusion of survival relevant covariates is essential to attain the scheduled power, since the stronger the covariate the more power is lost when omitting the covariate information.

The implementation of an optional sample size recalculation is of general relevance in clinical trials for various reasons. A well-discussed reason is the availability of only limited or uncertain planning information when the drug is new or when medical practice changes rapidly. Another reason has an economical and trial management background when the sponsor is not prepared for a large study because of financial restrictions for supporting investigator initiated trials or when the cost benefit ratio is unfavourable in case of a rare disease. The option for a mid-time or early term 'expense budgeting' to evaluate the feasibility of the study appears therefore attractive and helps at least to start the trial. Also structural limitations due to hierarchies in the decision process within a company may favour such a procedure.

Another relevant factor to design a study with the option of sample size reassessment is the desire to be prepared for unexpected results for primary and secondary endpoints during the study. At study start assumptions may not have been justified for nuisance parameters, in particular, regarding the estimate of variances, treatment effects or relationships between primary and secondary endpoints. In a wider sense one may also want to be prepared for the arrival of new external information which may necessitate a change of the treatment or even the whole concept. Also safety data may affect the planning relevant effect size in the light of the risk benefit ratio of the new drug and such lead to reconsidering the design.

The concept of adaptive designs is not as new as it seems today [1] and goes back to seminal work of Peter Bauer in the late 1980s when he fostered repeated and sequential testing for clinical trials research without violation of the error principles of statistical testing. However, methods for time-to-event endpoints have been developed only more recently [9, 12, 15, 28] and only few methodological papers addressed the highly relevant sample size reassessment [3, 23]. Two approaches are at present under development, which deal with the difficult question of how to handle censoring when patients at the interim analysis are event-free and will be followed up beyond the adaptation. One is based on the more classical approach of independent increments in a group sequential setting of [28], see also [35] and [15]. Another one is based on the patient-wise separation where parts of the data are omitted at the analysis. Therefore, [14] and [16] preferred a method based on the conditional error approach in contrast to [22] who made a conservative assumption to guarantee the type 1 error, when a substantial subset of observed event times is not used. The method of [16] assumes in addition a fixed preplanned time for the interim analysis. A one-sample adaptive design was discussed recently by Schmidt et al. [30] who compare the independent increment method with the patient-wise separation. In view of these developments our approaches for the two-sample problem combine elements of group sequential testing with the adaptive concept and can be seen as variants of the independent increment method when applying the CRP principle of [28] for the special case of overall small sample sizes in trials on rare cancer.

However, important caveats should be recognized when adapting the design of a clinical study. Sample size recalculation should not be a means to compensate drop-outs or to collect beyond the veil more data, e.g., for drug safety information, nor can it be an excuse for sloppy planning.

Details of the sample size recalculation and implementation in the study course should be prespecified in the trial protocol. Thus, planning and conducting adaptive designs will often involve earlier and more intense communication and discussion with the approving authorities. Trial simulations are an important tool to understand adaptive designs, especially when little is known in advance about the drugs to be investigated. The FDA encourages the use of simulations for adaptive design trials in their guidances [10, 11] stating that 'computer simulations can play a crucial role in adaptive designs and can provide the operating characteristics of the study design under different scenarios' and that 'for drug development programs where there is little prior experience with the product, drug class, patient population or other critical characteristics, clinical trial simulations can be performed with a range of potential values for relevant parameters encompassing the uncertainty in current knowledge'. Planning and simulating adaptive trial designs is facilitated by available commercial and free software, for which brief reviews may be found in [34], and [1]. It is obvious that adaptive design needs considerably more biostatistical thinking, biostatistical expert knowledge, preparation time for simulations and communication with the principal investigator, study group leaders or authorities than conventional, 'well-understood' trial designs.

Appendix: Numerical Results

The following tables present characteristic parameters for the two-stage group sequential designs with two-sided hypothesis and the possibility to stop early for futility (Table 3) and detailed listings of numerical results of the simulation studies underlying the development of an adaptive group sequential design for an early clinical comparative trial in rare tumour disease (Tables 4, 5, 6, and 7).

Table 3 Design parameters for balanced two-stage group sequential designs with two-sided hypotheses, possibility to stop early for futility and β-spending according to the γ-family

α	β	t'_1	d_{fix}	$\gamma = -4$					$\gamma = -2$					$\gamma = -1$				
				b_1	b_2	Δ	d_{max}	N_{max}	b_1	b_2	Δ	d_{max}	N_{max}	b_1	b_2	Δ	d_{max}	N_{max}
0.05	0.20	0.3	153.7	0.236	24.335	2.813	154.9	187.3	0.740	24.408	2.836	157.5	190.4	1.201	24.483	2.858	159.9	193.3
		0.4		0.710	24.318	2.814	155.0	187.4	1.870	24.354	2.837	157.6	190.5	2.808	24.382	2.858	159.9	193.3
		0.5		1.855	24.285	2.813	154.9	187.3	4.037	24.282	2.833	157.1	190.0	5.470	24.281	2.855	159.6	192.9
0.10	0.20	0.3	121.1	0.166	18.128	2.498	122.2	147.7	0.514	18.175	2.523	124.6	150.7	0.834	18.235	2.547	127.0	153.6
		0.4		0.454	18.105	2.501	122.5	148.1	1.207	18.118	2.526	124.9	151.0	1.850	18.126	2.549	127.2	153.8
		0.5		1.107	18.072	2.501	122.5	148.1	2.523	18.041	2.523	124.6	150.7	3.546	18.000	2.544	126.7	153.2
0.20	0.20	0.3	88.3	0.113	12.045	2.135	89.2	107.9	0.344	12.068	2.161	91.4	110.5	0.551	12.091	2.186	93.6	113.1
		0.4		0.274	12.024	2.138	89.5	108.2	0.748	12.011	2.168	92.0	111.3	1.146	11.999	2.194	94.3	113.9
		0.5		0.625	11.988	2.140	89.7	108.4	1.467	11.919	2.167	91.9	111.2	2.113	11.852	2.190	93.9	113.5
0.30	0.20	0.3	69.1	0.083	8.591	1.885	69.6	84.1	0.262	8.587	1.912	71.6	86.5	0.421	8.602	1.940	73.7	89.1
		0.4		0.200	8.537	1.884	69.5	84.0	0.544	8.526	1.921	72.3	87.4	0.837	8.486	1.949	74.4	89.9
		0.5		0.436	8.527	1.893	70.2	84.8	1.022	8.428	1.924	72.5	87.6	1.484	8.344	1.950	74.5	90.0
0.30	0.30	0.3	47.7	0.090	7.115	1.561	47.7	57.7	0.273	7.063	1.582	49.0	59.2	0.440	7.045	1.605	50.4	61.0
		0.4		0.200	7.059	1.561	47.7	57.7	0.534	7.010	1.589	49.4	59.8	0.814	6.952	1.613	50.9	61.6
		0.5		0.402	7.029	1.565	48.0	58.0	0.943	6.923	1.590	49.5	59.8	1.370	6.826	1.612	50.9	61.5

Given are for $\gamma = -4, -2, -1$, varying α and β and varying information times the acceptance and rejection boundaries b_1 and b_2, the characteristic effect $\Delta = \sigma_Z \theta_a \sqrt{d_{max}}$, the maximum number of events d_{max} for the group sequential design, the number of events d_{fix} for the fixed sample size design and the maximum sample size N_{max} for the group sequential design for a clinical relevant effect $\theta_a = \log(\frac{11}{7})$, exponential parameter of random censoring $\phi = 0.005$ and event rate $\psi = 0.827$. These values have been determined iteratively and numerically by the modified algorithm of Chang et al. [5]

Table 4 Simulation results for group sequential strategies GSD1 and GSD2 without adaptation under the null hypothesis $\theta = 0$

α	β	t_1^I	GSD1						GSD2						
			α_{est} [%]	$CI_{\alpha,low}$ [%]	$CI_{\alpha,up}$ [%]	$stop_{early}$ [%]	ADN	ASN	α_{est} [%]	$CI_{\alpha,low}$ [%]	$CI_{\alpha,up}$ [%]	$stop_{early}$ [%]	$\overline{n_2}$	ADN	ASN
0.05	0.20	0.3	0.3	0.1	0.9	2.4	78.9	101.8	3.9	2.8	5.3	2.4	20.0	148.3	184.1
		0.4	0.8	0.3	1.6	6.0	93.8	117.5	6.4	5.0	8.1	6.0	17.2	148.9	180.8
		0.5	1.4	0.8	2.3	18.1	104.7	129.0	4.2	3.0	5.6	18.1	14.6	142.7	171.2
0.10	0.20	0.3	3.4	2.5	4.5	1.9	73.0	95.8	9.0	7.6	10.6	1.9	11.8	120.1	146.3
		0.4	4.5	3.5	5.7	3.7	81.1	104.5	11.1	9.5	12.9	3.7	13.0	119.5	145.1
		0.5	6.1	4.9	7.5	10.4	91.1	115.0	10.9	9.3	12.7	10.4	10.9	118.0	141.2
0.20	0.20	0.3	15.8	14.3	17.4	1.4	72.6	95.3	21.3	19.6	23.1	1.4	4.7	91.2	107.1
		0.4	16.0	14.5	17.6	4.1	71.9	94.6	21.2	19.5	22.9	4.1	6.4	88.7	105.9
		0.5	19.1	17.5	20.8	7.3	76.0	98.3	21.7	20.0	23.5	7.3	7.0	88.1	104.9
0.30	0.20	0.3	29.4	27.9	31.0	1.5	69.2	83.2	30.5	29.0	32.1	1.5	2.7	72.4	83.3
		0.4	27.0	25.5	28.5	2.6	68.7	82.9	28.8	27.3	30.4	2.6	2.9	71.7	83.0
		0.5	30.3	28.8	31.8	6.4	68.0	83.0	30.8	29.3	32.4	6.4	3.5	70.9	83.1
0.30	0.30	0.3	30.1	28.6	31.7	2.0	50.4	57.4	29.5	28.0	31.1	2.0	1.3	50.5	57.4
		0.4	30.4	28.9	32.0	3.4	49.8	57.3	30.4	28.9	32.0	3.4	1.2	50.4	57.3
		0.5	28.4	26.9	30.0	7.0	48.9	56.9	29.2	27.7	30.8	7.0	1.1	49.9	56.9

Sample size calculation is done under $\theta_a = \log(11/7)$ and β-spending according to the γ-family with $\gamma = -4$. Average type I error rate α_{est}, corresponding confidence interval [$CI_{\alpha,low}$, $CI_{\alpha,up}$], proportion of early stops $stop_{early}$, average reestimated accrual rate $\overline{n_2}$ in stage 2 (GSD2), average total number of observed events ADN and average sample number ASN for 1000 simulation iterations are given. In GSD1 the accrual rate in stage 2 is fixed at $n_2 = 2$

Table 5 Simulation results for adaptive designs with sample size recalculation under the null hypothesis $\theta = 0$

α	β	$\beta_{cond,2}$	t_1^I	α_{est} [%]	$CI_{\alpha,low}$ [%]	$CI_{\alpha,up}$ [%]	$stop_{early}$ [%]	$\overline{n_2}$	ADN	ASN
0.05	0.20	0.19	0.3	5.3	4.0	6.9	2.4	33.8	209.7	265.6
		0.19	0.4	3.9	2.8	5.3	6.0	36.1	228.8	287.5
		0.18	0.5	4.4	3.2	5.9	18.1	37.6	227.6	284.6
0.10	0.20	0.19	0.3	10.7	9.1	12.4	1.9	19.7	163.5	203.6
		0.19	0.4	10.9	9.3	12.7	3.7	26.8	178.9	224.4
		0.18	0.5	10.0	8.5	11.7	10.4	28.3	188.3	235.0
0.20	0.20	0.19	0.3	20.8	19.1	22.5	1.4	7.7	121.0	144.0
		0.19	0.4	19.3	17.7	21.0	4.1	13.3	128.1	157.3
		0.18	0.5	22.2	20.5	24.0	7.3	19.0	137.9	171.6
0.30	0.30	0.19	0.3	29.0	27.5	30.6	2.0	3.6	90.1	104.2
		0.18	0.4	27.7	26.2	29.2	3.4	4.3	94.7	110.7
		0.17	0.5	27.8	26.3	29.3	7.0	5.6	100.2	118.6

Sample size calculation is done under $\theta_a = \log(11/7)$ and β-spending according to the γ-family with $\gamma = -4$. Conditional power $\beta_{cond,2}$ is calculated for overall conditional power of 0.8. Average type I error rate α_{est}, corresponding confidence interval $[CI_{\alpha,low}, CI_{\alpha,up}]$, proportion of early stops $stop_{early}$, average reestimated accrual rate $\overline{n_2}$, average total number of observed events ADN and average sample number ASN for 1000 simulation iterations are given

The tables showing the simulation results list, in particular, the parameters characterizing the underlying initial group sequential trial: the true treatment effect θ, the trial parameters α, β and the information time t_1^I of the interim analysis. Furthermore, for the adaptive strategy (Tables 5 and 7) the conditional type II error $\beta_{cond,2}$ for stage 2 is given to reach conditional overall power $1 - \beta_{cond} = 0.8$, see formula (2.5). The tables list the numerical results of the (α, β)-combinations which are depicted in the paper.

Table 6 Simulation results for group sequential strategies GSD1 and GSD2 without adaptation under different alternatives θ

θ	β	α	t_1'	GSD1				GSD2				
				$power_{est}$ [%]	$stop_{early}$ [%]	ADN	ASN	$power_{est}$ [%]	$stop_{early}$ [%]	$\overline{n}_2$	ADN	ASN
$\log\left(\frac{10}{7}\right)$	0.20	0.05	0.3	9.4	1.7	78.3	105.6	49.7	1.7	19.5	140.2	185.0
			0.4	17.1	3.4	94.1	122.5	53.0	3.4	16.5	143.4	183.6
			0.5	30.3	5.6	108.5	137.8	56.3	5.6	13.8	146.2	182.3
	0.20	0.10	0.3	23.5	0.9	70.4	97.4	56.8	0.9	13.1	113.0	147.2
			0.4	29.6	2.4	81.2	108.8	55.8	2.4	12.4	114.4	146.2
			0.5	40.2	4.5	92.6	121.1	58.3	4.5	10.2	116.6	145.2
	0.20	0.20	0.3	51.1	1.7	68.4	95.2	63.8	1.7	5.1	85.7	107.0
			0.4	53.6	1.7	69.8	96.3	66.1	1.7	7.0	84.9	107.2
			0.5	59.8	3.2	76.5	102.4	65.5	3.2	6.5	86.0	106.7
	0.20	0.30	0.3	62.9	1.4	65.6	83.2	66.8	1.4	2.8	68.8	83.4
			0.4	66.4	2.2	65.5	83.1	68.7	2.2	3.1	68.3	83.2
			0.5	64.2	3.2	66.0	84.0	65.7	3.2	3.6	68.4	84.1
	0.30	0.30	0.3	58.7	2.0	47.9	57.5	59.2	2.0	1.3	48.4	57.5
			0.4	60.8	2.8	47.7	57.4	61.5	2.8	1.2	48.7	57.4
			0.5	60.3	5.0	47.1	57.3	60.9	5.0	1.0	48.6	57.3
$\log\left(\frac{11}{7}\right)$	0.20	0.05	0.3	16.9	0.7	78.4	107.1	68.9	0.7	19.3	138.9	186.2
			0.4	30.5	2.0	94.3	124.2	73.5	2.0	16.3	142.3	185.0
			0.5	46.8	2.3	109.3	140.4	72.9	2.3	13.6	146.8	185.1
	0.20	0.10	0.3	36.0	0.6	69.8	98.0	73.9	0.6	13.3	111.0	147.4
			0.4	45.6	1.4	81.4	110.3	74.7	1.4	12.2	113.3	146.9
			0.5	58.3	2.9	92.9	122.8	74.8	2.9	10.0	116.0	146.2

(continued)

Table 6 (continued)

θ	α	β	t_1^l	GSD1 $power_{est}$ [%]	$stopP_{early}$ [%]	ADN	ASN	GSD2 $power_{est}$ [%]	$stopP_{early}$ [%]	$\overline{n}_2$	ADN	ASN
	0.20	0.20	0.3	64.1	1.0	67.7	95.6	77.5	1.0	5.2	84.5	107.4
			0.4	68.3	1.2	69.2	96.9	80.3	1.2	7.1	83.8	107.4
			0.5	72.7	1.8	76.4	103.5	78.8	1.8	6.3	85.3	107.3
	0.30	0.20	0.3	76.7	0.9	64.7	83.4	78.9	0.9	2.8	67.8	83.6
			0.4	78.4	2.0	64.4	83.2	80.3	2.0	3.1	67.2	83.3
			0.5	76.5	2.5	65.4	84.2	76.7	2.5	3.7	67.7	84.3
	0.30	0.30	0.3	68.0	1.9	47.1	57.5	68.6	1.9	1.3	47.7	57.5
			0.4	69.6	2.6	46.9	57.5	71.1	2.6	1.1	48.1	57.5
			0.5	70.7	3.3	46.6	57.5	72.1	3.3	0.9	48.2	57.5
$\log\left(\frac{12}{7}\right)$	0.05	0.20	0.3	26.9	0.2	78.4	108.3	83.5	0.2	19.1	137.1	186.8
			0.4	47.7	0.6	94.5	125.9	86.5	0.6	16.1	141.5	186.4
			0.5	64.6	0.8	109.6	142.2	86.6	0.8	13.4	145.9	186.3
	0.10	0.20	0.3	46.9	0.6	69.1	98.7	84.5	0.6	13.4	109.0	147.5
			0.4	59.7	1.2	81.1	111.4	86.2	1.2	12.0	111.7	147.1
			0.5	72.4	1.5	93.3	124.5	87.1	1.5	9.8	115.4	147.1
	0.20	0.20	0.3	75.6	0.6	66.6	95.8	85.1	0.6	5.3	83.2	107.6
			0.4	77.8	0.7	68.6	97.6	87.8	0.7	7.2	82.7	107.7
			0.5	83.6	1.3	76.3	104.2	87.7	1.3	6.1	84.5	107.5
	0.30	0.20	0.3	85.3	0.9	63.6	83.5	87.3	0.9	2.8	66.7	83.6
			0.4	85.7	0.7	63.8	83.7	87.9	0.7	3.1	66.6	83.7
			0.5	84.7	2.1	64.6	84.4	87.3	2.1	3.7	66.9	84.5

$\log\left(\frac{14}{7}\right)$		θ_a	ASN	$stop_{early}$	ADN	$power_{est}$	ASN	$stop_{early}$	$\bar{n}_2$	ADN	$power_{est}$
0.30	0.30	0.3	76.6	1.5	46.5	57.6	77.8	1.5	1.3	47.2	57.6
		0.4	79.3	2.4	46.1	57.6	80.5	2.4	1.1	47.4	57.5
		0.5	78.1	2.0	46.2	57.8	80.3	2.0	0.9	47.9	57.8
0.05	0.20	0.3	52.7	0.4	77.6	110.1	96.5	0.4	18.8	132.9	186.5
		0.4	74.9	0.2	94.1	128.2	97.8	0.2	15.8	137.9	186.8
		0.5	87.0	0.3	109.4	144.9	98.1	0.3	13.0	142.7	186.8
0.10	0.20	0.3	69.3	0.3	68.3	100.0	96.2	0.3	13.6	105.8	147.8
		0.4	83.5	0.2	80.8	113.7	96.2	0.2	11.7	109.3	147.9
		0.5	90.2	0.2	93.3	127.2	96.6	0.2	9.5	113.6	147.9
0.20	0.20	0.3	89.7	0.5	64.6	95.9	95.7	0.5	5.6	80.3	107.7
		0.4	91.1	0.2	67.6	98.8	95.9	0.2	7.2	80.4	107.9
		0.5	94.1	0.1	76.0	105.5	96.1	0.1	5.8	83.0	108.0
0.30	0.20	0.3	94.5	0.5	62.0	83.7	95.5	0.5	2.8	65.0	83.8
		0.4	94.7	0.6	62.0	83.7	95.1	0.6	3.2	64.6	83.8
		0.5	94.3	1.2	63.4	84.6	95.6	1.2	3.6	65.3	84.7
0.30	0.30	0.3	88.8	1.0	45.3	57.7	89.5	1.0	1.3	46.1	57.7
		0.4	88.7	1.1	45.1	57.8	88.7	1.1	1.1	46.6	57.7
		0.5	88.8	2.1	44.6	57.8	90.3	2.1	0.9	46.6	57.7

Sample size calculation is done under $\theta_a = \log(11/7)$ and β-spending according to the γ-family with $\gamma = -4$. Estimated power $power_{est}$, average reestimated accrual rate $\bar{n}_2$ in stage 2 (GSD2), proportion of early stops $stop_{early}$, average total number of observed events ADN and average sample number ASN for 1000 simulation iterations are given. In GSD1 the accrual rate in stage 2 is fixed at $n_2 = 2$

Table 7 Simulation results for adaptive designs with sample size recalculation under different alternatives θ

θ	α	β	$\beta_{cond,2}$	t_1^I	$power_{est}$ [%]	$stop_{early}$ [%]	$\overline{n_2}$	ADN	ASN
$\log\left(\frac{10}{7}\right)$	0.05	0.20	0.19	0.3	62.8	1.7	30.5	185.0	250.0
			0.19	0.4	60.2	3.4	28.7	192.2	254.1
			0.18	0.5	60.1	5.6	26.5	195.8	254.1
	0.10	0.20	0.19	0.3	63.4	0.9	20.9	147.9	198.0
			0.19	0.4	59.8	2.4	21.9	152.6	201.7
			0.18	0.5	62.9	4.5	20.6	157.4	204.5
	0.20	0.20	0.19	0.3	64.5	1.7	8.0	108.6	138.5
			0.19	0.4	68.4	1.7	13.0	111.9	146.4
			0.18	0.5	71.0	3.2	13.8	115.4	149.4
	0.30	0.30	0.19	0.3	68.6	2.0	3.7	85.7	105.1
			0.18	0.4	67.5	2.8	4.3	87.3	108.0
			0.17	0.5	69.3	5.0	5.3	89.3	112.1
$\log\left(\frac{11}{7}\right)$	0.05	0.20	0.19	0.3	78.0	0.7	28.2	174.5	239.2
			0.19	0.4	78.4	2.0	24.8	176.0	234.7
			0.18	0.5	80.2	2.3	21.6	177.9	231.9
	0.10	0.20	0.19	0.3	77.9	0.6	20.0	139.4	190.1
			0.19	0.4	76.9	1.4	19.4	142.2	189.7
			0.18	0.5	77.0	2.9	17.3	144.5	188.8
	0.20	0.20	0.19	0.3	79.2	1.0	7.8	104.0	135.0
			0.19	0.4	79.4	1.2	11.8	104.3	137.9
			0.18	0.5	79.4	1.8	11.7	107.0	139.5
	0.30	0.30	0.19	0.3	81.9	1.9	3.7	83.5	103.8
			0.18	0.4	79.6	2.6	4.1	83.3	104.8
			0.17	0.5	80.3	3.3	5.0	84.8	107.7
$\log\left(\frac{12}{7}\right)$	0.05	0.20	0.19	0.3	89.0	0.2	25.7	163.3	226.2
			0.19	0.4	90.0	0.6	21.1	161.0	216.0
			0.18	0.5	88.9	0.8	17.0	160.0	208.0
	0.10	0.20	0.19	0.3	87.8	0.6	18.8	130.8	181.0
			0.19	0.4	89.0	1.2	16.9	130.7	176.0
			0.18	0.5	87.4	1.5	14.4	132.9	174.0
	0.20	0.20	0.19	0.3	89.5	0.6	7.7	99.5	131.2
			0.19	0.4	87.0	0.7	10.7	97.3	130.1
			0.18	0.5	87.0	1.3	9.8	98.7	129.1
	0.30	0.30	0.19	0.3	89.4	1.5	3.6	81.0	102.5
			0.18	0.4	86.2	2.4	3.9	78.9	100.6
			0.17	0.5	89.2	2.0	4.6	80.0	103.1

(continued)

Table 7 (continued)

θ	α	β	$\beta_{cond,2}$	t_1^I	$power_{est}$ [%]	$stop_{early}$ [%]	$\overline{n_2}$	ADN	ASN
$\log\left(\frac{14}{7}\right)$	0.05	0.20	0.19	0.3	96.7	0.4	21.0	141.1	199.9
			0.19	0.4	95.2	0.2	14.7	133.9	180.2
			0.18	0.5	94.0	0.3	11.2	136.3	176.2
	0.10	0.20	0.19	0.3	95.5	0.3	16.1	115.9	163.7
			0.19	0.4	94.2	0.2	12.5	112.4	152.6
			0.18	0.5	94.3	0.2	10.1	116.1	151.9
	0.20	0.20	0.19	0.3	94.5	0.5	7.0	89.3	121.1
			0.19	0.4	93.7	0.2	8.5	85.7	116.3
			0.18	0.5	95.0	0.1	7.2	88.4	116.5
	0.30	0.30	0.19	0.3	96.1	1.0	3.5	76.2	99.2
			0.18	0.4	93.1	1.1	3.5	72.2	94.3
			0.17	0.5	93.4	2.1	3.8	69.5	91.5

Sample size calculation is done under $\theta_a = \log(11/7)$ and β-spending according to the γ-family with $\gamma = -4$. Conditional power $\beta_{cond,2}$ is calculated for overall conditional power 0.8. Estimated power $power_{est}$, proportion of early stops $stop_{early}$, average reestimated accrual rate $\overline{n_2}$, average total number of observed events ADN and average sample number ASN for 1000 simulation iterations are given

References

1. Bauer, P., Bretz, F., Dragalin, V., König, F., & Wassmer, G. (2016). Twenty-five years of confirmatory adaptive designs: opportunities and pitfalls. *Statistics in Medicine, 35*, 325–347.
2. Bauer, P., & Köhne, K. (1994). Evaluation of experiments with adaptive interim analyses. *Biometrics, 50*, 1029–1041.
3. Bauer, P., & Posch, M. (2004). Letter to the editor. Modification of the sample size and the schedule of interim analyses in survival trials based on data inspections, by H. Müller, H.-H. Schäfer, Statistics in Medicine 2001; 20:3741–3751. *Statistics in Medicine, 23*, 1333–1335.
4. Brannath, W., Posch, M., & Bauer, P. (2002). Recursive combination tests. *Journal of the American Statistical Association, 97*, 236–244.
5. Chang, M. N., Hwang, I. K., & Shih, W. J. (1998). Group sequential designs using both type I and type II error probability spending functions. *Communications in Statistics, Theory and Methods, A27(6)*, 1323–1339.
6. Cox, D. R. (1972). Regression models and life tables (with discussion). *Journal of the Royal Statistical Society, B34*, 187–220.
7. Cox, D. R. (1975). Partial likelihood. *Biometrika, 62(2)*, 269–276.
8. Denne, J. S. (2001). Sample size recalculation using conditional power. *Statistics in Medicine, 20*, 2645–2660.
9. Desseaux, K., & Porcher, R. (2007). Flexible two-stage design with sample size reassessment for survival trials. *Statistics in Medicine, 26(27)*, 5002–5013.
10. FDA. (2010). Adaptive design clinical trials for drugs and biologics – Draft guidance. *U.S. Department of Health and Human Services, Food and Drug Administration.* http://www.fda. gov/Drugs/GuidanceComplianceRegulatoryInformation/Guidances/ucm121568.htm

11. FDA. (2016). Adaptive designs for medical device clinical studies – Guidance for industry and food and drug administration staff. *U.S. Department of Health and Human Services, Food and Drug Administration.* https://www.fda.gov/downloads/medicaldevices/deviceregulationandguidance/guidancedocuments/ucm446729.pdf

12. Gu, M., & Ying, Z. (1995). Group sequential methods for survival data using partial likelihood score processes with covariate adjustment. *Statistica Sinica, 5,* 793–804.

13. Hwang, I. K., Shih, W. J., & De Cani, J. S. (1990). Group sequential designs using a family of type I error probability spending functions. *Statistics in Medicine, 9,* 1439–1445.

14. Irle, S., & Schäfer, H. (2012). Interim design modifications in time-to-event studies. *Journal of the American Statistical Association, 107,* 341–348.

15. Jahn-Eimermacher, A., & Ingel, K. (2009). Adaptive trial design: A general methodology for censored time to event data. *Contemporary Clinical Trials, 30,* 171–177.

16. Jenkins, M., Stone, A., & Jennison, C. (2011). An adaptive seamless phase II/III design for oncology trials with subpopulation selection using correlated survival endpoints. *Pharmaceutical Statistics, 10(4),* 347–356.

17. Jennison, C., & Turnbull, B. W. (2000). Group sequential methods with applications to clinical trials. London: Chapman and Hall/CRC.

18. Lachin, J. M., & Foulkes, M. A. (1986). Evaluation of sample size and power for analyses of survival with allowance for nonuniform patient entry, losses to follow-up, noncompliance and stratification. *Biometrics, 42,* 507–519.

19. Lan, K. K. G., & DeMets, D. L. (1983). Discrete sequential boundaries for clinical trials. *Biometrika, 70(3),* 659–663.

20. Lan, K. K. G., & DeMets, D. L. (1989). Group sequential procedures: Calendar versus information time. *Statistics in Medicine, 8,* 1191–1198.

21. Lehmacher, W., & Wassmer, G. (1999). Adaptive sample size calculations in group sequential trials. *Biometrics, 55,* 1286–1290.

22. Magirr, D., Jaki, T., Koenig, F., & Posch, M. (2014). Adaptive survival trials. https://arxiv.org/abs/1405.1569

23. Mehta, C. R., & Pocock, S. J. (2011). Adaptive increase in sample size when interim results are promising: A practical guide with examples. *Statistics in Medicine, 30(28),* 3267–3284.

24. Müller, H. H., & Schäfer, H. (2004). A general statistical principle for changing a design any time during the course of a trial. *Statistics in Medicine, 23,* 2497–2508.

25. Project team, R. The R Project for Statistical Computing. https://www.r-project.org

26. Rubinstein, L. V., Gail, M. H., & Santner, T. J. (1981). Planning the duration of a comparative clinical trial with loss to follow-up and a period of continued observation. *Journal of Chronic Diseases, 34,* 469–479.

27. Rudser, K. D., & Emerson, S. S. (2008). Implementing type I and type II error spending for two-sided group sequential designs. *Contemporary Clinical Trials, 29,* 351–358.

28. Schäfer, H., & Müller, H. H. (2001). Modification of the sample size and the schedule of interim analyses in survival trials based on data inspections. *Statistics in Medicine, 20,* 3741–3751.

29. Schmidinger, M., & Zielinski, C. C. (2009). Novel agents for renal cell carcinoma require novel selection paradigms to optimise first-line therapy. *Cancer Treatment Reviews, 35,* 289–296.

30. Schmidt, R., Faldum, A., & Kwiecien, R. (2018). Adaptive designs of the one-sample logrank test. *Biometrics, 74,* 529–537. https://doi.org/10.1111/biom.12776

31. Schrader, A. J., Olbert, P. J., Hegele, A., Varga, Z., & Hofmann, R. (2008). Metastatic non-clear cell renal cell carcinoma: current therapeutic options. *BJU International, 101,* 1343–1345.

32. Tsiatis, A. (1981). A large sample study of Cox's regression model. *The Annals of Statistics, 9(1),* 93–108.

33. Tsiatis, A., Rosner, G. L., & Tritchler, D. L. (1985). Group sequential tests with censored survival data adjusting for covariates. *Biometrika, 72(2),* 365–373.

34. Tymofyeyev, Y. (2014). A review of available software and capabilities for adaptive designs. In W. He, J. Pinheiro, & O. M. Kuznetsova (Eds.), *Practical considerations for adaptive trial design and implementation* (pp. 139–155). New York: Springer.
35. Wassmer, G. (2006). Planning and analyzing adaptive group sequential survival trials. *Biometrical Journal, 48,* 714–729.
36. Wunder, C., Kopp-Schneider, A., & Edler, L. (2012). An adaptive group sequential phase II design to compare treatments for survival endpoints in rare patient entities. *Journal of Biopharmaceutical Statistics, 22(2),* 294–311.

Parameter Estimation for Multivariate Nonlinear Stochastic Differential Equation Models: A Comparison Study

Wei Gu, Hulin Wu, and Hongqi Xue

Abstract Statistical methods have been proposed to estimate parameters in multivariate stochastic differential equations (SDEs) from discrete observations. In this paper, we propose a method to improve the performance of the local linearization method proposed by Shoji and Ozaki (Biometrika 85:240–243, 1998), i.e., to avoid the ill-conditioned problem in the computational algorithm. Simulation studies are performed to compare the new method to three other methods, the benchmark Euler method and methods due to Pedersen (1995) and to Hurn et al. (2003). Our results show that the new method performs the best when the sample size is large and the methods proposed by Pedersen and Hurn et al. perform better when the sample size is small. These results provide useful guidance for practitioners.

Keywords Diffusion process · Euler–Maruyama scheme · Simulated maximum likelihood methods · Kernel density estimator · Local linearization method

1 Introduction

In past decades, diffusion models described by stochastic differential equations (SDE) have been used in many fields, such as economics (Black and Scholes [4]; Cox et al. [7]) and biology (Gillespie [11]; Arkin et al. [3]; Dalal et al.

W. Gu
School of Statistics and Mathematics, Zhongnan University of Economics and Law, Wuhan, Hubei, People's Republic of China

H. Wu (✉)
University of Texas Health Sciences Center, Houston, Texas, USA
e-mail: Hulin.Wu@uth.tmc.edu

H. Xue
Department of Biostatistics and Computational Biology, University of Rochester, New York, NY, USA
e-mail: Hongqi_Xue@URMC.Rochester.edu

© Springer Nature Switzerland AG 2020
A. Almudevar et al. (eds.), *Statistical Modeling for Biological Systems*,
https://doi.org/10.1007/978-3-030-34675-1_13

245

[8]). Parameters of SDE models are often unknown in practical applications, and it is challenging to efficiently estimate these parameters in SDE models from experimental data. This paper focuses on maximum likelihood estimation methods of the unknown parameters in multivariate nonlinear SDE models, the key step is to obtain the transition density function of diffusions, which is usually unknown and its explicit form is not available except for some special cases. Recently, parameter estimation methods have been proposed based on approximation of the transition density function (Pedersen [20]; Brandt and Santa-Claras [5]; Durham and Gallant [9]). Comprehensive reviews about other parameter estimation methods for SDE can be found in the literature (Nielsen et al. [18]; Sørensen [28]; Jimenez et al. [14]; Aït-sahalia [1]; Hurn and Jeisman [12]).

Four popular parameter estimation methods for SDE models are considered here, which are easy to implement. The first approach is the Euler method, which is regarded as the benchmark of other methods. The second one is known as the simulated maximum likelihood (SML) method proposed by Pedersen [20] and Brandt and Santa-Claras [5] in which the transition density is approximated by a Monte Carlo integral method. The third approach is also the simulated maximum likelihood method considered by Hurn et al. [13], but this method is based on the kernel density estimator. The fourth one is called the *improved local linearization* (ILL) scheme based on Shoji and Ozaki [24, 25], in which the stochastic differential equations are discretized by the *local linearization* (LL) scheme. The ILL method is mainly considered in this paper.

In this paper, first we extend the Hurn et al. [13] method to deal with multidimensional SDEs by assuming diffusions to be pairwise independent. In addition, we propose a new approach to improve the performance of Shoji and Ozaki's method [25] to deal with the potential singular or ill-conditioned Jacobian matrix of the drift coefficient. We also extend their method to estimate the parameters of SDEs with nonlinear drift and diffusion coefficients. We employ the Lamperti transformation to obtain a new SDE with unit diffusion coefficient, and for this new SDE, our ILL method still works even when the drift coefficient has a singular or ill-conditioned Jacobian matrix. This is because we propose computing the integrals using the Jacobian matrix exponentials, instead of evaluating the inverse of the matrix.

The paper is organized as follows. In Sect. 2, we briefly introduce the three estimation methods, including the Euler method, the SML method proposed by Pedersen [20] and Brandt and Santa-Claras [5], and the SML method based on the kernel density estimator (Hurn et al. [13]). We propose the ILL scheme of Shoji and Ozaki [25] in Sect. 3. We evaluate the four estimation methods via Monte Carlo simulation studies in Sect. 4. A brief discussion is given in Sect. 5.

2 Estimation Methods for Nonlinear SDEs

Consider a diffusion process $X_t \in \mathcal{R}^d$ that satisfies the following nonlinear stochastic differential equation:

$$dX_t = f(X_t; \theta)dt + \sigma(X_t; \theta)dW_t, \quad X_{t_0} = x_0, \quad t \geq 0, \tag{2.1}$$

and it has a componentwise form

$$dX_t^l = f^l(X_t; \theta)dt + \sum_{j=1}^{d} \sigma_{lj}(X_t; \theta)dW_t^j, \quad l = 1, \cdots, d, \tag{2.2}$$

where W_t is a d-dimensional vector of independent Brownian motions, $f : \mathcal{R}^d \times \Theta \to \mathcal{R}^d$ and $\sigma : \mathcal{R}^d \times \Theta \to \mathcal{R}^{d \times d}$ are known functions depending on an unknown parameter vector $\theta \in \Theta \subseteq \mathcal{R}^p$. Assume that the diffusion process X is observed at discrete time points $0 = t_0 < t_1 < \cdots < t_n$, that is $X^{obs} = (x_0, x_1, x_2, \cdots, x_n)$. The initial value of X at t_0 is denoted by $X_{t_0} = x_0$, and the sampling time interval $\Delta = t_i - t_{i-1}$ is considered as a constant for simplicity of presentation.

Let $p(x, t \mid y, s; \theta)$ be the transition probability density of the diffusion X ($s \leq t$). From the Markov property of the diffusion X, we can obtain the following likelihood function conditional on the initial value x_0 if the transition density p is known, that is

$$L_n(\theta) = \prod_{i=1}^{n} p(x_i, t_i \mid x_{i-1}, t_{i-1}; \theta). \tag{2.3}$$

An estimate of θ can be obtained by maximizing the likelihood function (2.3), which is equivalent to

$$\hat{\theta} = \arg\min_{\theta} \left\{ -\sum_{i=1}^{n} \log p(x_i, t_i \mid x_{i-1}, t_{i-1}; \theta) \right\}. \tag{2.4}$$

However, the transition densities of X are usually unknown in practice. Various numerical approaches are proposed to approximate p based on consecutive observations of X. We review some of these approaches in the following subsections.

2.1 Euler Method

The Euler method is the simplest method we consider, which can serve as a comparison benchmark for evaluating other estimation approaches. Assume that

$f(x_{i-1}; \theta)$ and $\sigma(x_{i-1}; \theta)$ are matrices which do not depend on time; then the Euler–Maruyama scheme can be used to discretize model (2.1) as,

$$x_i = x_{i-1} + f(x_{i-1}; \theta)\Delta + \sigma(x_{i-1}; \theta)\sqrt{\Delta}\varepsilon_i \tag{2.5}$$

where ε_i is a vector of independent standard normal random variates, so that we have the following transition densities between x_{i-1} and x_i:

$$p^{Euler}(x_i, t_i \mid x_{i-1}, t_{i-1}; \theta)$$
$$= ((2\pi)^d |V_{i-1}|)^{-1/2} \exp(-\frac{1}{2}(x_i - m_{i-1})' V_{i-1}^{-1}(x_i - m_{i-1})), \tag{2.6}$$

where $m_{i-1} = x_{i-1} + f(x_{i-1}; \theta)\Delta$ and $V_{i-1} = \sigma(x_{i-1}; \theta)\sigma^T(x_{i-1}; \theta)\Delta$, $|V(\cdot)|$ and $V^{-1}(\cdot)$ denote the determinant and inverse of V, respectively.

2.2 Simulated Maximum Likelihood Method

In the following, we consider the simulated maximum likelihood method proposed by Pedersen [20]. First divide the time interval $[t_{i-1}, t_i]$ into N equally spaced subintervals, that is $t_{i-1} = \tau_{i,0} < \tau_{i,1} < \cdots < \tau_{i,N} = t_i$ with the step size $\delta = \Delta/N$. We discretize model (2.1) using the Euler–Maruyama scheme to obtain the approximations $(\hat{x}_{i,n})_{n=1,\cdots,N-1}$, which are Gaussian processes (Kloeden and Platen [16]).

By the Markov property of the process X and the Chapman–Kolmogorov equation, we have the following approximation to the transition density between two consecutive observations:

$$p^{Pedersen}(x_i, t_i \mid x_{i-1}, t_{i-1}; \theta)$$
$$= \int_{\mathcal{R}^{d(N-1)}} \prod_{n=1}^{N} p^{(1)}(\xi_n, \tau_{i,n} \mid \xi_{n-1}, \tau_{i,n-1}; \theta) d\xi_1 \cdots d\xi_{N-1}$$
$$= E[p^{(1)}(x_i, t_i \mid \hat{x}_{i,N-1}, \tau_{i,N-1}; \theta)] \tag{2.7}$$

with $\xi_0 = x_{i-1}$ and $\xi_N = x_i$, where $p^{(1)}(\xi_n, \tau_{i,n} \mid \xi_{n-1}, \tau_{i,n-1}; \theta)$ is the Gaussian transition density with mean $\xi_{n-1} + f(\xi_{n-1}; \theta)\Delta$ and variance $\sigma(\xi_{n-1}; \theta)\sigma^T(\xi_{n-1}; \theta)\Delta$; and $E[p^{(1)}(x, t \mid y, s; \theta)]$ is the expectation with respect to $P_{y,s,\theta}$, the joint probability distribution of (y, s) at true value θ. Then apply the Monte Carlo integral method to approximate $p^{Pedersen}(x_i, t_i \mid x_{i-1}, t_{i-1}; \theta)$ as

$$p_M^{Pedersen}(x_i, t_i \mid x_{i-1}, t_{i-1}; \theta)$$

$$= \frac{1}{M} \sum_{m=1}^{M} p^{(1)}(x_i, t_i \mid \hat{x}_{i,N-1}^{(m)}, \tau_{i,N-1}; \theta) \tag{2.8}$$

by choosing the integer M sufficiently large, where $\hat{x}_{i,N-1}^{(m)}$ are independent approximated values using the Euler–Maruyama scheme with step size δ and initial value x_{i-1}. Pedersen [20] proved that $L_n^{(N)}(\theta) = \prod_{i=1}^{n} p_M^{Pedersen}(x_i, t_i \mid x_{i-1}, t_{i-1}; \theta)$ converges to $L_n(\theta)$ in (2.3) in probability as $N \to \infty$. In practice, we use $p_M^{Pedersen}$ to approximate $p^{Pedersen}$ with an error order $O((1/N)(1/\sqrt{M}))$. To choose M and N, Durham and Gallant [9] suggest that N ranging 5–10 and M ranging from 2500 to 5000 are sufficient for a four-dimensional diffusion process.

Hurn et al. [13] considered an alternative approach to approximate the transition densities of the diffusion process based on the kernel density estimator. However, for fixed x and θ, the transition densities of the diffusion process defined by (2.1) can be seen as the solution to a *forward Kolmogorov equation*

$$\frac{\partial}{\partial t} p(x, t \mid y, s; \theta)$$

$$= -\sum_{l=1}^{d} \frac{\partial}{\partial y_l} (f^l(y; \theta) p(x, t \mid y, s; \theta))$$

$$+ \frac{1}{2} \sum_{l=1}^{d} \sum_{j=1}^{d} \frac{\partial^2}{\partial y_j \partial y_l} (b^{lj}(y; \theta) p(x, t \mid y, s; \theta)) \tag{2.9}$$

with the initial value $p(x, s \mid y, s; \theta) = \delta(x - y)$, where $\delta(x - y)$ is the Dirac delta function, and $b(x; \theta) \equiv \sigma(x; \theta)\sigma^T(x; \theta)$. Poulsen [22] used the Crank–Nicolson finite difference method to solve (2.9) for each observation $(x_i)_{i=1,\cdots,n}$.

Hurn et al. [13] proposed a two-step procedure to approximate

$$p^l(x_i^l, t_i \mid x_{i-1}^l, t_{i-1}; \theta),$$

the transition density function of the l-th component X^l between consecutive observations x_{i-1}^l and x_i^l:

Step 1 Divide $[t_{i-1}, t_i]$ into N equally spaced subintervals with a fixed size

$$\delta = \Delta / N.$$

Then use the Euler–Maruyama, Milstein, or stochastic Runge–Kutta scheme to discretize (2.1) by taking x_{i-1}^l as the starting value. We can obtain the approximation $\hat{x}_i^l$ of x_i^l. Repeating this numerical approximation M times, we have $(\hat{x}_i^{l,m})_{m=1,\cdots,M}$.

Step 2 Apply nonparametric smoothing approaches to approximate

$$p^l(x_i^l, t_i \mid x_{i-1}^l, t_{i-1}; \theta).$$

In this paper, we adopt the kernel smoothing approach to obtain the density estimator:

$$p^l(x_i^l, t_i \mid x_{i-1}^l, t_{i-1}; \theta) = \frac{1}{M h_{i,l}} \sum_{m=1}^{M} K\left(\frac{x_i^l - \hat{x}_i^{l,m}}{h_{i,l}}\right),$$

where $h_{i,l}$ is the kernel bandwidth at time t_i, and $K(\cdot)$ is a symmetric non-negative kernel function. The choice of $K(\cdot)$ can be a normal density kernel, Epanechnikov kernel, or some other kernel (Fan and Gijbels [10]). We use the normal density kernel in this paper with bandwidth

$$h_{i,l} = (4/(d+2))^{1/(d+4)} s_{i,l} M^{-1/(d+4)}, \quad i = 1, \cdots, n; \ l = 1, \cdots, d,$$

where $s_{i,l}$ is the standard deviation of $(\hat{x}_i^{l,m})_{m=1,\cdots,M}$ (Scott [23]; Silverman [26]).

For the multivariate diffusion processes (Singer [27]), assuming $(X_t^l)_{l=1,\cdots,d}$ to be independent, then the joint likelihood can be written as

$$p^{Hurn}(x_i, t_i \mid x_{i-1}, t_{i-1}; \theta) = \prod_{l=1}^{d} p^l(x_i^l, t_i \mid x_{i-1}^l, t_{i-1}; \theta). \tag{2.10}$$

In reality, the independence assumption may not hold. In this case, this method may not be well justified, although a pseudo-likelihood argument can be used. However, Hurn et al. [13] has pointed out that the simulated maximum likelihood method based on the kernel density estimator is quite robust. Its performance is relatively insensitive to the choice of kernel, bandwidth, discretization step size, and discretization method as well as other assumptions.

3   Improved Local Linearization (ILL) Method

In this section, we first apply the Lamperti transformation (Aït-Sahalia [2]) to convert the general SDE (2.1) to the SDE with a unit diffusion coefficient so that we can employ the local linearization approach proposed by Shoji and Ozaki [24, 25] to approximate the drift term of the SDE. We also propose a method to avoid a potential problem due to the singular or ill-conditioned Jacobian matrix in evaluating the transition density.

Under the assumption that $\sigma(\cdot)$ in the SDE (2.1) is reducible, first we introduce a definition and a proposition given by Aït-Sahalia [2].

Definition 3.1 (Aït-Sahalia [2]) The diffusion X is said to be reducible to unit diffusion (or reducible, in short) if and only if there exists a one-to-one transformation of the diffusion X into a diffusion Y whose diffusion matrix σ_Y is the identity matrix. That is, there exists an invertible function $\gamma(x)$, infinitely differentiable in X on the domain $\Im_X$, such that $Y_t \equiv \gamma(X_t)$ satisfies the SDE

$$dY_t = \mu_Y(Y_t)dt + dW_t$$

on the domain $\Im_Y$.

Proposition 3.1 (Aït-Sahalia [2]) *The diffusion X is reducible if and only if*

$$\sum_{l=1}^{d} \frac{\partial \sigma_{i,k}(x)}{\partial x_l} \sigma_{l,j}(x) = \sum_{l=1}^{d} \frac{\partial \sigma_{i,j}(x)}{\partial x_l} \sigma_{l,k}(x) \tag{3.1}$$

for each triplet $(i, j, k) = 1, \cdots, d$ such that $k > j$. If σ is nonsingular, then the condition can be expressed as

$$\frac{\partial [\sigma^{-1}]_{i,j}(x)}{\partial x_k} = \frac{\partial [\sigma^{-1}]_{i,k}(x)}{\partial x_j}. \tag{3.2}$$

If $\sigma(x; \theta)$ in the SDE (2.1) satisfies the above proposition, we can apply the Lamperti transformation $\nabla \gamma(X_t; \theta) = \sigma^{-1}(X_t; \theta)$ to model (2.1), where $\nabla \gamma(x) = [\partial \gamma_i(x)/\partial x_l]_{i=1,\cdots,d; l=1,\cdots,d}$ denotes the Jacobian matrix of γ, $\sigma(\cdot)$ is $d \times d$ coefficient matrix of SDE (2.1), and $\sigma^{-1}(\cdot)$ is the inverse of $\sigma(\cdot)$. Letting $Y_t \equiv \gamma(X_t; \theta)$, we have

$$dY_t = g(Y_t; \theta)dt + dW_t. \tag{3.3}$$

We assume the new drift coefficient to be a d-valued smooth function which satisfies the Lipschitz condition and has non-zero Jacobian matrix. For one-dimensional SDE, the Lamperti transformation can be written as

$$Y_t \equiv \gamma(X_t; \theta) = \int^{X_t} \sigma^{-1}(u; \theta)du$$

and

$$g(y; \theta) = \frac{f(\gamma^{-1}(y; \theta); \theta)}{\sigma(\gamma^{-1}(y; \theta); \theta)} - \frac{1}{2} \frac{\partial \sigma}{\partial x}(\gamma^{-1}(y; \theta); \theta).$$

In the following discussion, we will focus on SDE (3.3). By the Itô formula, the drift function g in SDE (3.3) can be linearized as

$$dg = Jdy + Mdt, \tag{3.4}$$

where

$$J = \frac{\partial(g_1, \cdots, g_d)}{\partial(y_1, \cdots, y_d)}, \quad M = [\frac{1}{2}tr(H_1), \cdots, \frac{1}{2}tr(H_d)]', \text{ and } H_l = (\frac{\partial^2 g_l}{\partial y_k \partial y_j})_{1 \le k, j \le d}.$$

We are approximating J and M by constants over $t \in [s, s + \Delta t]$, that is, $J_t = J_s$ and $M_t = M_s$, where J_s and M_s are evaluated at y_s. SDE (3.3) has the following linear form:

$$dy_t = (J_s y_t + M_s t + g(y_s) - J_s y_s - M_s s)dt + dW_t. \tag{3.5}$$

Note that the linear SDE (3.5) has the following closed-form solution (Jimenez et al. [15]):

$$y_t = y_s + \Phi(y_s; \Delta t) + \xi(y_s; \Delta t), \tag{3.6}$$

where $\xi(y_s; \Delta t) = \int_s^t \exp(J_s(t - u))dW(u)$ is a stochastic integral with mean 0 and covariance

$$V_s = \int_s^t \exp\{J_s(t - u)\} \exp\{J_s^T (t - u)\}du, \tag{3.7}$$

J_s^T denoting the transpose of matrix J_s, and

$$\Phi(y_s; \Delta t) = r_0(J_s, h)g(y_s) + (hr_0(J_s, h) - r_1(J_s, h))M_s, \tag{3.8}$$

where

$$r_i(M, a) = \int_0^a \exp(Mu)u^i du, \quad i = 0, 1, \tag{3.9}$$

a is a positive number and M is a square matrix.

If $\Phi(\cdot)$ and V_s are obtained, the transition density function of the diffusion process Y satisfying the SDE (3.3) can be evaluated by

$$p^{Shoji}(y_t, t \mid y_s, s; \theta)$$
$$= ((2\pi)^n |V_s|)^{-1/2} \exp\left(-\frac{1}{2}(y_t - E_s)^T V_s^{-1}(y_t - E_s)\right), \tag{3.10}$$

where $E_s = y_s + \Phi(y_s; \Delta t)$. If we denote $v(x) \equiv \sigma(x)\sigma^T(x)$, the log-transition density of the diffusion process X for model (2.1) is given by the Jacobian formula

$$\log(p^{Shoji}(x_t, t \mid x_s, s; \theta))$$
$$= -\frac{1}{2}\log(|v(x)|) + \log(p^{Shoji}(y_t, t \mid y_s, s; \theta)). \tag{3.11}$$

It is critical to develop methods for numerically evaluating $\Phi(\cdot)$ and V_s in the above procedure. Shoji and Ozaki [25] suggested that $\Phi(\cdot)$ could be obtained from the following expression:

$$\Phi(y_s; \Delta t)$$
$$= J_s^{-1}(\exp(J_s(t-s)) - I)g(y_s)$$
$$+ (J_s^{-1})^2(\exp(J_s(t-s)) - I - J_s(t-s))M_s \tag{3.12}$$

and V_s can be obtained by solving the following matrix equation:

$$J_s V_s + V_s J_s^T = \exp(J_s(t-u)) \exp(J_s^T(t-u)) - I. \tag{3.13}$$

However, notice that the above method for evaluating $\Phi(\cdot)$ and V_s from formulas (3.12) and (3.13) may fail if the Jacobian matrix J_s is singular or ill-conditioned. To solve this problem, we suggest using the following theorem to evaluate the integrals (3.7) and (3.8) (Van Loan [30]; Carbonell et al. [6]).
setcounterthm0

Theorem 3.1 (Van Loan [30]) *Let $n_1, n_2, n_3,$ and n_4 be the dimensions of the square matrices $A_1, A_2, A_3,$ and $A_4,$ and set $m = n_1 + n_2 + n_3 + n_4$. If the $m \times m$ block triangular matrix C is defined by*

$$C = \begin{pmatrix} A_1 & B_1 & C_1 & D_1 \\ 0 & A_2 & B_2 & C_2 \\ 0 & 0 & A_3 & B_3 \\ 0 & 0 & 0 & A_4 \end{pmatrix}$$

then for $t \geq 0$

$$\exp(Ct) = \begin{pmatrix} F_1(t) & G_1(t) & H_1(t) & K_1(t) \\ 0 & F_2(t) & G_2(t) & H_2(t) \\ 0 & 0 & F_3(t) & G_3(t) \\ 0 & 0 & 0 & F_4(t) \end{pmatrix},$$

where

$$F_j(t) = e^{A_j t}, \quad j = 1, 2, 3, 4$$

$$G_j(t) = \int_0^t e^{A_j(t-s)} B_j e^{A_{j+1}s} ds, \quad j = 1, 2, 3$$

$$H_j(t) = \int_0^t e^{A_j(t-s)} C_j e^{A_{j+2}s} ds$$

$$+ \int_0^t \int_0^s e^{A_j(t-s)} B_j e^{A_{j+1}(s-r)} B_{j+1} e^{A_{j+2}r} dr ds, \quad j = 1, 2,$$

$$K_1(t) = \int_0^t e^{A_1(t-s)} D_1 e^{A_4 s} ds$$

$$+ \int_0^t \int_0^s e^{A_1(t-s)} [C_1 e^{A_3(s-r)} B_3 + B_1 e^{A_2(s-r)} C_2] e^{A_4 r} dr ds$$

$$+ \int_0^t \int_0^s \int_0^r e^{A_1(t-s)} B_1 e^{A_2(s-r)} B_2 e^{A_3(r-w)} B_3 e^{A_4 w} dw dr ds.$$

First, we construct the following block matrices M_1, M_2, and M_3 to obtain r_0, r_1 in (3.8) and V_s in (3.7), respectively,

$$M_1 = \begin{pmatrix} 0 & I_d \\ 0 & J_s \end{pmatrix}, M_2 = \begin{pmatrix} -J_s & I_d & 0 \\ 0 & 0 & I_d \\ 0 & 0 & 0 \end{pmatrix}, M_3 = \begin{pmatrix} J_s & I_d \\ 0 & -J_s^T \end{pmatrix},$$

where I_d is the d-dimensional identity matrix. Then, from Theorem 3.1, we obtain the following:

Corollary 3.1 r_0 *can be obtained in the following block matrix:*

$$\begin{pmatrix} F_1^{r_0}(J_s; h) & r_0(J_s; h) \\ 0 & F_2^{r_0}(J_s; h) \end{pmatrix} = \exp(h M_1),$$

$r_1 = e^{J_s h} H_1^{r_1}(J_s; h)$ *is obtained from the following block matrix:*

$$\begin{pmatrix} F_1^{r_1}(J_s; h) & G_1^{r_1}(J_s; h) & H_1^{r_1}(J_s; h) \\ 0 & F_2^{r_1}(J_s; h) & G_2^{r_1}(J_s; h) \\ 0 & 0 & F_3^{r_1}(J_s; h) \end{pmatrix} = \exp(h M_2),$$

and $V_s = G_1^{V_s}(J_s; h) F_1^{V_s}(J_s; h)^T$ *is obtained from the following block matrix:*

$$\begin{pmatrix} F_1^{V_s}(J_s; h) & G_1^{V_s}(J_s; h) \\ 0 & F_2^{V_s}(J_s; h) \end{pmatrix} = \exp(h M_3).$$

Proof r_0 and r_1 are the direct results of the Van Loan theorem. We can obtain the result for V_s by applying the rules of integral calculus to (3.7):

$$V_s = \int_s^t \exp(J_s(t-u)) \exp(J_s^T(t-u))du = \int_0^h \exp(J_s(h-u)) \exp(J_s^T(h-u))du.$$

□

Now we only need to evaluate the matrix exponentials using appropriate numerical algorithms, such as Padé approximations, the Schur decomposition, or Krylov subspace method (see Moler and Van Loan [17] for a review), which can be found in MATLAB for implementation.

4 Numerical Example: HIV Dynamic Model

In past decades, ordinary differential equation (ODE) models have been widely used to describe the interactions between HIV virus and immune cellular response (see Perelson and Nelson [21]; Nowak and May [19]; Tan and Wu [29] for detailed reviews). One of the popular deterministic HIV dynamic models can be written as (Nowak and May [19]),

$$\begin{cases} dx_1(t) = (\lambda - \rho x_1(t) - \eta x_1(t)x_3(t))dt, \\ dx_2(t) = (\eta x_1(t)x_3(t) - \delta x_2(t))dt, \\ dx_3(t) = (N_1 \delta x_2(t) - c x_3(t))dt, \end{cases} \tag{4.1}$$

where x_1, x_2, and x_3 denote the concentrations of uninfected target T cells, the infected cells, and the free virus, respectively. Parameters λ and ρ represent the proliferation rate and death rate of uninfected target cells, η denotes the infection rate, N_1 denotes the number of virions produced from each infected cell during its life time, δ and c are the death or clearance rates of infected cells and free virions, respectively. However, this ODE model does not account for the stochastic feature of the biological process. Here we consider a SDE counterpart of this ODE model, i.e.,

$$\begin{cases} dx_1(t) = (\lambda - \rho x_1(t) - \eta x_1(t)x_3(t))dt + \sigma_1 x_1(t)dW_1(t), \\ dx_2(t) = (\eta x_1(t)x_3(t) - \delta x_2(t))dt + \sigma_2 x_2(t)dW_2(t), \\ dx_3(t) = (N_1 \delta x_2(t) - c x_3(t))dt + \sigma_3 x_3(t)dW_3(t), \end{cases} \tag{4.2}$$

where σ_1, σ_2, and σ_3 are diffusion parameters. Similar to Dalal et al. [8], we multiply the noise by $x_i(t)$ in model (4.2). In this section, we perform Monte Carlo simulation studies to evaluate the proposed methods for estimation of the unknown parameters in the HIV SDE model (4.2).

The following parameter values and initial conditions of SDE (4.2) are used: $t \in [0, 10]$, $x_1(0) = 600$, $x_2(0) = 30$, $x_3(0) = 1e + 5$, $\rho = 0.108$, $\delta = 0.5$, $\eta = 9.5e - 6$, $\lambda = 36$, $N_1 = 1000$, $C = 3$, $\sigma_1 = 0.1$, $\sigma_2 = 0.1$, and $\sigma_3 = 0.1$. To generate simulation data from SDE (4.2), we first discretize the SDE (4.2) using a Milstein scheme with a step size $h_g = 0.001$ (Kloeden and Platen [16]). Assume that the observations are taken at equally sampled time intervals $h_o = 0.02$ and 0.1, which result in the sample size $n = 500$ and 100, respectively. These two sampling schedules allow us to evaluate the performance of the four methods for different sample sizes. To reduce the computational burden for the simulation study, we fixed parameters ρ, δ, and η as their true values and applied the four methods to estimate the parameters $(\lambda, N_1, C, \sigma_1, \sigma_2, \sigma_3)$. Note that for Pedersen's and Hurn et al.'s methods, we selected $M = 2000$ and $N = 5$.

For evaluating the performance of different estimation methods, we define the average relative estimation error (ARE) as

$$ARE = \frac{1}{R} \sum_{r=1}^{R} \frac{|\hat{\theta}_r - \theta|}{\theta} \times 100\%,$$

where $\hat{\theta}_r$ is the estimate of the parameter θ from the r-th simulation data set and $R=200$ is the total number of simulation runs.

We report the AREs for the estimates of the parameters $(\lambda, N_1, C, \sigma_1, \sigma_2, \sigma_3)$ in Table 1. In addition, we also list the computational time of each method to evaluate the computational cost. We use the computational time of Euler's method as the standard unit and express the computational cost of other methods as fold change compared to Euler's method. From Table 1, we can see that, for a small sample size ($n = 100$), Pedersen and Hurn et al.'s methods perform reasonably well for all parameters, while the Euler and ILL methods perform poorly. However, when the sample size is large ($n = 500$), all four methods perform similarly, in particular the proposed ILL method performs the best among the four methods for all parameters. In addition, Hurn et al.'s method performs the worst almost for all parameters among the four methods for the large sample size case although the magnitude of the differences is not large. We also notice that Pedersen and Hurn et al.'s methods need to generate a large amount of Monte Carlo simulations which is computationally much more expensive than the Euler and ILL methods, and among which Euler's method is lowest and the Hurn et al.'s method is highest in computational cost.

5 Conclusion

In this paper, we evaluated four different estimation methods for multivariate nonlinear SDE models. In particular, we have proposed a method to avoid the singular or ill-conditioned problem of the Jacobian matrix for the method proposed by Shoji and Ozaki [25]. We employed an HIV dynamic SDE model to assess the performance of four estimation methods by Monte Carlo simulations. We found out that, for a small sample size, the Euler and ILL methods perform poorly and should not be used. In this case, either Pedersen or Hurn et al.'s method can be used and a high computational cost has to be paid. When the sample size is large enough, the

Table 1 Numerical simulation results of HIV dynamic model (4.2)

	λ ARE(%)	N_1 ARE(%)	C ARE(%)	σ_1 ARE(%)	σ_2 ARE(%)	σ_3 ARE(%)	Relative time
$n = 100$							
Euler	13.70	17.66	16.79	6.35	50.32	14.05	1.00
Pedersen	13.10	6.78	6.31	5.66	5.59	6.93	34.93
Hurn et al.	15.46	6.30	5.93	7.68	6.55	8.37	47.28
ILL	14.38	19.79	18.61	13.09	9.44	14.83	8.44
$n = 500$							
Euler	13.29	6.23	5.75	2.68	2.79	3.03	1.00
Pedersen	13.70	6.04	5.56	2.54	3.04	2.25	44.40
Hurn et al.	14.92	6.59	6.01	3.40	3.36	2.69	58.04
ILL	13.22	5.84	5.40	2.58	2.78	1.96	8.06

The ARE is calculated based on 200 simulation runs and the computational time (cost) is normalized to that of Euler's method. The sampling time intervals are taken as $h_o = 0.1$ and 0.02 with corresponding sample sizes of 100 and 500, respectively

ILL method performs the best and the computational cost is relatively low, and thus is recommended for practical use. Note that for a small sample size, further research is needed to stabilize the computational algorithm of the Euler or ILL method so that the estimates can be improved. Here we did not consider the measurement error in the observation data. Some of these methods can be extended to accommodate the measurement error for SDE models.

Acknowledgements This research was supported by the NIAID/NIH grants HHSN272201000055C and AI087135, and by two University of Rochester CTSI (UL1RR024160) pilot awards from the National Center for Research Resources of NIH.

References

1. Aït-Sahalia, Y. (2007). Estimating continuous-time models with discretely sampled data. In R. Blundell, P. Torsten, & W. K. Newey (Eds.), *Advances in Economics and Econometrics, Theory and Applications, Ninth World Congress. Econometric society monographs.* New York: Cambridge University Press.
2. Aït-Sahalia, Y. (2008). Closed-form likelihood expansions for multivariate diffusions. *Annals of Statistics, 36*, 906–937.
3. Arkin, A., Ross, J., & McAdams, H. H. (1998). Stochastic kinetic analysis of developmental pathway bifurcation in phage lambda-infected Escherichia coli cells. *Genetics, 149*, 633–648.
4. Black, F., & Scholes, M. S. (1973). The pricing of options and corporate liabilities. *Journal of Political Economy, 81*, 637–659.
5. Brandt, M., & Santa-Clara, P. (2002). Simulated likelihood estimation of diffusions with an application to exchange rate dynamics in incomplete markets. *Journal of Financial Economics, 63*, 161–210.
6. Carbonell, F., Jimenez, J. C., & Pedroso, L. M. (2008). Computing multiple integrals involving matrix exponentials. *Journal of Computational and Applied Mathematics, 213*, 300–305.

7. Cox, J. C., Ingersoll, J. E., & Ross, S. A. (1985). A theory of the term structure of interest rates. *Econometrica, 53*, 385–407.
8. Dalal, N., Greenhalgh, D., & Mao, X. (2008). A stochastic model for internal HIV dynamics. *Journal of Mathematical Analysis and Applications, 341*, 1084–1101.
9. Durham, G., & Gallant, A. (2002). Numerical techniques for maximum likelihood estimation of continuous-time diffusion processes. *Journal of Business & Economic Statistics, 20*, 297–316.
10. Fan, J., & Gijbels, I. (1996). *Local polynomial modelling and its applications*. London: Chapman and Hall.
11. Gillespie, D. T. (1992). A rigorous derivation of the chemical master equation. *Physica A: Statistical Mechanics and its Applications, 188*, 404–425.
12. Hurn, A. S., & Jeisman, J. (2007). Seeing the wood for the trees: A critical evaluation of methods to estimate the parameters of stochastic differential equations. *Journal of Financial Economics, 5*, 390–455.
13. Hurn, A. S., Lindsay, K. A., & Martin, V. I. (2003). On the efficacy of simulated maximum likelihood for estimating the parameters of stochastic differential equations. *Journal of Time Series Analysis, 24*, 43–63.
14. Jimenez, J. C., Biscay, R. J., & Ozaki, T. (2006). Inference methods for discretely observed continuous-time stochastic volatility models: A commented overview. *Asia-Pacific Financial Markets, 12*, 109–141.
15. Jimenez, J. C., Shoji, I., & Ozaki, T. (1999). Simulation of stochastic differential equations through the local linearization method. A comparative study. *Journal of Statistical Physics, 94*, 587–602.
16. Kloeden, P., & Platen, E. (1992). *Numerical solution of stochastic differential equations*. Berlin: Springer.
17. Moler, C. B., & Van Loan, C. F. (2003). Nineteen dubious methods for computing the matrix exponential, twenty-five years later. *SIAM Review, 45*, 3–49.
18. Nielsen, J. N., Madsen, H., & Young, P. C. (2000). Parameter estimation in stochastic differential equations: An overview. *Annual Reviews in Control, 24*, 83–94.
19. Nowak, M. A., & May, R. M. (2000). *Virus dynamics: Mathematical principles of immunology and virology*. Oxford: Oxford University Press.
20. Pedersen, A. R. (1995). A new approach to maximum likelihood estimation for stochastic differential equations based on discrete observations. *Scandinavian Journal of Statistics, 22*, 55–71.
21. Perelson, A. S., & Nelson, P. W. (1999). Mathematical analysis of HIV-1 dynamics in vivo. *SIAM Review, 41*, 3–44.
22. Poulsen, R. (1999). Approximate maximum likelihood estimation of discretely observed diffusion processes. Working Paper report No. 29, Center for Analytical Finance, Aarhus.
23. Scott, D. W. (1992). *Multivariate density estimation: Theory, practice and visualisation*. New York: Wiley.
24. Shoji, I., & Ozaki, T. (1997). Comparative study of estimation methods for continuous time stochastic processes. *Journal of Time Series Analysis, 18*, 485–506.
25. Shoji, I., & Ozaki, T. (1998). A statistical method of estimation and simulation for systems of stochastic differential equations. *Biometrika, 85*, 240–243.
26. Silverman, B. W. (1986). *Density estimation for statistics and data analysis*. London: Chapman and Hall.
27. Singer, H. (2002). Parameter estimation of nonlinear stochastic differential equations: Simulated maximum likelihood versus extended Kalman filter and Itô-Taylor expansion. *Journal of Computational and Graphical Statistics, 11*, 972–995.
28. Sørensen, H. (2004). Parametric inference for diffusion processes observed at discrete points in time: A survey. *International Statistical Review, 72*, 337–354.
29. Tan, W. Y., & Wu, H. (2005). Deterministic and stochastic models of AIDS epidemics and HIV infections with intervention. Singapore: World Scientific.
30. Van Loan, C. F. (1978). Computing integrals involving the matrix exponential. *IEEE Transactions on Automatic Control, 23*, 395–404.

On Frailties, Archimedean Copulas and Semi-Invariance Under Truncation

David Oakes

Abstract Definitions and basic properties of bivariate Archimedean copula models for survival data are reviewed with an emphasis on their motivation via frailties. I present some new characterization results for Archimedean copula models based on a notion I call semi-invariance under truncation.

1 Introduction

In this paper, dedicated to the memory of Andrei Yakovlev, I present a brief primer on multivariate frailty models and copulas. My interest in this area, which relates in some degree to Andrei's work on cure models, was first stimulated by a remarkable short paper of Clayton [5] and enhanced by a subsequent paper of Genest and MacKay [11]. I have fond memories of discussing some of this work with Andrei.

Much of this paper is review, and necessarily very selective in view of the substantial literature on the topic, but the later sections include some results that I believe to be new, particularly the notion of "semi-invariance" and its use in characterizing the class of bivariate Archimedean copula models. I conclude with some brief speculations about higher-dimensional copulas.

2 Basics

We consider first a continuous non-negative survival time T with survivor function $B(t) = \mathrm{pr}(T > t)$, density $-B'(t)$, and hazard function

D. Oakes (✉)
Department of Biostatistics and Computational Biology, University of Rochester Medical Center, Rochester, NY, USA
e-mail: David_Oakes@URMC.Rochester.edu

© Springer Nature Switzerland AG 2020
A. Almudevar et al. (eds.), *Statistical Modeling for Biological Systems*,
https://doi.org/10.1007/978-3-030-34675-1_14

259

$$b(t) = \lim_{\Delta \to 0+} \frac{1}{\Delta} \mathrm{pr}(T \le t + \Delta \mid T > t) = -\frac{B'(t)}{B(t)}.$$

A simple form of Cox's [6] proportional hazards model asserts that if T has the aforementioned distribution under certain baseline conditions, denoted by $x = 0$, where x is a vector of covariates, then the hazard function of T for a general value of x is $b(t, x) = \theta b(t)$, where $\theta = \exp(\beta^{\top} x)$, for some vector β of coefficients. The corresponding survivor function is

$$B(t, x) = \exp\left\{-\int_0^t b(u, x) du\right\} = \exp\left\{-\theta \int_0^t b(u) du\right\} = B(t)^{\theta}.$$

We now explore the consequences of replacing the fixed multiplier θ in this model by a random variable, denoted by W and called a frailty. While the x's in Cox's model generally denote observed covariates, the frailty is typically unobserved.

Consider the univariate model with non-negative frailties drawn independently from a common distribution which need not be continuous and may even have an atom at zero. The previous equation, now written $B(t \mid w) = B(t)^w$, gives the unconditional survivor function of T as

$$\mathrm{pr}(T > t) = S(t) = E\{B(t)^W\} = E \exp[-W\{-\log B(t)\}] = p(s), \qquad (2.1)$$

say, where $p(s) = E \exp(-sW)$ is the Laplace transform of the distribution of W and $s = -\log B(t) = \int_0^t b(u) du$ is the integrated hazard function for the baseline distribution of T. It is clear that we cannot hope to recover information about the two unknown functions $p(s)$ and $B(t)$ solely from information on the observable survivor function $S(t)$.

Suppose however that we can observe two random variables T_1 and T_2 which each follow a proportional hazards model on the same frailty variable W. That is, we now assume that T_1 and T_2 are conditionally independent given W, and

$$\mathrm{pr}(T_1 > t_1 \mid W = w) = \{B_1(t_1)\}^w, \quad \mathrm{pr}(T_2 > t_2 \mid W = w) = \{B_2(t_2)\}^w.$$

Here, for $j = 1, 2$, the $B_j(t)$ are the survivor functions for the conditional distributions of T_j given $W = 1$, and we will write $b_j(t)$ for the corresponding hazard functions. The observable joint survivor function of T_1 and T_2 is then

$$\begin{aligned}
S(t_1, t_2) = \mathrm{pr}(T_1 > t_1, T_2 > t_2) &= E\{\mathrm{pr}(T_1 > t_1, T_2 > t_2 \mid W)\} \qquad (2.2) \\
&= E\{\mathrm{pr}(T_1 > t_1 \mid W)\mathrm{pr}(T_2 > t_2 \mid W)\} = E\{B_1(t_1)^W B_2(t_2)^W\} \\
&= E \exp[-W\{-\log B_1(t_1) - \log B_2(t_2)\}] \\
&= p(s_1 + s_2),
\end{aligned}$$

where now the $s_j = -\log B_j(t_j) = \int_0^t b_j(u)du$ for $j = 1, 2$ are the integrated hazards for the two baseline distributions.

We can express the joint survivor function in terms of the marginals $S_1(t_1)$ and $S_2(t_2)$ by setting $t_2 = 0$ and $t_1 = 0$ in succession, giving

$$S_1(t_1) = S(t_1, 0) = p(s_1), \quad S_2(t_2) = S(0, t_2) = p(s_2),$$

so that

$$S(t_1, t_2) = p[q\{S_1(t_1)\} + q\{S_2(t_2)\}], \qquad (2.3)$$

where $q(v)$ is the inverse function to $p(s)$. Note that since $p(s)$ is a Laplace transform it is concave and decreasing, so the inverse function always exists and is also concave and decreasing, with $q(1) = 0$. If W has an atom at zero, $\text{pr}(W = 0) = a$, say then $p(\infty) = \exp(-a) = q_0$ say and $q(v)$ is defined only on the interval $(q_0, 1]$, otherwise $q(v)$ is defined on the entire interval $(0, 1]$.

The representation (2.3) is an example of an *Archimedean copula*. See, for example, Nelsen [19], a key reference on this topic. Bivariate copulas are bivariate distribution functions with uniform marginals. Their importance in statistics arises from Sklar's theorem which states that any continuous bivariate distribution function $F(t_1, t_2)$ has a unique representation

$$F(t_1, t_2) = \tilde{C}\{F_1(t_1), F_2(t_2)\}, \qquad (2.4)$$

where $\tilde{C}(u_1, u_2)$ is a copula and $F_1(t_1) = F(t_1, \infty)$ and $F_2(t_2) = F(\infty, t_2)$ are the marginal distributions. (If either marginal is discontinuous, the representation still exists but is not generally unique, see, for example, Genest and Nešlehová [12]. In the present paper we avoid such complications by requiring that joint distributions be absolutely continuous.) In survival analysis we typically work with the associated copula $C(u_1, u_2) = \tilde{C}(1 - u_1, 1 - u_2) - 1 + u_1 + u_2$ (note however that this is also a bivariate distribution function!) giving the corresponding representation

$$S(t_1, t_2) = C\{S_1(t_2), S_2(t_2)\} \qquad (2.5)$$

of the joint survivor function in terms of the marginals. For reasons that I find somewhat mysterious (see the discussion in Nelsen [19]), the term Archimedean is used for the particular class of copulas with the quasi-additive structure

$$C(u_1, u_2) = p\{q(u_1) + q(u_2)\} \qquad (2.6)$$

seen in (2.3). Archimedean copulas are more general than bivariate frailty models because the frailty derivation requires that $p(s)$ be a Laplace transform, and thus completely monotone (i.e., its derivatives must alternate in sign), see Feller [8], whereas (2.3) defines a bona fide bivariate survivor function so long as $p(0) = 1$, $p'(s) \leq 0$, and $p''(s) \geq 0$.

Copula representations allow the form of the marginal distributions to be specified separately from that of the dependence structure between them. Nonparametric measures of association such as Kendall's tau and Spearman's rho depend on $S(t_1, t_1)$ only through $C(u_1, u_2)$. To put the point another way, if modified variables $T'_j = g_j(T_j)$ are defined by continuous increasing transformations $g_j(t)$ of the original T_j then the copula representations (2.4) and (2.5) for the transformed variables (T'_1, T'_2) have the same copula functions C and $\tilde{C}$ as those for (T_1, T_2). However this separation between the form of the marginals and the dependence structure does not in general imply statistical orthogonality, in terms of Fisher information. Detailed examination of this point is beyond the scope of this paper; suffice it to say that the general theory of inference in copula models is rich and challenging.

For calculations regarding Archimedean copula models it is often convenient to work with the transformed variables $q\{S_j(T_1)\}$ ($j = 1, 2$), which have the simple joint density

$$f(s_1, s_2) = p''(s_1 + s_2).$$ (2.7)

3 Uniqueness

As mentioned above, the univariate frailty model (2.1) does not allow recovery of the two functions $p(s)$ and $B(t)$ from the single function $S(t)$. However the situation is very different for the bivariate frailty model (2.2) or (2.3).

Theorem 3.1 *The representation (2.3) is unique up to a scale factor in* $p(s)$.

Proof (Oakes [21]; see also Aczél [2]) We give an explicit formula for the inverse function $q(v)$. To do this we introduce the *cross-ratio* function

$$\theta(v) = \theta\{S(t_1, t_2)\} = \theta^*(t_1, t_2) = \frac{h(t_1 \mid T_2 = t_2)}{h(t_1 \mid T_2 > t_2)}.$$

Here the numerator and denominator are the hazard functions for the conditional survivor function of T_1 given $T_2 = t_2$, namely $-S^{(11)}(t_1, t_2)/S^{(01)}(t_1, t_2)$, and of T_1 given $T_2 > t_2$, namely $-S^{(10)}(t_1, t_2)/S^{(00)}(t_1, t_2)$. Here, and in the sequel, the orders of differentiations in the first and second argument of a function of two variables are indicated by superscripts, so that, for example, $S^{(01)}(t_1, t_2) = \partial S(t_1, t_2)/\partial t_2$. We obtain

$$\theta^*(t_1, t_2) = \frac{S^{(11)}(t_1, t_2)S^{(00)}(t_1, t_2)}{S^{(01)}(t_1, t_2)S^{(10)}(t_1, t_2)}.$$ (3.1)

This expression also shows that $\theta^*(t_1, t_2)$ is invariant under exchange of T_1 and T_2.

The cross-ratio function $\theta^*(t_1, t_2)$ is defined for any continuous bivariate survivor function $S(t_1, t_2)$ and is easily seen to depend only on the associated copula $C(u_1, u_2)$, with

$$\theta^*(t_1, t_2) = \frac{C^{(11)}(u_1, u_2)C^{(00)}(u_1, u_2)}{C^{(01)}(u_1, u_2)C^{(10)}(u_1, u_2)}.$$

When the Archimedean copula representation (2.3) holds we have

$$\theta^*(t_1, t_2) = \frac{p''(s)p(s)}{p'(s)^2},$$

where $s_j = q\{S_j(t)\}$ and $s = s_1 + s_2$. Note that terms $q'\{S_1(t_1)\}$, $q'\{S_2(t_2)\}$, $S_1'(t_1)$, $S_2'(t_2)$ occur in both numerator and denominator and so cancel. Writing $s = q(v)$, we see that $v = p(s) = S(t_1, t_2)$, so, as asserted, $\theta^*(t_1, t_2)$ is a function of (t_1, t_2) only through $v = S(t_1, t_2)$. So we have

$$\theta(v) = \frac{p''(s)p(s)}{p'(s)^2}.$$

As it stands this equation does not seem useful because the arguments on the left and right side differ. However, since $v = p(s)$, we have $s = q(v)$, so that

$$\theta(v) = \frac{p''\{q(v)\}p\{q(v)\}}{p'\{q(v)^2\}}.$$

Also, by the inverse function theorem,

$$p'\{q(v)\} = 1/q'(v), \quad p''\{q(v)\} = -q''(v)/q'(v)^3,$$

so that

$$\theta(v) = -\frac{vq''(v)}{q'(v)}. \tag{3.2}$$

This equation can be rearranged and integrated twice to give

$$q(v) = \int_{z=v}^{1} \exp\left\{ \int_{y=z}^{1-\kappa} \frac{\theta(y)}{y} dy \right\} dz. \tag{3.3}$$

Here κ ($0 < \kappa < 1$) functions as one constant of integration, corresponding to an arbitrary scale factor in the distribution of W. Clearly, in the basic representation (2.2) we may replace W by cW with Laplace transform $p_c(s) = p(cs)$ and $q_c(v) = c^{-1}q(v)$. The second constant of integration is eliminated by the condition $q(1) = 0$.

There is an interesting connection with the theory of exponential families, distributions of a random variable Y with density of the form $f_Y(y) = \exp\{\eta y - b(\eta) + c(y)\}$. For such distributions $\mu = E(Y) = b'(\theta)$, $\text{var}(Y) = b''(\theta)$. The variance of Y, written as a function $V(\mu)$ of μ, is called the variance function (McCullagh and Nelder [17]). The problem of determining an exponential family, if it exists, with a given variance function leads to equations similar to (3.2). The equivalence between these two problems is not surprising. It is easily seen that

$$\theta(v) = 1 + \frac{\text{var}(W \mid T_1 > t_1, T_2 > t_2)}{\{E(W \mid T_1 > t_1, T_2 > t_2)\}^2},$$

so that $\theta(v)$ equals one plus the squared coefficient of variation of the distribution of W among pairs with $T_1 > t_1, T_2 > t_2$. $\square$

Reverting to copula models, Oakes [21] further showed that the condition that $\theta^*(t_1, t_2)$ depends on (t_1, t_2) only through $S(t_1, t_2)$, i.e., that level curves of the survivor function correspond to level curves of $\theta^*(t_1, t_2)$, characterizes the class of bivariate survivor functions with an Archimedean copula representation. Scheffé [24] proved the same mathematical result in a different context. We recapitulate the proof here as the result will be important later.

Theorem 3.2 *(Oakes [21]; see also Aczél [2]; Scheffé [24]) Suppose that $S(t_1, t_2)$ is an absolutely continuous bivariate survivor function whose cross-ratio function $\theta^*(t_1, t_2)$ satisfies $\theta^*(t_1, t_2) = \theta\{S(t_1, t_2)\}$. Then $S(t_1, t_2)$ satisfies the Archimedean representation (2) with the function $p(s)$ determined up to a scale parameter by the expression (3.3) for its inverse function $q(v)$.*

Proof The defining equation (3.1) with $\theta^*(t_1, t_2) = \theta\{S(t_1, t_2)\}$ may be written as

$$\frac{\partial}{\partial t_1}\left[\log\{-S^{(01)}(t_1, t_2)\}\right] = \frac{\partial}{\partial t_1}\left\{\int_{S(t_1,t_2)} \frac{\theta(v)}{v} dv\right\}.$$

Integration between 0 and t_1, exponentiating and rearranging gives

$$q'\{S(t_1, t_2)\}S^{(01)}(t_1, t_2) = q'\{S(0, t_2)\}S^{(01)}(0, t_2),$$

where the function $q(v)$ is now defined by (3.3). The right side of this equation is functionally independent of t_1 and the left side is $\partial/\partial t_2 [q\{S(t_1, t_2)\}]$. Integration with respect to t_2 with the boundary condition $q(1) = 0$ gives $q\{S(t_1, t_2)\} = q\{S(t_1, 0)\} + q\{S(0, t_2)\}$, equivalent to (2.3). $\square$

Genest and MacKay [11] present a different criterion for a copula to be Archimedean, based on reasoning presented by Abel [1]. I take the liberty of appending an elementary proof based on a standard technique for solving first order partial differential equations.

Theorem 3.3 *The copula* $C(u_1, u_2)$ *is Archimedean if and only if there is a function* $f(u)$ *such that*

$$f(u_1)C^{(01)}(u_1, u_2) = f(u_2)C^{(10)}(u_1, u_2). \tag{3.4}$$

Proof Necessity follows by direct differentiation of (2.6). We find that $f(u) = q'(u)$. To prove sufficiency, note that the partial differential equation (3.4) is a Lagrange equation (Piaggio [23]) with subsidiary equations

$$\frac{du_2}{f(u_1)} = -\frac{du_1}{f(u_2)} = \frac{dC(u_1, u_1)}{0}.$$

Two particular solutions of these equations, valid for any a and b, are $q(u_1) + q(u_2) = a$, where now $q(u)$ is defined to be $\int^u f(\tilde{u})d\tilde{u}$, and $C(u_1, u_2) = b$. The general solution of (3.4) is then $g\{C, q(u_1) + q(u_2)\} = 0$ for an arbitrary bivariate function $g(a, b)$ or equivalently

$$C(u_1, u_2) = h\{q(u_1) + q(u_2)\}$$

for an arbitrary univariate function h. That $h(s) = p(s)$ follows from the boundary conditions $C(u_1, 1) = u_1, C(1, u_2) = u_2$. □

Note It seems possible to relax the condition (3.4) to allow different functions $f_1(u_1)$ and $f_2(u_2)$. For then the general solution has the form $C(u_1, u_2) = h\{q_1(u_1) + q_2(u_2)\}$ and the fact that $q_1 = q_2 = h^{-1}$ follows as before from the fact that $C(u_1, u_2)$, being a copula, has uniform marginals.

4 Some Examples

We now consider some specific examples, generated by specific forms for the Laplace transform $p(s)$. Some, but not all, of these families permit extensions of the parameter space in which $p(s)$ is no longer a Laplace transform but (2.3) still represents a valid joint survivor function. Such extensions are needed for (2.3) to represent negative dependence between T_1 and T_2. For example, Clayton's model permits such an extension but Hougaard's model does not.

4.1 Clayton's Model—Gamma Distributed Frailties

Clayton [5] took as his starting point the equation $\theta^*(t_1, t_2) = \theta$ and deduced an expression for the joint survivor function $S(t_1, t_2)$ in terms of the constant θ and two arbitrary functions of t_1 and t_2. He also gave a separate derivation using gamma distributed random effects, a forerunner of our formulation (2.2). Oakes [20, 21] recast Clayton's expression for $S(t_1, t_2)$ in the form

$$S(t_1, t_2) = \left\{ S_1(t_1)^{1-\theta} + S_2(t_2)^{1-\theta} - 1 \right\}^{1/(1-\theta)},$$

a special case of (2.3) with $p(s) = 1/(1+s)^{1/(\theta-1)}$, the Laplace transform of a gamma distribution with index $\eta = 1/(\theta - 1)$. Genest and MacKay [11] noted the connection with the copula formulation (2.2). The constancy of $\theta(v)$ reflects the fact that the squared coefficient of variation of the gamma distribution with index η is $1/\eta$, whatever the scale parameter. The limiting case of $\theta = 1$ yields a degenerate distribution for W giving independence between T_1 and T_2. The model can be extended to the case $\theta < 1$ by taking the positive part of the previous expression for $S(t_1, t_2)$. The frailty interpretation does not apply in this case, but the model is still valid and its defining property $\theta^*(t_1, t_2) = \theta$ still holds.

Clayton's model has some interesting invariance properties under truncation which we will discuss later. The simple forms of both the density and Laplace transform of the gamma distribution facilitate parametric and semiparametric inference. However the constancy of $\theta(v)$ which implies, roughly speaking, as strong a pattern of dependence in the bivariate tail of the distribution (i.e., for t_1, t_2 both large) as in the original distribution, may not always be realistic.

4.2   Inverse Gaussian Model

The inverse Gaussian density

$$k(w) = \frac{\kappa}{(2\pi\rho w^3)^{1/2}} \exp\left(-\frac{\rho w}{2} + \kappa - \frac{\kappa^2}{2\rho w} \right), \quad (\kappa > 0, \rho > 0)$$

has Laplace transform

$$p(s) = \exp\left[-\kappa \left\{ 1 - \left(1 + \frac{2s}{\rho} \right)^{\frac{1}{2}} \right\} \right].$$

In this parameterization $E(W) = \kappa/\rho$ and $\text{var}(W) = \kappa/\rho^2$. Here

$$\theta(v) = 1 + \frac{1}{\eta - \log v},$$

which decreases from $1 + 1/\eta$ to 1 as v decreases from 1 to 0.

4.3 Logarithmic Series Distribution: Frank's Model

Fisher's logarithmic series distribution, based on the series expansion of $-\log(1 - \alpha)$, has

$$pr(W = k) = -\frac{1}{\log(1 - \alpha)}\frac{\alpha^k}{k} \quad (k = 1, 2, \ldots).$$

The Laplace transform is

$$p(s) = \frac{\log(1 - \alpha e^{-s})}{\log(1 - \alpha)}.$$

This model is notable primarily for a mathematical curiosity—in this case (and only this case) the survival copula $C(u_1, u_2)$ has the same mathematical form as $\tilde{C}(u_1, u_2) = C(1 - u_1, 1 - u_2) - 1 + u_1 + u_2$. See Frank [9] and Genest [10].

4.4 Poisson Model

The Poisson distribution with mean λ has Laplace transform $p(s) = \exp\{-\lambda(1 - e^{-s})\}$. Since $W = 0$ with positive probability, this model allows the possibility of infinite survival; however, this must occur for both components or neither. The cross-ratio function is

$$\theta(v) = 1 + \frac{1}{\lambda + \log v}.$$

The possibility of infinite survival means that $v > \exp(-\lambda)$, so that $\theta(v)$ is well defined for all relevant values of v. As $v \to \exp(-\lambda)$, $\theta(v) \to \infty$. This example is of interest because unlike most frailty distributions, it leads to increasing strength of dependence as $(t_1, t_2) \to \infty$. The model is unlikely to be realistic for most situations. It can be made more so at some cost in complexity of calculation by considering a frailty variable with displaced Poisson distribution $W' = W + \delta$ say, where $\delta > 0$. This removes the possibility of infinite survival. It can be shown that for this model, $\theta(v)$ initially increases to a unique maximum as v decreases from one, but ultimately decreases to unity as $v \to 0$.

4.5 Positive Stable Model

The positive stable distributions, that is distributions with Laplace transform

$$p(s) = E \exp(-sW) = \exp(-s^\alpha), \quad (0 < \alpha < 1), \tag{4.1}$$

are familiar to probabilists; see, for example, Feller [8]. The name arises from the fact that if $W_1, \ldots, W_k$ are independent identically distributed random variables each with the distribution determined by (4.1), then

$$W = k^{-1/\alpha} \sum_1^k W_j,$$

has the same distribution, a fact easily verified by the calculation of its Laplace transform. That (4.1) is the Laplace transform of a non-negative random variable follows from the fact that it is a completely monotone function as is shown by Feller. In general there is no explicit form for the density or distribution function. Zolotarev [27] provided a fairly simple recipe for simulating positive stable distributions. Devroye [7] is a recent accessible reference. Williams [26] showed that positive stable distributions with $\alpha = 1/m$, where m is an integer, can be expressed as certain products of reciprocals of independent chi-squared random variables.

Clearly the mean and higher moments of W must be infinite. However the moments of $\log W$ are finite and can be calculated explicitly. In particular

$$E(\log W) = \gamma(\phi - 1),$$

where $\gamma \approx 0.57726$ is Euler's constant, and

$$\text{var}(\log W) = \frac{\pi^2}{6}(\phi^2 - 1).$$

The negative moments $E(W^{-k})$ also exist for all positive integers k and can be calculated easily. These calculations can be accomplished by a neat mathematical trick. Suppose that Y follows a unit exponential distribution, that W follows the positive stable distribution with Laplace transform (4.1), and that Y and W are independent. Consider the random variable $T = Y/W$. Its survivor function is

$$\text{pr}(T > t) = \text{pr}(Y/W > t) = E\left[\text{pr}(Y/W > t \mid W)\right] = E\left[\text{pr}(Y > tW \mid W)\right]$$
$$= E\left[\exp(-tW)\right] = \exp(-t^\alpha).$$

The fourth equality arises by substituting tW into the survivor function for the unit exponential distribution, the last equality follows directly from (4.1). The final expression shows that T has a Weibull distribution with index α. The moments of $\log W$ can be calculated from the relation $\log T = \log Y - \log W$ and the independence of Y and W.

The copula model (2.3) gives a bivariate survivor function of the form

$$S(t_1, t_2) = \exp\left(-\left[\{-\log S_1(t_1)\}^\phi + \{-\log S_2(t_2)\}^\phi\right]^\alpha\right),$$

where we write $\phi = 1/\alpha$ to simplify notation. This class of bivariate distributions appears to have been first considered by Gumbel [13] in the context of extreme value theory. The derivation via frailties was given by Hougaard [14]. We obtain a simple form when the marginal distributions are unit exponential,

$$S(t) = \exp\{-(t_1^\phi + t_2^\phi)^\alpha\}. \tag{4.2}$$

For the positive stable frailty model

$$\theta(v) = 1 + \frac{1-\alpha}{-\alpha \log v}.$$

This is infinite at $v = 1$ but decreases to unity as $v \to 0$.

4.6 Exterior Power Families

Suppose that $p(s)$ is the Laplace transform of an infinitely divisible distribution, so that for every η, $p(s, \eta) = p(s)^\eta$ is the Laplace transform of a non-negative random variable. Then as η varies we obtain a single parameter family of frailties. The corresponding family of survival distributions is closed under pairwise minima, i.e., if $(T_1^{(1)}, T_2^{(1)})$ and $(T_1^{(2)}, T_2^{(2)})$ are independent copies of (T_1, T_2), following the model with $p(s)^\eta$, and $T_j^{(m)} = \min(T_j^{(1)}, T_j^{(2)})$ for $j = 1, 2$, then $(T_1^{(m)}, T_2^{(m)})$ follows the same model with η replaced by 2η. All the previous families, except the logarithmic series distribution, are closed under this operation and so clearly satisfy the infinite divisibility condition.

4.7 Interior Power Families

If W is a random variable with Laplace transform $p(s)$ and W_α is a positive stable random variable with index α, $0 < \alpha \le 1$, independent of W, then the Laplace transform of $W^\phi W$ is

$$p_\alpha(s) = E\left\{\exp(-sW^\phi W)\right\} = E\left\{E\left\{\exp(-sW^\phi W) \mid W\right\}\right\}$$
$$= E\left\{\exp\{-(sW^\phi)^\alpha\}\right\} = E\left\{\exp(-s^\alpha W)\right\} = p(s^\alpha).$$

This may be called an interior power family. Clearly the inverse functions satisfy $q_\alpha(v) = q(v)^\phi$.

5   The Kendall Distribution and Kendall's Tau

It is well known that for a continuous univariate survival time T with survivor function $S(t) = \text{pr}(T > t)$ the probability integral transform $S(T)$ follows a uniform distribution over $(0, 1)$. No such general result can hold for the distribution of $V = S(T_1, T_2)$ where (T_1, T_2) has bivariate survivor function $S(t_1, t_2)$. This can be seen by considering the special cases $T_1 = T_2$, in which case V does have the uniform distribution, and T_1, T_2 independent, for which $S(T_1, T_2) = S_1(T_1)S_2(T_2)$ will have the distribution of the product of two independent uniforms, which has density $f(v) = -\log(v)$ $(0 < v < 1)$. In general there appears to be no simple explicit form for the distribution of V.

However for Archimedean copulas the distribution function $K(v) = \text{pr}(V < v) = \text{pr}\{S(T_1, T_2) < v\}$ can be calculated explicitly. We obtain

$$K(v) = v - \frac{q(v)}{q'(v)},$$

with density

$$k(v) = \frac{q''(v)q(v)}{q'(v)^2},$$

eerily reminiscent of the expression for the cross-ratio function in terms of $p(u)$ and its derivatives. To derive these results we may use the fact that V and

$$U = \frac{q\{S_1(T_1)\}}{q\{S_1(T_1) + q\{S_2(T_2)\}}$$

are independent random variables, with U having a uniform distribution over $(0, 1)$. This follows immediately from the fact that the reduced density (2.7) depends on (s_1, s_2) only through $s_1 + s_2$.

The population version of Kendall's tau is defined as

$$\tau = E\left\{\text{sign}\left(T_1^{(1)} - T_1^{(2)}\right)\left(T_2^{(1)} - T_2^{(2)}\right)\right\}.$$

Here $(T_1^{(j)}, T_2^{(j)})$, $(j = 1, 2)$ denote independent copies of (T_1, T_2). Using the reduced density (2.7) we can calculate

$$\tau = 2\left\{\int up(u)p''(u)du - \int up'(u)^2du\right\} = 4\int up(u)p''(u)du - 1$$

$$= 1 - 4\int up'(u)^2du.$$

In terms of the inverse function $q(v)$ we have

$$\tau = 1 - 4 \int \frac{q(v)}{q'(v)} dv.$$

It is easily seen that $\tau = 4E(V) - 1$, explaining the name "Kendall distribution."
Most of the examples given above yield explicit expressions for $K(v)$ and for τ. For Clayton's model with $\theta > 1$ we find that

$$\tau = \frac{\theta - 1}{\theta + 1}$$

and

$$K(v) = (\theta v - v^\theta)/(\theta - 1),$$

$$k(v) = \theta(1 - v^{\theta - 1})/(\theta - 1),$$

with the limiting forms $K(v) = v - v \log(v)$ and $k(v) = -\log v$ as $\theta \to 1$.
For the Hougaard model

$$\tau = 1 - \alpha$$

and

$$k(v) = 1 - \alpha + \alpha \log v.$$

This last result is noteworthy because it expresses the density for a general value of α as a simple mixture of the uniform density $k_0(v) = 1$, that would apply in the degenerate case as $\alpha \to 0$, and the density of the product of two uniforms that would apply in the case of independence, as $\alpha \to 1$. Inverting the transformation $(T_1, T_2) \to (U, V)$ gives the very simple representation of the standard form (4.2)

$$T_1 = U^\alpha Z, \quad T_2 = (1 - U)^\alpha Z,$$

where U is uniform $(0, 1)$ and $Z = -\log V$ follows the mixed gamma distribution with density $f_Z(z) = (1 - \alpha + \alpha z)e^{-z}$, a representation first noted by Lee [15]. This representation provides an easy way to simulate random variables with this joint density and is useful for some calculations.

6   Truncation Invariance and Semi-Invariance

We revert to the basic copula representation (2.5) of the survivor function of a continuous bivariate survival time (T_1, T_2). We consider the class of distributions obtained from (2.5) by bivariate left truncation, that is, by conditioning on $T_1 > x_1$,

$T_2 > x_2$ for some point $x = (x_1, x_2)$. Left truncation is an important concept in epidemiologic applications. We obtain the class of survivor functions

$$S_x(t_1, t_2) = \frac{S(t_1, t_2)}{S(x_1, x_2)}, \quad (t_1 > x_1, t_2 > x_2).$$

The marginals of this distribution are

$$S_{1,x}(t_1) = \frac{S(t_1, x_2)}{S(x_1, x_2)} \quad \text{and} \quad S_{2,x} = \frac{S(x_1, t_2)}{S(x_1, x_2)}.$$

Of course in general these are not identical to the marginals $S_j(t_j)/S_j(x_j)$ obtained by conditioning on a single variable ($j = 1, 2$).

Sklar's theorem implies the existence and uniqueness of a copula $C_x(u_1, u_2)$, which will in general depend on x, such that

$$S_x(t_1, t_2) = C_x\{S_{1,x}(t_1, t_2), S_{2,x}(t_1, t_2)\},$$

that is

$$\frac{S(t_1, t_2)}{S(x_1, x_2)} = C_x\left\{ \frac{S(t_1, x_2)}{S(x_1, x_2)}, \frac{S(t_2, x_1)}{S(x_1, x_2)} \right\}. \tag{6.1}$$

Suppose now that $C(u_1, u_2)$ has the Archimedean form (2.6). Then, writing $y = S(x_1, x_2)$, direct calculation shows that

$$S_x(t_1, t_2) = p_x[q_x\{S_{1,x}(t_1)\} + q_x\{S_{2,x}(t_2)\}],$$

where $p_x(s) = p\{s + q(y)\}$ and $q_x(v) = q(yv) - q(y)$. Under the bivariate frailty model (2.2), when $p(s)$ is the Laplace transform of the distribution of W, $p_x(s)$ is the Laplace transform of the conditional distribution of W given $T_1 > t_1, T_2 > t_2$.

We see that $S_x(t_1, t_2)$ also has the form of an Archimedean copula model and that the functions $p_x(s)$ and $q_x(v)$ depend on the truncation point $x = (x_1, x_2)$ only through y. Moreover, for Clayton's model, $p_x(s) = p\{s/(1 + y)\}$ and $q_x(s) = (1 + y)q(s)$, so that the resulting copula does not depend on the truncation point.

Now let S be an absolutely continuous joint survivor function with the general copula representation (2.5). We shall say that a S or its generating copula C is *semi-invariant* under left truncation if the former property holds, i.e., if C_x depends on x only through $S(x_1, x_2)$, and *invariant* if the latter holds, i.e., $C_x = C$ for all (x_1, x_2). The latter condition implies that the dependence structure of the truncated distribution $S_x(t_1, t_2)$ does not depend on x. The former relaxes this condition, but still requires that truncation points (x_1, x_2) equally far from the origin (as defined by $S(x_1, x_2)$) yield the same dependence structure for the truncated distribution.

We now show that the properties of semi-invariance and invariance under left truncation Characterize, respectively, Archimedean copulas and Clayton's copula.

The general argument closely follows that of Oakes [22]; however, the first part of the theorem does not appear to have been stated previously.

Theorem 6.1 *The copula C is semi-invariant if and only if it is Archimedean, invariant if and only if it has the form of Clayton's copula.*

Proof Sufficiency was proved above. To show necessity, let $S(t_1, t_2)$ be an absolutely continuous joint survivor function generated by a copula $C(u_1, u_2)$. Suppose that $t_j = x_j + dx_j$ $(j = 1, 2)$ and write $u_j = S_j(x_j)$, so that $u_j + du_j = S_j(x_j + dx_j)$. Suppressing the arguments (u_1, u_2) of $C(u_1, u_2)$ for notational simplicity we obtain on expanding the left side of (6.1),

$$1 + \frac{C^{(10)}}{C}du_1 + \frac{C^{(01)}}{C}du_2 + \frac{1}{2}\frac{C^{(20)}}{C}du_1^2 + \frac{1}{2}\frac{C^{(02)}}{C}du_2^2 + \frac{C^{(11)}}{C}du_1 du_2.$$

The right side of (6.1) is

$$C_x\left(1 + \frac{C^{(10)}}{C}du_1 + \frac{1}{2}\frac{C^{(20)}}{C}du_1^2, 1 + \frac{C^{(01)}}{C}du_2 + \frac{1}{2}\frac{C^{(02)}}{C}du_2^2\right).$$

This may be further expanded since

$$C_x^{(10)}(1, 1) = C_x^{(01)}(1, 1) = 1, \quad \text{and } C_x^{(20)}(1, 1) = C_x^{(02)}(1, 1) = 0,$$

because, C_x being a copula, must have the properties of a bivariate distribution function with uniform marginals. This further expansion yields

$$1 + \frac{C^{(10)}}{C}du_1 + \frac{1}{2}\frac{C^{(01)}}{C}du_1^2 + \frac{C^{(01)}}{C}du_2 + \frac{1}{2}\frac{C^{(10)}}{C}du_2^2 + \theta_x\frac{C^{(10)}}{C}\frac{C^{(01)}}{C}du_1 du_2,$$

where $\theta_x = C_x^{(11)}(1, 1)$.

The zero and first order terms on the two sides of this equation are equal for any copula C. For equality of the second order terms we must have

$$\frac{C^{(11)}}{C} = \theta_x \frac{C^{(10)}}{C}\frac{C^{(01)}}{C},$$

so that

$$\theta_x = \frac{C^{(11)}C}{C^{(10)}C^{(01)}}.$$

However by Theorem 3.2, invariance and semi-invariance require, respectively, that θ_x be x-free and that θ_x depend on x only through $S(x_1, x_2)$. Since $\theta_x = \theta^*(x_1, x_2)$ as defined in (3.1), the result follows immediately. $\qquad\square$

7 The Kendall Distribution Under Truncation

Since the Kendall distribution $K_x(v)$ (say) for the truncated survivor function $S_x(t_1, t_2)$ is determined by the copula $C_x(u_1, u_2)$, it follows immediately that $K_x(v)$ is invariant (respectively semi-invariant) for Clayton (respectively general Archimedean) copulas. Wang and Oakes [25] presented, among other results, a general expression for $K_x(v)$ for Archimedean copula models and mentioned its semi-invariance property. Earlier, Manatunga and Oakes [16] had derived a general expression for τ_x, Kendall's tau for truncated bivariate frailty models, mentioned its semi-invariance property for these models and showed that τ_x as a function of $S(x_1, x_2)$ characterizes the frailty distribution. Their results extend immediately to absolutely continuous Archimedean copula models.

In this section we show that semi-invariance of $K_x(v)$ characterizes Archimedean copula models among absolutely continuous bivariate copula models. Whether the characterization holds under the weaker condition of semi-invariance of τ_x appears to be an open question at this time.

Theorem 7.1 *Let* (U_1, U_2) *have an absolutely continuous joint distribution* $C(u_1, u_2)$ *with uniform marginals and set* $V = C(U_1, U_2)$. *Suppose that the conditional distribution function of* V *given* $(U_1 \leq x_1, U_2 \leq x_2)$ *is semi-invariant, i.e., depends on* (x_1, x_2) *only through* $C(x_1, x_2)$. *Then* $C(u_1, u_2)$ *is Archimedean.*

Note We have chosen to work directly with the copula here, which reverses the direction of the conditioning and also the direction of some of the inequalities.

Proof It is easily seen that $C(u_1, u_2)$ is semi-invariant if and only if

$$I(x_1, x_2) = \int_{u_1=0}^{x_1} \int_{u_2=0}^{x_2} 1\{C(u_1, u_2) \leq v\} C^{(11)}(u_1, u_2) du_1 du_2$$

is semi-invariant (as a function of (x_1, x_2) for fixed v). Semi-invariance of I always holds if $v \geq C(x_1, x_2)$, when $I(x_1, x_2) = C(x_1, x_2)$, so consider the case that $v < C(x_1, x_2)$, which implies that $x_1 > v$ and $x_2 > v$. In this case let a_2 and a_1 be the solutions of $C(a_2, x_2) = v$ and of $C(x_1, a_1) = v$, respectively, and note that, for fixed v, each a_j ($j = 1, 2$) is a function of (x_1, x_2) only through x_j.

We have

$$\frac{\partial I}{\partial x_1} = \int_{u_2=0}^{x_2} 1\{C(x_1, u_2) \leq v\} C^{(11)}(x_1, u_2) du_2$$

$$= \int_{u_2=0}^{a_1} C^{(11)}(x_1, u_2) du_2$$

$$= C^{(10)}(x_1, a_1)$$

and similarly

$$\frac{\partial I}{\partial x_2} = C^{(01)}(a_2, x_2).$$

Semi-invariance of I requires that its differential along a level curve of $C(x_1, x_2)$ is zero, i.e., that

$$\frac{\partial I}{\partial x_1}dx_1 + \frac{\partial I}{\partial x_2}dx_2 = 0,$$

provided

$$C^{(10)}(x_1, x_2)dx_1 + C^{(01)}(x_1, x_2)dx_2 = 0.$$

For fixed v, and $x_1, x_2 \geq v$, this has the form

$$f_1(x_1)C^{(01)}(x_1, x_2) = f_2(x_2)C^{(10)}(x_1, x_2).$$

This is a Lagrange equation and, following the general line of reasoning in the proof of Theorem 3.3, we find that its general solution is of the form

$$C(x_1, x_2) = h\{g_1(x_1) + g_2(x_2)\},$$

for certain functions f, g and h. These functions may also depend on v, and the solution is valid at least for $x_1, x_2 \geq v$. Without loss of generality, take $g_1(1) = g_2(1) = 0$. Since C has uniform marginals we must have $x_1 = h\{g_1(x_1)\}$ and $x_2 = h\{g_2(x_2)\}$ at least in $x_1, x_2 \geq v$.

Switching notation, this gives the familiar Archimedean form,

$$C(u_1, u_2) = p\{q(u_1) + q(u_2)\}. \tag{7.1}$$

We still need to address the dependence on v. Although the function p and its inverse q may depend on v, the copula $C(u_1, u_2)$ does not. The uniqueness result (Theorem 3.2) shows that p and q are unique up to a scale factor whenever they are defined. Also, since a solution valid for $x_1, x_2 \geq v$ is *a fortiori* valid for $x_1, x_2 \geq v'$ for any $v' \geq v$, by considering a sequence $v_i \to 0$ we can extend (7.1) to the range $0 < x_1, x_2 \leq 1$ and by continuity to $0 \leq x_1, x_2 \leq 1$. So $C(u_1, u_2)$ is Archimedean and the result is proved. □

Corollary *If $K_x(v)$ is invariant, C is Clayton's copula.*

8 Higher Dimensions

The copula representations of a general survivor function including the special case of an Archimedean copula extend naturally to higher dimensions, viz.

$$S(t_1, \ldots, t_m) = C\{S_1(t_1), \ldots, S_m(t_m)\},$$

where $C(u_1, \ldots, u_m)$ is a multivariate distribution with uniform marginals, and

$$C(u_1, \ldots, u_m) = p\{q(u_1) + \ldots + q(u_m)\}, \tag{8.1}$$

respectively. As in the two-dimensional case, (8.1) may be generated by a frailty model with $p(s)$ denoting the Laplace transform of the distribution of W. In one respect the generalization simplifies the theory, if (8.1) is to hold for arbitrary m (with the same functions $p(s)$ and $q(v)$) then all derivatives of $p(s)$ must alternate in sign (since the m-dimensional density of C must be non-negative) so that $p(s)$ must be completely monotone and therefore a Laplace transform.

Equation (8.1) is highly restrictive; for example, it requires exchangeability of the components of C. Several authors have proposed generalizations of (8.1) that allow more flexibility. For example, Bandeen-Roche and Liang [4] propose a model for multivariate survival data subject to multiple levels of clustering, using a recursive definition in which the clustering at each level is represented by different functions p_j, q_j. Their model has the attractive property of allowing bivariate marginals to follow possibly different Archimedean copula models, allowing stronger dependence (say) for closely related than for distantly related individuals. However it is notationally complex and requires external specification of the clustering structure.

For many years I was bothered by the question of whether there is a natural extension of Clayton's model to higher dimensions, other than the trivial (8.1). It is clearly possible to define higher order cross-ratio functions, i.e., generalizing $\theta^*(t_1, t_2)$, and requiring constancy of these cross-ratio functions over m-dimensional time should yield a m-dimensional joint distribution with non-exchangeable dependence structure but still invariant under m-dimensional left truncation. But does such a distribution exist? The essentially negative answer to this question was given by Ahmadi Javid [3]. He showed that in the class of absolutely continuous and twice differentiable copulas, products of algebraically independent multivariate Clayton copulas and standard uniform distributions are the only ones possessing truncation invariance. The proof is elementary, building on the result of Oakes [22] for the bivariate case, but (unlike this case) long and intricate.

The notion of semi-invariance extends in an obvious way to higher dimensional copulas. It remains to be seen whether this extension is fruitful, for example, whether Theorems 6.1 and 7.1 hold in higher dimensions.

The representation (V, U) discussed in Sect. 5 extends naturally to higher dimensional Archimedean copulas. Specifically, if we write $V = S(T_1, \ldots, T_m)$ and set $U_j = q\{S_j(T_j)\}/[q\{S_1(T_1)\} + \ldots + q\{S_m(T_m)\}], (j = 1, \ldots, m)$, we find that

$(U_1, \ldots, U_m)$ is independent of V and has a uniform distribution over the simplex $u_1 + \ldots + u_m = 1$. This property is the basis of a recent elegant treatment of Archimedean copulas by McNeil and Nešlehová [18] using Williamson transforms.

References

1. Abel, N. H. (1826). Recherche des fonctions de deux quantités variables indépendantes x et y, telles que $f(x, y)$, qui ont la propriétée que $f(z, f(x, y))$ est une fonction symétrique de z, x et y. *Journal fur die Reine und Angewandte Mathematik, 1*, 11–15.
2. Aczél, J. (1950). Einige aus funktionalgleichungen zweier veränderlichen ableitbare differentialgleichungen. *Acta Scientiarum Mathematicarum (Szeged), 13*, 179–189.
3. Ahmadi Javid, A. (2009). Copulas with truncation-invariance property. *Communications in Statistics - Theory and Methods, 38*, 3756–3771.
4. Bandeen-Roche, K. J., & Liang, K. Y. (1996). Modelling failure-time associations in data with multiple levels of clustering. *Biometrika, 83*, 29–39.
5. Clayton, D. G. (1978). A model for association in bivariate life tables and its application in epidemiological studies of familial tendency in chronic disease incidence. *Biometrika, 65*, 141–151.
6. Cox, D. R. (1972). Regression models and life tables (with discussion). *Journal of the Royal Statistical Society, Series B: Statistical Methodology, 34*, 187–220.
7. Devroye, L. (2009). On exact simulation algorithms for some distributions related to Jacobi theta functions. *Statistics & Probability Letters, 79*, 2251–2259.
8. Feller, W. (1971). *An introduction to probability theory and its applications, Vol. II.* New Delhi: Wiley.
9. Frank, M. J. (1979). On the simultaneous associativity of $F(x, y)$ and $x + y - F(x, y)$. *Aequationes Mathematicae, 19*, 194–226.
10. Genest, C. (1987). Frank's family of bivariate distributions. *Biometrika, 74*, 549–555.
11. Genest, C., & MacKay, R. J. (1986). Copules Archimédiennes et familles de lois bidimensionnelles dont les marges sont données. *The Canadian Journal of Statistics, 14*, 145–159.
12. Genest, C., & Nešlehová, J. (2007). A primer on copulas for count data. *Astrophysical Bulletin, 37*, 475–515.
13. Gumbel, E. J. (1960). Bivariate exponential distributions. *Journal of the American Statistical Association, 55*, 698–707.
14. Hougaard, P. (1986). A class of multivariate failure time distributions. *Biometrika, 73*, 671–678.
15. Lee, L. (1979). Multivariate distributions having Weibull properties. *Journal of Multivariate Analysis, 9*, 267–277.
16. Manatunga, A. K., & Oakes, D. (1996). A measure of association for bivariate frailty distributions. *Journal of Multivariate Analysis, 56*, 60–74.
17. McCullagh, P., & Nelder, J. A. (1989). *Generalized linear models*, 2nd edn. London: Chapman & Hall/CRC.
18. McNeil, A. J., & Nešlehová, J. (2009). Multivariate Archimedean copulas, d-monotone functions and ℓ_1-norm symmetric distributions. *The Annals of Statistics, 37*, 3059–3097.
19. Nelsen, R. B. (2006). *An introduction to copulas*, 2nd edn. New York: Springer.
20. Oakes, D. (1982). A model for association in bivariate survival data. *Journal of the Royal Statistical Society, Series B: Statistical Methodology, 44*, 414–422.
21. Oakes, D. (1989). Bivariate survival models induced by frailties. *Journal of the American Statistical Association, 84*, 487–493.
22. Oakes, D. (2005). On the preservation of copula structure under truncation. *The Canadian Journal of Statistics, 33*, 465–468.

23. Piaggio, H. T. H. (1962). *An elementary treatise on differential equations and their applications*. London: G. Bell & Sons.
24. Scheffé, H. (1959). *The analysis of variance*. New York: Wiley.
25. Wang, A., & Oakes, D. (2008). Some properties of the Kendall distribution in bivariate Archimedean copula models under censoring. *Statistics & Probability Letters, 78*, 2578–2583.
26. Williams, E. J. (1977). Some representations of stable random variables as products. *Biometrika, 64*, 167–169.
27. Zolotarev, V. M. (1966). On representation of stable laws by integrals. *Translation in Mathematical Statistics, 6*, 84–88.

Part II
Short Communications

The Generalized ANOVA: A Classic Song Sung with Modern Lyrics

Hui Zhang and Xin Tu

Abstract The widely used analysis of variance (ANOVA) suffers from a series of flaws that not only raise questions about conclusions drawn from its use, but also undercut its many potential applications to modern clinical and observational research. In this paper, we propose a *generalized* ANOVA model to address the limitations of this popular approach so that it can be applied to many immediate as well as potential applications ranging from an age-old technical issue in applying ANOVA to cutting-edge methodological challenges. By integrating the classic theory of U-statistics, we develop distribution-free inference for this new class of models to address missing data for longitudinal clinical trials and cohort studies.

Keywords Count response · Missing data · Overdispersion

1 Introduction

The analysis of variance (ANOVA) model is the most popular approach for comparing two or more sample means. However, the variance homogeneity, a fundamental assumption underlying the validity of inference by ANOVA, is almost always taken for granted. The lack of reliable tests for this assumption frequently contributes to flawed research and spurious findings. All available methods for testing variance homogeneity are *ad hoc* and in particular, none addresses missing data arising in longitudinal studies (see Duran [2] for a review of methods). As longitudinal study designs have become the standard benchmark rather than the exception in modern clinical trials and observational studies, this methodological gap in the current literature must be filled immediately to facilitate modern statistical applications.

H. Zhang (✉)
Feinberg School of Medicine, North Western University Chicago, IL, USA
e-mail: hzhang@northwestern.edu

X. Tu
Division of Biostatistics and Bioinformatics, UC San Diego School of Medicine, CA, USA
e-mail: x2tu@ucsd.edu

© Springer Nature Switzerland AG 2020
A. Almudevar et al. (eds.), *Statistical Modeling for Biological Systems*,
https://doi.org/10.1007/978-3-030-34675-1_15

281

Assessing variance homogeneity is of critical importance not only for valid inference using ANOVA, but also for our paradigm shift from executing tightly controlled efficacy trials involving patients with a single, well-defined disorder to conducting community-based studies implemented across diverse health service settings and broad populations. In the latter effectiveness trials, the primary goal is not to establish the superiority of one treatment over another, but rather to determine whether an efficacious treatment can have any measurable, beneficial effect when implemented in more broad population base (Clarke [1]; Hogarty et al. [4]; Taube et al. [13]). Thus, the high degree of internal validity associated with efficacy studies becomes threatened in effectiveness research due to the lack of rigid inclusion/exclusion criteria, decreased incentives for study participation, and lessened patient control and medication compliance. As a result, treatment outcomes from an effectiveness study are typically more variable than those from a similar efficacy trial. *Test of primary outcome variance difference is an important approach to valuate the effectiveness, in addition to the internal liability.*

Since all popular statistical methods are based on modeling the mean response, they cannot be applied to select optimal treatments when differences only lie in the variance. Non-parametric methods are no exception, either. For example, although the popular two-sample Wilcoxon (or Mann–Whitney–Wilcoxon) rank sum test posits no parametric distribution such as normality, it does assume that the distributions of the two samples are identical except for a location shift (Mann and Whitney [9]; Randles and Wolfe [10]; Wilcoxon [14]). Such "a non-parametric" assumption is not unique to the Wilcoxon test, and in fact, most non-parametric methods are developed based on this assumption, including the Kruskal–Wallis non-parametric approach to the analysis of variance (ANOVA) models (Kruskal and Wallis [6]). Further, none of these approaches applies to longitudinal studies with missing data.

In this paper, we present a new model to extend the classic ANOVA to concurrently modeling the mean and variance within a longitudinal data setting, and examine its performance in the presence of missing data using simulations. This generalized ANOVA (GANOVA) provides a systematic approach to address the aforementioned limitations of ANOVA plus some other important modeling issues often arising in psychosocial research. More details of this approach can be found in a technical report (Zhang et al. [15]).

2 Generalized ANOVA for Mean and Variance

When comparing the group means across multiple groups using ANOVA, the F test relies on the assumption of variance homogeneity for valid inference about the null hypothesis of equal group means. When this assumption fails, the normal-based F test no longer provides valid inference. Below, we develop an approach to model the group mean and variance simultaneously, which, in particular, allows us to test variance homogeneity. In addition, this approach does not assume normality or any parametric distribution, providing robust inference.

We define a distribution-free *generalized* ANOVA as follows:

$$E\left(y_{ki}\right) = \mu_k, \quad 1 \leq i \leq n, \quad 1 \leq k \leq K. \tag{2.1}$$

To add a component to the above for modeling the variance, consider first a relatively simple case with $K = 2$ and the function:

$$h_{k2ij} = h_2\left(y_{ki}, y_{kj}\right) = \frac{1}{2}\left(y_{ki} - y_{kj}\right)^2, \quad 1 \leq i < j \leq n, \quad 1 \leq k \leq K. \tag{2.2}$$

It is readily checked that $E\left(h_2\left(y_{ki}, y_{kj}\right)\right) = \sigma_k^2$. Thus, by combining (2.1) and (2.2), we derive a model for both mean and variance:

$$E\left(\mathbf{h}_{k\mathbf{i}}\right) = E\left(\mathbf{h}\left(y_{ki}, y_{kj}\right)\right) = \boldsymbol{\theta}_k, \quad \mathbf{i} = (i, j) \in C_2^{n_k}, \quad 1 \leq k \leq K, \tag{2.3}$$

where C_q^n denotes the set of $\binom{n}{q}$ combinations of q distinct elements $\left(i_1, \ldots, i_q\right)$ from the integer set $\{1, \ldots, n\}$, and

$$\mathbf{h}_{k\mathbf{i}} = \mathbf{h}\left(y_{ki}, y_{kj}\right) = (h_{k1\mathbf{i}}, h_{k2\mathbf{i}})^\top, \quad \boldsymbol{\theta}_k = \left(\mu_k, \sigma_k^2\right)^\top,$$

$$h_{k1\mathbf{i}} = h_1\left(y_{ki}, y_{kj}\right) = \frac{1}{2}\left(y_{ki} + y_{kj}\right), \quad h_{k2\mathbf{i}} = h_2\left(y_{ki}, y_{kj}\right) = \frac{1}{2}\left(y_{ki} - y_{kj}\right)^2.$$

3 Inference for Generalized ANOVA

The bivariate function, $\mathbf{h}\left(y_{ki}, y_{kj}\right)$, is a one-sample, two-argument vector-valued U-statistic, enabling us to use the U-statistics theory to facilitate inference about the GANOVA in (2.3) (Hoeffding [3]; Serfling [12]; Kowalski and Tu [5]). Specifically, consider the estimate of $\boldsymbol{\theta}_k$ defined by:

$$\widehat{\boldsymbol{\theta}}_k = \binom{n}{2}^{-1} \sum_{(i,j) \in C_2^{n_k}} \mathbf{h}\left(y_{ki}, y_{kj}\right) \quad 1 \leq k \leq K. \tag{3.1}$$

It is readily checked that $\widehat{\theta}_k$ is an unbiased estimate of θ_k, i.e., $E\left(\widehat{\theta}_k\right) = \theta_k$. However, the classic law of large numbers (LLN) and central limit theorem (CLT) cannot be applied to establish the asymptotic behavior of this estimate since it is defined by a dependent sum. By utilizing the theory of U-statistics, we can show that $\widehat{\theta}_k$ is consistent and asymptotically normal, and derive Wald-type test statistic for testing the null. *To estimate the asymptotic variance of $\widehat{\theta}_k$, i.e., the variance matrix of $(\widehat{\mu}_k, \widehat{\sigma}_k^2)$, μ_{k3} and $\mu_{k4} - \sigma_k^4$ will be estimated for $cov(\widehat{\mu}_k, \widehat{\sigma}_k^2)$ and $var(\widehat{\mu}_k, \widehat{\sigma}_k^2)$,* *respectively, where* μ_{k3} *and* μ_{k4} *are the third- and fourth-central moments of response (see Zhang et al. [15] for details).*

4 Extension to a Longitudinal Data Setting with Missing Data

To extend GANOVA to a longitudinal data setting, consider a study with K groups, n_k subjects within each kth group, and M assessments. To incorporate the additional time dimension, let

$$\mathbf{y}_{ki} = (y_{ki1}, \ldots, y_{kiM})^\top, \quad \boldsymbol{\theta}_{km} = (\theta_{k1m}, \theta_{k2m})^\top = \left(\mu_{km}, \sigma_{km}^2\right)^\top, \quad (4.1)$$

$$\boldsymbol{\theta}_k = \left(\boldsymbol{\theta}_{k1}^\top, \ldots, \boldsymbol{\theta}_{kM}^\top\right)^\top, \quad \mathbf{h}_{ki} = \mathbf{h}\left(\mathbf{y}_{ki}, \mathbf{y}_{kj}\right) = \left(\mathbf{h}_{ki1}^\top, \ldots, \mathbf{h}_{kiM}^\top\right)^\top,$$

$$\mathbf{i} = (i, j) \in C_2^{n_k}, \quad 1 \leq k < K, \quad 1 \leq m < M.$$

The GANOVA (2.3) now has the following form:

$$E\left(\mathbf{h}_{ki}\right) = E\left(\mathbf{h}\left(\mathbf{y}_{ki}, \mathbf{y}_{kj}\right)\right) = \boldsymbol{\theta}_k, \quad \mathbf{i} = (i, j) \in C_2^{n_k}, \quad 1 \leq k \leq K. \quad (4.2)$$

Thus, it has the same general form as its cross-sectional counterpart except for the extra components in the response and parameters to account for the added time dimension. Inference about $\boldsymbol{\theta}_k$ and any (smooth) function of them can be obtained by applying the theory of multivariate U-statistics (see Kowalski and Tu [5] for details).

Missing data is a common occurrence in longitudinal studies. For convenience, we assume that all subjects are observed at the baseline $t = 1$ and missing responses only occur at post-baseline visits. One approach to dealing with missing data is to simply remove the subjects missing one or more post-baseline outcomes and perform inference based on the remaining subjects. Such a complete-data approach not only reduces power, but more importantly yields biased estimates in many applications.

To address this issue so that GANOVA provides valid inference under the missing at random (MAR) assumption, a popular mechanism fitting most applications (Little and Rubin [7]), define a set of missing (or rather, observed) data indicators for each ith subject:

$$r_{kim} = \begin{cases} 1 \text{ if } y_{kim} \text{ is observed} \\ 0 \text{ if } y_{kim} \text{ is missing} \end{cases}, \quad \mathbf{r}_{ki} = (r_{ki1}, \ldots, r_{kiM})^\top.$$

Let $\pi_{kim} = \Pr(r_{kim} = 1 \mid \mathbf{y}_{ki})$ denote the probability of observing the response y_{kim} given $\mathbf{y}_{ki}$. Under monotone missing data pattern (MMDP) and MAR, π_{kim}

only depends on the observed y_{kit} prior to time m so that we may model and estimate π_{kim} using logistic regression based on the observed data (Robins et al. [11]; Kowalski and Tu [5]; Lu et al. [8]). We then revise the estimates $\widehat{\theta}_k$ in (3.1) by using inverse probability weighted (IPW) kernels:

$$\mathbf{g}_{kim} = \mathbf{g}\left(y_{kim}, y_{kjm}; r_{kim}, r_{kjm}\right) = \frac{r_{kim}r_{kjm}}{\pi_{kim}\pi_{kjm}}\mathbf{h}_m\left(y_{kim}, y_{kjm}\right),$$

$$\mathbf{g}_{ki} = \mathbf{g}\left(\mathbf{y}_{ki}, \mathbf{y}_{kj}; \mathbf{r}_{ki}, \mathbf{r}_{kj}\right) = \left(\mathbf{g}_{ki1}^\mathsf{T}, \ldots, \mathbf{g}_{kiM}^\mathsf{T}\right)^\mathsf{T}, \quad 1 \leq k \leq K.$$

Since

$$E\left(\mathbf{g}_{kim}\right) = E\left(\frac{r_{kim}r_{kjm}}{\pi_{kim}\pi_{kjm}}\mathbf{h}_{kim}\right)$$

$$= E\left[E\left(\pi_{kim}^{-1}\pi_{kjm}^{-1}r_{kim}r_{kjm}\mathbf{h}_{kim} \mid \mathbf{y}_{ki}, \mathbf{y}_{kj}\right)\right] = \theta_{km},$$

the new estimates are unbiased. Inference is similarly extended to the IPW-based $\widehat{\theta}_k$. See Zhang et al. [15] for details.

5 Simulation

Consider the problem of overdispersion in modeling a count response using a negative binomial (NB) model. We assumed one group and simulated data from a three-variate model:

$$\mathbf{y}_i = (y_{i1}, y_{i2}, y_{i3})^\mathsf{T} \sim NB\left(\boldsymbol{\mu} = (\mu_1, \mu_2, \mu_3)^\mathsf{T}, \iota\right), \quad \alpha > 0, \quad (5.1)$$

$$\mu_m = 5, \quad \iota = 0.05 \text{ or } 0.1, \quad 1 \leq m \leq 3,$$

where $NB(\boldsymbol{\mu}, \iota)$ denotes a NB model with marginal mean $E(y_{im}) = \mu_m$ and variance $Var(y_{im}) = \sigma_m^2 = \mu_m(1 + \iota\mu_m)$. The correlations between different time points are $corr(y_{i1}, y_{i2}) = corr(y_{i1}, y_{i3}) = corr(y_{i2}, y_{i3}) = 0.5$. This model defaults to a multivariate Poisson with mean $\boldsymbol{\mu}$, when $\iota = 0$. Otherwise, the dispersion parameter ι indicates the amount of extra-Poisson variation. The variance $\sigma_m^2 = 6.25$ (7.5) for $\iota = 0.05$ (0.1).

We simulated the missing response under MAR and MMDP so that the probability of the first missing y_{it} depends on its immediate predecessor, and yields about 10 and 20% missing responses at $m = 2$ and 3, respectively. We considered the following null:

$$H_0 : \mu_m = \sigma_m^2, \quad 1 \leq m \leq 3, \quad \text{or} \quad H_0 : K\theta = \mathbf{0},$$

Table 1 Estimates of mean and variance over each of three assessments along with their standard errors as well as test statistics and power estimates for testing the null of no overdispersion for the simulation study

Estimates of mean μ_m and variance σ_m^2 over time and power for detecting overdispersion under NB(μ, ι)

| Sample | | Mean \| Variance | | | | | | $H_0 : \mu_i = \sigma_i^2,\ i = 1, 2, 3$ |
size	ι	Time 1		Time 2		Time 3		Power (type I error= 0.05)
50	0.05	4.98	6.23	4.95	6.18	4.94	6.17	0.134
	0.1	4.98	7.49	4.94	7.40	4.94	7.41	0.381
100	0.05	4.99	6.23	4.99	6.23	5.00	6.28	0.257
	0.1	4.98	7.47	4.98	7.47	5.00	7.53	0.786
200	0.05	4.98	6.23	5.00	6.28	5.00	6.27	0.572
	0.1	4.98	7.46	5.00	7.54	5.00	7.54	0.991
300	0.05	4.99	6.26	5.00	6.26	5.00	6.26	0.816
	0.1	4.99	7.51	5.00	7.51	5.00	7.53	1

$$K = \begin{pmatrix} 1 & -1 & 0 & 0 & 0 & 0 \\ 0 & 0 & 1 & -1 & 0 & 0 \\ 0 & 0 & 0 & 0 & 1 & -1 \end{pmatrix}, \quad \theta = \left(\theta_1^\top, \theta_2^\top, \theta_3^\top \right)^\top, \quad \theta_m = \left(\mu_m, \sigma_m^2 \right)^\top.$$

Since the Wald statistic, $W = n\widehat{\theta}^\top K^\top (K \widehat{\Sigma} K^\top)^{-1} K \widehat{\theta}$, has an asymptotic central χ_3^2 distribution under H_0, we estimated power by calculating the empirical power, $\widehat{\varphi} = (1/1000) \sum_{j=1}^{1000} I_{\{W_j \geq q_{0.95}\}}$, where W_j denotes the test statistic W from the jth of 1000 MC replication constructed based on the asymptotic distribution of W and $q_{0.95}$ is the 95th percentile of χ_3^2.

Shown in Table 1 are the estimated mean and variance of the simulated $\mathbf{y}_i$, and power estimates for the two values of the dispersion parameter ι. The estimated mean and variance were quite close to their respective true values across all sample sizes in both cases of ι, with type I error set at 0.05. Although power was low for small sample sizes, it grew rapidly with increasing sample size as ι changed to 0.1. Since $\iota = 0.1$ still represents a small amount of overdispersion in the response, e.g., $\sigma_m^2 / \mu_m = 7.5/5 = 1.5$, the approach seems to be quite powerful to detect overdispersion.

6 Discussion

We have developed a new class of models to concurrently model the mean and variance of an outcome and utilized the theory of U-statistics to provide robust inference for such a generalized ANOVA within a longitudinal data setting. This new class of GANOVA has many immediate as well as potential applications ranging from addressing an age-old technical issue in applying the classic ANOVA to cutting-edge methodological challenges arising from the emerging effectiveness research paradigm.

Acknowledgements The authors sincerely thank Dr. W. Jack Hall and Ms. Cheryl Bliss-Clark at the University of Rochester for their help to improve the presentation of the manuscript. This paper was also supported by the ASA Best Student Paper Award and ENAR Distinguished Student Paper Award to be presented at the 2009 JSM in Washington and 2010 ENAR Spring Meeting in New Orleans, respectively.

References

1. Clarke, G. N. (1995). Improving the transition from basic efficacy research to effectiveness studies: Methodological issues and procedures. *Journal of Consulting and Clinical Psychology, 63*, 718–725.
2. Duran, B. S. (1976). A survey of nonparametric tests for scale. *Communications in Statistics - Theory and Methods, 5*, 1287–1312.
3. Hoeffding, W. (1948). A class of statistics with asymptotically normal distribution. *Annals of Mathematical Statistics, 19*, 293–325.
4. Hogarty, G. E., Schooler, N. R., & Baker, R. W. (1997). Efficacy versus effectiveness. *Psychiatric Services, 48*, 1107.
5. Kowalski, J., & Tu, X. M. (2007). *Modern applied U-statistics.* New York: Wiley.
6. Kruskal, W. H., & Wallis, W. A. (1952). Use of ranks in one-criterion variance analysis. *Journal of the American Statistical Association, 47*, 583–621.
7. Little, R. J. A., & Rubin, D. B. (1987). *Statistical analysis with missing data.* New York: Wiley.
8. Lu, N., Tang, W., He, H., Yu, Q., Crits-Christoph, P., Hui, Z., et al. (2009). On the impact of parametric assumptions and robust alternatives for longitudinal data analysis. *Biometrical Journal, 51*, 627–643.
9. Mann, H. B., & Whitney, D. R. (1947). On a test of whether one of two random variables is stochastically larger than the other. *Annals of Mathematical Statistics, 18*, 50–60.
10. Randles, R. H., & Wolfe, D. A. (1979). *Introduction to the theory of nonparametric statistics.* New York: Wiley.
11. Robins, J. M., Rotnitzky, A., & Zhao, L. P. (1995). Analysis of semiparametric regression models for repeated outcomes in the presence of missing data. *Journal of the American Statistical Association, 90*, 106–121.
12. Serfling, R. J. (1980). *Approximation theorems of mathematical statistics.* New York: Wiley.
13. Taube, C. A., Mechanic, D., & Hohmann, A. A. (1989). *The future of mental health services research.* Washington: U.S. Department of Health and Human Services.
14. Wilcoxon, F. (1945). Individual comparisons by ranking methods. *Biometrics, 1*, 80–83.
15. Zhang, H., & Tu, X. M. (2011). *Generalized ANOVA for concurrently modeling mean and variance within a longitudinal data setting.* Technical Report, Department of Biostatistics and Computational Biology, University of Rochester.

Analyzing Gene Pathways from Microarrays to Sequencing Platforms

Jeffrey Miecznikowski, Dan Wang, Xing Ren, Jianmin Wang, and Song Liu

Abstract Genetic microarrays have been the primary technology for quantitative transcriptome analysis since the mid-1990s. Via statistical testing methodology developed for microarray data, researchers can study genes and gene pathways involved in a disease. Recently a new technology known as RNA-seq has been developed to quantitatively study the transcriptome. This new technology can also study genes and gene pathways, although the statistical methodology used for microarrays must be adapted to this new platform. In this manuscript, we discuss methods of gene pathway analysis in microarrays and next generation sequencing and their advantages over standard "gene by gene" testing schemes.

Keywords Pathway analysis · Microarrays · RNA-Seq · GSEA · GSA · Multiple testing

1 Introduction

Microarrays and other high throughput genetic platforms can monitor the expression of thousands of genes within an organism. Using a statistical test for each gene, a researcher can conduct "gene by gene" testing to determine a subset of "interesting" genes, that is, genes that are differentially expressed between two conditions (e.g.,

J. Miecznikowski (✉)
SUNY University at Buffalo, Department of Biostatistics, Buffalo, NY, USA

Roswell Park Comprehensive Cancer Center, Buffalo, NY, USA
e-mail: jcm38@buffalo.edu

X. Ren
SUNY University at Buffalo, Department of Biostatistics, Buffalo, NY, USA
e-mail: xingren@buffalo.edu

D. Wang · J. Wang · S. Liu
Roswell Park Comprehensive Cancer Center, Buffalo, NY, USA
e-mail: Dan.Wang@roswellpark.org; Jianmin.Wang@roswellpark.org;
Song.Liu@roswellpark.org

© Springer Nature Switzerland AG 2020
A. Almudevar et al. (eds.), *Statistical Modeling for Biological Systems*,
https://doi.org/10.1007/978-3-030-34675-1_16

289

cancer vs healthy), or genes that are significantly correlated with an outcome (e.g., survival) or phenotype. Gene by gene testing is limited by the multiple testing challenge of controlling a suitable Type I error rate to limit the number of false positives. Further, in many experiments, there can be large discrepancies in reproducibility due to the noise present in microarrays, e.g., cross hybridization, background subtraction, and normalization across arrays.

In search of results using a smaller number of tests with greater reproducibility, researchers developed methods to assess differential expression in predetermined biological sets of genes, for example, all the genes in the calcium signaling pathway. These methods are collectively referred to as gene pathway analysis or gene class testing [1]. There are numerous methods designed for pathway analysis using microarray data.

Recently, next generation sequencing technologies, especially RNA-seq, have replaced microarray experiments. RNA-seq allows for massively parallel sequencing with improved efficiency over microarrays. Researchers can also perform "gene by gene" testing in RNA-seq experiments; however, the statistical tests must be changed since RNA-seq is count data (discrete) while microarray data are considered continuous measures of expression. In light of this difference and others, researchers are actively developing new gene set/pathway analysis algorithms.

In the following sections, we discuss methods for pathway analysis in microarrays highlighting a meta-analysis case study designed using gene pathway analysis. We also highlight gene pathway analysis methods available for next generation sequencing and conclude with a discussion on the similarities and differences in analyzing gene pathways in traditional microarrays and next generation sequencing techniques.

2   Pathway Methods for Microarrays

With a univariate test to determine significance for each gene, e.g., a two sample t test, a set or group of genes can be assessed for significance using a contingency table with a Fisher's exact test or chi-squared test to assess overrepresentation of the number of differentially expressed genes. These methods assume a cut point or threshold to determine significance and also assume independence between the genes.

Gene set enrichment analysis (GSEA) was one of the first methods designed to address the weaknesses in the contingency table based methods [20]. The major elements of GSEA include calculation of an enrichment score and estimation of significance level. In short, the genes are ranked according to their correlation with the outcome (e.g., p-values sorted from smallest to largest). An enrichment score is produced for each pathway to determine the degree to which the gene set under consideration is overrepresented at extremes of the ranked list, see (2.1) and (2.2).

Significance of the enrichment statistic is computed by estimating the nominal p-value of the statistic via a null distribution obtained from permuting the sample labels.

Mathematically, we rank order N genes to form the list $L = \{g_1, \ldots, g_N\}$ according to correlation, $r(g_j) = r_j$ of their expression profiles with the outcome C. Next, we evaluate the fraction of genes in set S ("hits") weighted by correlation and the fraction of genes not in S ("misses") present up to a given position i in L. For each gene set S, we compute

$$P_{hit}(S, i) = \sum_{g_j \in S, \, j \leq i} \frac{|r_j|^p}{N_R}, \text{ where } N_R = \sum_{g_j \in S} |r_j|^p, \qquad (2.1)$$

and

$$P_{miss}(S, i) = \sum_{g_j \text{ not in } S, \, j \leq i} \frac{1}{(N - N_R)}. \qquad (2.2)$$

The enrichment score (ES) for set S is the maximum deviation from zero, $P_{hit} - P_{miss}$, over all values i from 1 to N. If $p = 0$, ES for set S reduces to standard Kolmogorov–Smirnov statistic, when $p = 1$ we are weighting by correlation with C normalized by the sum of the correlations over all genes in S.

Note, the GSEA algorithm does not consider the enrichment statistic for a pathway against an enrichment statistic for a randomly chosen set of genes. An improvement over GSEA is the gene set analysis (GSA) algorithm that uses an improved statistic and tests for significance against a null distribution determined by both permutation of the sample labels and randomly chosen sets of statistics (genes).

2.1   The GSA Method

In short, GSA offers two potential improvements to GSEA, namely the maxmean statistic for summarizing gene sets and restandardization for more accurate inferences [2]. The restandardization process of GSA consists of a randomization step and a permutation step. The randomization step standardizes the maxmean statistic with respect to its randomized mean and standard deviation; then, the permutation step computes the p-value for the statistic from a permutation distribution. The following subsections detail how significance or "enrichment" is determined for gene sets.

2.1.1 The Maxmean Statistic

Commonly, we are interested in comparing genes between two conditions A and B. The two sample t test is a common test to use in this setting. We let $z_i = \Phi^{-1}(t_i)$ where t_i represents the two sample t statistic for gene i and Φ is the cumulative distribution function (CDF) for a standard normal random variable.

Let $\{z_1, z_2, \ldots, z_m\}$ denote the gene scores in gene set S with m genes. We desire an enrichment statistic for each pathway that has power against both shift and scale alternatives. Following the method in [2], for each gene set S, we obtain a summary statistic S for S as follows,

$$S = \max\{\bar{s}_S^{(+)}, \bar{s}_S^{(-)}\}, \tag{2.3}$$

where

$$\bar{s}_S^{(+)} = \sum_{i=1}^{m} \frac{s^{(+)}(z_i)}{m} \text{ and } \bar{s}_S^{(-)} = \sum_{i=1}^{m} \frac{s^{(-)}(z_i)}{m}, \tag{2.4}$$

and

$$s^{(+)}(z) = \max\{z, 0\} \text{ and } s^{(-)}(z) = -\min\{z, 0\}. \tag{2.5}$$

We obtain S for each gene pathway under consideration. We then standardize S by its randomized mean and standard deviation obtaining,

$$S' = (S - mean_s)/stdev_s, \tag{2.6}$$

where $mean_s$ and $stdev_s$ are the sample mean and sample standard deviation, respectively, of S from all gene pathways under consideration. We determine significance for this statistic by deriving its null distribution according to a permutation scheme.

2.1.2 Determining the Univariate p-Value

To determine the univariate significance for gene pathway S, we permute the sample labels of the gene expression profile $B = 2000$ times, and recompute score S' for pathway S on each permuted dataset, yielding permutations values $S'^{*1}, S'^{*2}, \ldots, S'^{*B}$. The estimated p-value for S is specified in either direction by:

$$p.hi = \frac{\sum_{j=1}^{B} I(S'^{*j} > S')}{B}, \tag{2.7}$$

$$p.lo = \frac{\sum_{j=1}^{B} I(S'^{*j} < S')}{B}, \tag{2.8}$$

where $I()$ denotes the indicator function, $p.hi$ denotes the percentage of scores higher than S', and $p.lo$ denotes the percentage of permuted scores lower than S'. Lastly, we compute the overall p-value for a pathway as the minimum between $p.lo$ and $p.hi$ as follows,

$$p\text{- value for pathway } S = \min\{p.lo, \ p.hi\}. \tag{2.9}$$

In the next section we apply a modified GSA algorithm to several microarray datasets designed to study genes involved in breast cancer.

3 Case Study

An estimated 12% of females in the USA are at risk of developing breast cancer in their lifetime and in 2008 breast cancer caused approximately 450,000 deaths worldwide. In this study we compare 5 breast cancer microarrays datasets where the results are fully presented in [13, 14]. For our analysis, the pathway database is compiled from the Kyoto Encyclopedia of Genes and Genomes (KEGG) [4] with the addition of curated pathways from the human protein reference database (HPRD) [15]. The combined KEGG and HPRD pathway database contains 232 human pathways that include metabolism, genetic information processing, environmental information processing, cellular processes, human diseases, and drug development. For each dataset we employed a modified GSA method to measure gene pathway correlation with overall survival after accounting for estrogen receptor (ER) status and tumor size. After controlling for multiple testing, we determined that the *cell cycle pathway*, *pyrimidine metabolism pathway*, and *biosynthesis of phenylpropanoids pathway* can each significantly stratify survival within subsets of patients across several of the datasets. This was in stark contrast to "gene by gene" testing which showed little, if any, reproducibility across the datasets. The case study concludes that in light of clinical variables, there are significant gene pathways in common across the datasets that can further stratify patients according to their survival.

4 Next Generation Sequencing Tests

Sequencing technologies essentially began with Sanger sequencing [18, 19]. Technical improvements have yielded quantitative methods based on tagged sequences including SAGE (serial analysis of gene expression) to MPSS (massively parallel signature sequencing). Next generation sequencing technologies, especially RNA-seq, allow for massively parallel sequencing with improved efficiency over microarrays. Details on the preparation of a sample for RNA sequencing, and preprocessing of the RNA-seq data including alignment and normalization can be found in [5–7, 9, 12, 17].

For RNA-seq technology, the input data (after alignment and normalization) for statistical analysis is a matrix $Y = [y_{ij}]$, where y_{ij} denotes the number of reads of transcript i in sample j. Let n_j be the total number of reads in sample j. For a robust statistical test of differential expression, a distribution should be specified for the y_{ij}. Commonly for comparison of two treatments A and B the hypotheses are $H_0 : q_{iA} = q_{iB}$ vs $H_1 : q_{iA} \neq q_{iB}$ where q_{iA} and q_{iB} denote the amount of transcript i in treatments A and B, respectively. A two sample t test is inappropriate because of the (discrete) count nature of the data. Given the specified model and scaling factor estimated, the standard procedures may employ a Wald test, score test [8], or the likelihood ratio test [11]. Due to the cost of sequencing, however, usually only a small number of samples are available. This suggests using exact test procedures rather than tests based on large sample approximations.

In [16], the authors develop an exact test for small sample estimation in the negative binomial model. Specifically, for a comparison between m_A samples in treatment A and m_B samples in treatment B, we define, $k_{iT} = \sum_{j \in T} y_{ij}$, $T = A, B$, where T denotes the set of sample indices in Treatment T. Thus k_{iT} is the sum of transcript i in treatment T and $k_{iS} = k_{iA} + k_{iB}$ is the overall sum. Given the negative binomial model, the probability of the event $k_{iA} = a$ and $k_{iB} = b$, denoted $p(a, b)$ can be calculated for any values a and b. Then the two sided p-value for the exact test is the probability of observing treatment sums more extreme than the observed k_{iA}, k_{iB}, conditional on the overall sum k_{iS}. In other words, the p-value for transcript i is given by the following,

$$
p_i = \frac{\displaystyle\sum_{\substack{a+b=k_{iS}, \\ p(a,b) \leq p(k_{iA}, k_{iB})}} p(a, b)}{\displaystyle\sum_{a+b=k_{iS}} p(a, b)}, \tag{4.1}
$$

where the denominator is the probability of observing the overall sum k_{iS}, and the numerator is the sum of probabilities less than or equal to $p(k_{iA}, k_{iB})$ given the overall sum k_{iS}.

The p-values defined in (4.1) can be adopted for testing with GSA by calculating a z-statistic via $z_i = \Phi^{-1}(p_i)$ where Φ represents the cumulative distribution function (CDF) for a standard normal random variable. Using the z-statistics, it is possible to implement the GSA algorithm defined in Sect. 2.1. It is future work for our group to explore the power and robustness properties of the GSA algorithm when using RNA-seq derived test statistics.

5 Discussion and Conclusions

There are over 80 different algorithms and software packages designed to study gene set/pathway analysis in microarrays with some algorithms adaptable for next generation sequence data. Relative to microarray gene pathway analysis methods, there is a limited number of methods suitable for RNA-seq datasets although active research has resulted in several new methods including [3, 21]. As previously discussed, the discrete count data rather than continuous intensity signal is one of the main differences between RNA-seq data and microarray data. Additionally, it is noted that RNA-seq is a competitive experiment in that transcripts or genes "compete" against each other to be counted [10]. This is not the case in microarray experiments. As discussed in [22], there exist biases that do not exist in microarray data. Namely, in RNA-seq experiments, long transcripts or highly expressed transcripts are more likely to be called differentially expressed resulting in a larger number of false positives in those situations. These aspects separating microarray data from RNA-seq data should be accounted for when adapting microarray based gene pathway methods for RNA-seq data.

References

1. Allison, D. B., Cui, X., Page, G. P., & Sabripour, M. (2006). Microarray data analysis: From disarray to consolidation and consensus. *Nature Reviews Genetics, 7*, 55–65.
2. Efron, B., & Tibshirani, R. (2007). On testing the significance of sets of genes. *The Annals of Applied Statistics, 1*, 107–129.
3. Hänzelmann, S., Castelo, R., & Guinney, J. (2013). GSVA: Gene set variation analysis for microarray and RNA-seq data. *BMC Bioinformatics, 14*, 7.
4. Kanehisa, M., & Goto, S. (2000). KEGG: Kyoto encyclopedia of genes and genomes. *Nucleic Acids Research, 28*, 27–30.
5. Langmead, B., Hansen, K. D., & Leek, J. T. (2010). Cloud-scale RNA-Sequencing differential expression analysis with Myrna. *Genome Biology, 11*, R83.
6. Li, H., & Durbin, R. (2009). Fast and accurate short read alignment with Burrows-Wheeler transform. *Bioinformatics, 25*, 1754–1760.
7. Li, H., Ruan, J., & Durbin, R. (2008). Mapping short DNA sequencing reads and calling variants using mapping quality scores. *Genome Research, 18*, 1851–1858.
8. Li, J., Witten, D. M., Johnstone, I. M., & Tibshirani, R. (2012). Normalization, testing, and false discovery rate estimation for RNA-Sequencing data. *Biostatistics, 13*, 523–538.
9. Li, R., Yu, C., Li, Y., Lam, T. W., Yiu, S. M., Kristiansen, K., et al. (2009). SOAP2: An improved ultrafast tool for short read alignment. *Bioinformatics, 25*, 1966–1967.
10. Mak, H. C., & Storey, J. D. (2011). Interview with nature biotechnology: New statistical methods for high-throughput sequencing. *Nature Biotechnology, 29*, 331–333.
11. Marioni, J. C., Mason, C. E., Mane, S. M., Stephens, M., & Gilad, Y. (2008). RNA-Seq: An assessment of technical reproducibility and comparison with gene expression arrays. *Genome Research, 18*, 1509–1517.
12. Miecznikowski, J. C., Liu, S., & Ren, X. (2012). Statistical modeling for differential transcriptome analysis using RNA-seq technology. *Journal of Solid Tumors, 2*, 33–44.

13. Miecznikowski, J. C., Wang, D., Gold, D. L., & Liu, S. (2012). Meta-analysis of high throughput oncology data. In R. Chakraborty, C. R. Rao, & P. K. Sen (Eds.), *Handbook of statistics: Bioinformatics in human health and heredity* (pp. 67–96). Amsterdam: North Holland.
14. Miecznikowski, J. C., Wang, D., Liu, S., Sucheston, L., & Gold, D. (2010). Comparative survival analysis of breast cancer microarray studies identifies important prognostic genetic pathways. *BMC Cancer, 10*, 573.
15. Mishra, G. R., Suresh, M., Kumaran, K., Kannabiran, N., Suresh, S., Bala, P., et al. (2006). Human protein reference database–2006 update. *Nucleic Acids Research, 34*, D411.
16. Robinson, M. D., & Smyth, G. K. (2008). Small-sample estimation of negative binomial dispersion, with applications to SAGE data. *Biostatistics, 9*, 321–332.
17. Rumble, S. M., Lacroute, P., Dalca, A. V., Fiume, M., Sidow, A., & Brudno, M. (2009). SHRiMP: Accurate mapping of short color-space reads. *PLoS Computational Biology, 5*, e1000386.
18. Sanger, F., & Coulson, A. R. (1975). A rapid method for determining sequences in DNA by primed synthesis with DNA polymerase. *Journal of Molecular Biology, 94*, 441–448.
19. Sanger, F., Nicklen, S., & Coulson, A. R. (1977). DNA sequencing with chain-terminating inhibitors. *Proceedings of the National Academy of Sciences of the United States of America, 74*, 5463–5467.
20. Subramanian, A., Tamayo, P., Mootha, V. K., Mukherjee, S., Ebert, B. L., Gillette, M. A., et al. (2005). Gene set enrichment analysis: A knowledge-based approach for interpreting genome-wide expression profiles. *Proceedings of the National Academy of Sciences of the United States of America, 102*, 15545–15550.
21. Varemo, L., Nielsen, J., & Nookaew, I. (2013). Enriching the gene set analysis of genome-wide data by incorporating directionality of gene expression and combining statistical hypotheses and methods. *Nucleic Acids Research, 1*, 14.
22. Young, M. D., Wakefield, M. J., Smyth, G. K., Oshlack, A., Young, M., Wakefield, M., et al. (2010). Gene ontology analysis for RNA-seq: Accounting for selection bias. *Genome Biology, 11*, R14.

A New Approach for Quantifying Uncertainty in Epidemiology

Elart von Collani

Abstract Epidemiology is the branch of science on which public health research is founded. This essay shall review some of the principles underlying current methodology, revealing some ambiguities and inconsistencies. A new approach is proposed, the *Bernoulli space*, which is a complete model of uncertainty in a given situation. Each part of the model is necessary and the entire model is sufficient for describing all relevant parts of uncertainty. Using the Bernoulli space two aims are achieved: (1) Reliable and accurate predictions are obtained as basis for the decision-making process; (2) A unique interpretation of the obtained experimental results is obtained.

Keywords Epidemiology · Inference

1 Preamble

Invited by Andrei Yakovlev, I delivered a seminar lecture on modeling uncertainty at the Huntsman Cancer Institute in Salt Lake City in March 2001. During the seminar, I proposed a new approach to quantify and model uncertainty. With this essay I aim at continuing the discussion that was started on that occasion.

2 Uncertainty and Epidemiology

The main aim of epidemiology is to plan and evaluate strategies to prevent illness and diseases. In order to accomplish the task of prevention, reliable and accurate predictions are necessary. They are difficult because the future development

E. von Collani (deceased)
Professor von Collani wrote this chapter while at University of Würzburg Sanderring 2, Würzburg, Germany

© Springer Nature Switzerland AG 2020
A. Almudevar et al. (eds.), *Statistical Modeling for Biological Systems*,
https://doi.org/10.1007/978-3-030-34675-1_17

is indeterminate and the same holds for the consequences of an implemented prevention strategy. Uncertainty about the future development constitutes therefore the main obstacle for a successful campaign against infectious diseases. (In order to avoid misunderstandings, the term uncertainty is used here exclusively with respect to the indeterminate future and not to the determinate past.)

Uncertainty about the future development has two sources, an internal one that is characteristic of man and an external one that is characteristic for the universe. Human ignorance is the internal source, while the external source is often called randomness. Randomness leads to the fact that a process, when repeated, will result in different outcomes. Dealing scientifically with infectious diseases means the development of a mathematical model that mirrors sufficiently well all relevant aspects of the given situation and thus can be used as a basis for making decisions. The most relevant issue concerning the future development is uncertainty and therefore modeling uncertainty in an appropriate way is crucial for enabling reliable and accurate predictions and for making correct decisions.

Unfortunately, science does not provide an appropriate model of "uncertainty." In lieu thereof many different models can be found in the scientific literature as well in applications. Actually, scientists traditionally keep away from uncertainty, but recently mathematicians have invented an ever increasing number of different approaches to uncertainty without clearly stating what uncertainty is. Recently, Hans Kuijper, a Dutch sinologist turned systemicist, expressed his helplessness facing all the uncertainty concepts with the following words:

> 'Crisp sets', 'fuzzy sets', 'rough sets', 'grey sets', 'fuzzy rough sets', 'rough fuzzy sets', 'fuzzy grey sets', 'grey fuzzy sets', 'rough grey sets', 'grey rough sets', and now 'affinity sets'. My goodness! Is there anybody around who can enlighten me, i.e., help me to see a clear pattern in this set of sets, allegedly providing powerful tools to model various kinds of uncertainty?

Beside the approaches mentioned by Kuijper there are many "traditional" approaches to uncertainty developed and adopted in the context of mathematical probability theory, such as the frequentist approach, the Bayesian approach, the propensity approach, and so on. These probabilistic approaches are not at all consistent with one another and lead to different methods and different results for the same problem. As a consequence, the results obtained are not uniquely interpretable, which necessarily leads to wrong discoveries and, subsequently, to wrong decisions. Recently, Xiao-Li Meng complains in The American Statistician [3] about too many false discoveries, misleading information, and misguided policies of astronomers, engineers, geophysicists, psychiatrists, and social scientists because of inappropriately taking account of uncertainty. Meng even proposes a "police of science!"

Any decision in epidemiology concerning an illness or a disease should be based on reliable and accurate predictions which are only possible if the actual situation with respect to the illness or disease is sufficiently well known and quantitatively modeled. Thus, the values of relevant quantities must be determined, which is done in statistics by estimation methods or, more generally, by measurement procedures, and again the problem of measurement uncertainty arises and has to be solved by means of mathematical models that describe the underlying uncertainty generated by randomness and ignorance sufficiently well.

Before the value of a quantity can be determined or measured, the quantity must be "quantified." Quantification means that the quantity is represented mathematically, i.e., by numbers, sets of numbers, variables, or functions. An appropriate quantification results in uniquely interpretable values, and only if such a quantification of a characteristic has been achieved, can one develop measurement procedures for determining the actual, but unknown value of the quantity. There are several issues that make an appropriate quantification of uncertainty of the future development more difficult than the quantification of most other attributes.

1. Uncertainty refers to something that will become visible only in future.
2. Uncertainty contradicts the prevailing thinking that is characterized by the assumption that the future development follows cause–effect relations.
3. There are two extremely different mechanisms that generate uncertainty (human ignorance and natural randomness) and each has to be quantified separately.

We will stress the need to model separately the two sources of uncertainty, randomness and ignorance, and discuss the consequences of failing to do so.

2.1 Quantification of Randomness

Any given situation in epidemiology refers to facts with respect to the population, the disease, and the diagnostic and therapeutic means. To know them is necessary, because these "initial conditions" largely determine the necessary means to be taken and influence the future development of the disease.

However, even if the initial conditions are completely known, it is impossible to predict the exact future development due to what is called "randomness." A necessary requirement for solving the prediction problem is an adequate quantification and a subsequent modeling of uncertainty about the indeterminate future development since uncertainty is the only aspect that prevents exact predictions.

Randomness refers to future events which may occur or may not occur. The uncertainty due to randomness was quantified more than 300 years ago by the Swiss theologian and mathematician Jakob Bernoulli (see [1, 2]). He explained the uncertainty about the occurrence of a future event by the "degree of certitude of its occurrence" and called this degree "probability." Jakob Bernoulli allocated some probability mass to each future event with the interpretation that this probability mass is a part of the certitude of occurrence. This "definition" of probability allows only one interpretation and, moreover, yields immediately the mathematical properties to be required by the mathematical expression representing the probability of a future event. Any impossible event represents the "natural zero" with respect to its probability, and an event that will occur with certainty must necessarily have probability 1. Moreover, combining two mutually exclusive events means to put together their probability masses which leads to the well-known additivity of the probability measure.

Unfortunately, Bernoulli's ideas and developments fell on deaf ears at the time, and ultimately slipped into oblivion. The result is that a variety of different and inconsistent interpretations of probability are used, and this fact prevents their unique interpretation.

In summary, randomness of future development that extends to all future events is quantified by what is known a probability measure or probability distribution.

2.2   Quantification of Ignorance

Besides randomness with respect to the future, ignorance about the initial conditions (that is, about the past), is the second source of uncertainty. If the actual initial conditions is given by the value d, then d is generally unknown, and has to be determined by a measurement procedure (estimation procedure) that is subject to randomness. Therefore the actual value d cannot be determined exactly. Instead, any (reliable) measurement procedure results in a set of possible values for the unknown quantity, while all values outside the set may be excluded.

Thus, we conclude that ignorance about the initial conditions is quantified by a set that covers all those values that cannot be excluded.

3   Ambiguity of Epidemiological Results

During the last years, the number of large-scale epidemiological studies has rapidly increased. However, after the completion of most of these studies, a controversy breaks out about the interpretation of the results obtained. Often, the various proposed interpretations contradict one another and not only the public is confused, but also the decision makers. Thus, necessary decisions are postponed or wrong decisions are made.

We will illustrate this point with the following case study. In 2003 an epidemiological study *Childhood Cancer in the Vicinity of Nuclear Power Plants* (KiKK)[1] was undertaken in Germany which was "intended to find out whether cancer in children under 5 years of age is more frequent in the immediate vicinity of nuclear power plants (NPP) than further away." It was carried out by the German Childhood Cancer Registry (Deutsches Kinderkrebsregister, DKKR), where the study design was developed in consultation with an expert committee assembled by the Federal

---

[1] http://www.bfs.de/en/bfs/druck/Ufoplan/4334_KIKK.html, accessed September 2009.

Office for Radiation Protection (BfS). The conclusions of this study were published at the end of 2007 and read as follows:

> The present study confirms that in Germany there is a correlation between the distance of the home from the nearest NPP at the time of diagnosis and the risk of developing cancer (respectively leukemia) before the 5th birthday. This study is not able to state which biological risk factors could explain this relationship. Exposure to ionizing radiation was neither measured nor modeled. Although previous results could be reproduced by the current study, the present status of radiobiologic and epidemiologic knowledge does not allow the conclusion that the ionizing radiation emitted by German NPPs during normal operation is the cause. This study cannot conclusively clarify whether confounders, selection, or randomness plays a role in the distance trend observed.[2]

The study and the conclusion were hardly published when a fierce discussion started about the correct interpretation. The first doubts about the results were expressed by the external advisory immediately after the report was published.[3] The board stated at the beginning of its evaluation that the study design complies with the state of the art of epidemiological science, and that most suggestions of the external panel of experts regarding analyses of the quality of data and results were realized. These analyses did not give any indications that the results were significantly distorted.

But, then the report of the advisory board continues among others with the following words:

1. The evaluation plan did not provide for the calculations on the attributive risk. Stating the risk that can be attributed to the distance of the home to a reactor and the population-related risk is indispensable for the communication of the results to politicians and the general public. In the case under consideration the calculations were not carried out correctly.
2. The authors write that "[...] due to the current radio-biological and radio-epidemiological knowledge the ionizing radiation emitted by German nuclear power plants in normal operation can basically not be interpreted as the cause." In contrast to the authors, all members of the external panel of experts are convinced that due to the particularly high radiation risk for small children and due to the insufficient data on emissions of power reactors, this interrelation can on no account be excluded. Furthermore, there are several epidemiological causality criteria in favor of such an interrelation. The task of science now is to find an explanatory approach for the difference between epidemiological and radio-biological evidence.
3. To explain the risk around nuclear power plants proved by them, the authors refer to so-called confounders, selection mechanisms not described in detail, or statistical chance. In view of the study results, the external panel considers all three explanatory approaches to be improbable.

---

[2] http://www.bfs.de/en/bfs/druck/Ufoplan/4334_KIKK_Zusamm.pdf, accessed September 2009.

[3] http://www.bfs.de/en/kerntechnik/kinderkrebs/Expertengremium.html, accessed September 2009.

Since then, the discussion continues about how the study results shall be interpreted and, of course, no decisions at all were made with respect to children living close to nuclear power plants. In the meantime, it seems useful to consider whether or not these ambiguities could be resolved by a less complicated, but more sound, model of uncertainty.

3.1 Statistics and Ambiguity

In epidemiology, statistical methods are widely used that are based on the concept of "probability." The question arises whether or not this fundamental concept meets the requirement of having a unique definition as a necessary condition of unambiguity.

A glance into statistical textbooks does not help much, as often no interpretation of the mathematical concept "probability" is given. A very good source for the interpretations of "probability" is the Stanford Encyclopedia of Philosophy,[4] which lists five main interpretations and a number of secondary ones indicating the ambiguity of the concept "probability." In statistics, the two most frequently used interpretations are the so-called frequency interpretation and the Bayesian (or subjective) interpretation:

1. The frequency interpretation is based on a sequence of repeatable experiments and explains the "probability of an outcome" as its relative frequency "in the long run" (infinite frequency interpretation).
2. The Bayesian interpretation refers to statements and explains probability as the individual's degree of belief in it.

Clearly, both interpretations are not at all consistent with one another. Moreover, according to the frequency interpretation probability needs an infinite sequence of experiments, which obviously does not exist. In other words, the frequency interpretation does not explain the concept of probability, but leads itself to irresolvable interpretation difficulties. The Bayesian interpretation, on the other hand, approves "belief" as a means to develop science and thereby puts it close to religion. As a result, the basic concept of statistics, that is, "probability," appears to have not one (objective) meaning but several (subjective) meanings that are not consistent with one another. There are some immediate consequences:

1. The results of any statistical analysis cannot be interpreted in a unique way and may therefore lead to disputes and errors.
2. The methods developed in statistics heavily depend on the chosen probability interpretation, so that for the same problem there are different methods yielding different results, with different interpretations.

---

[4]http://plato.stanford.edu.

3. In most cases, the models used in making a statistical analysis violate reality in an obvious way. Therefore, it is in general impossible to judge the relevance of any obtained result.

We will continue our inquiry by considering a seminal epidemiological problem, that of comparing two fixed but unknown probabilities.

3.2 Epidemiological Measurements

Based on the probability measure, a number of derived measures are used in epidemiology. At least some of these measures were historically derived in the realm of gambling. One of the most often used quantities in epidemiology is the odds ratio, where the odds[5] in favor of an event E are defined by means of probability. Let $p = P(E)$ be the probability of the event E, then the odds $o(E)$ of the event E are defined as

$$o(E) = \frac{p}{1 - p}.$$

The "odds" in favor of an event are also called "relative frequencies" and cause much confusion as they are often mixed up with the probability of the event in question, which in epidemiology is also called the "chance" of the event. Evidently, for $p < 1$, the probability and the odds are equivalent quantities as

$$p = \frac{o(E)}{1 + o(E)}.$$

However, comparing the two concepts "probability" and "odds," it turns out that the "odds" is the essentially more complex quantity. According to Bernoulli, the probability quantifies the degree of certitude of the occurrence of the event and hence is restricted to a bounded range given by the unit interval, while in contrast the odds are unbounded. Given the equivalence of the two quantities, the question naturally arises as to why both are needed by epidemiologists. One answer to this question lies in the use of the "odds ratio." Consider the event E, but with two different initial conditions resulting in two different probabilities $p_1 = P_1(E)$ and $p_2 = P_2(E)$. Then the odds ratio of the event E is defined as

$$or(E) = \frac{p_1/(1 - p_1)}{p_2/(1 - p_2)}.$$

---

[5]The odds may be defined with respect to a (future) event or a proposition. The term "odds" in favor or against an event is used in some games of chance to indicate the amount of winnings that will be paid if the event actually occurs.

This odds ratio is commonly used for comparing two probabilities p_1 and p_2. Comparing two real numbers is generally done without any further transformation, because the task is simple, well understood, and exercised already in elementary school. By calculating the odds, (i.e., a "relative quantity"), the understanding becomes more difficult as any relative quantity is more complex than an absolute one. But, no information about the situation is lost by the transformation from probability to odds.

By calculating the odds ratio, however, a clear understanding becomes almost impossible and, even worse, the better part of the available information is lost, as the actual values p_1 and p_2 cannot be regained from the value of the corresponding odds ratio. The implications of this will be illustrated in the next section.

4   A Stochastic Model of Uncertainty

We have argued that human uncertainty about a future development can have only two sources, randomness, which concerns uncertainty of future development, and ignorance, which concerns initial conditions in the past. In order to enable a description of a future development, consideration must be limited to what is of specific interest. This is done by specifying the future aspect of interest and quantifying it by a "random variable" represented by X.

Predictions are possible, because the future development is more or less strongly influenced by the past initial conditions. Therefore, for modeling the transition from the initial conditions to the future outcome of the random variable X, the initial conditions must be described by a variable, say D, called the deterministic variable. Ignorance is quantified by a probability distribution, determined by certain (distribution) parameters, the values of which are fixed by the given situation (or facts). Let the deterministic variable D have the actual value d.

4.1   The Bernoulli Space

The pair of variables (X, D) determines a model describing the transition from past to future, introduced in [4, 5] and is called the "Bernoulli space," denoted $\mathbb{B}_{X,D}$, defined by the three components:

$$\mathbb{B}_{X,D} = (\mathcal{D}, \mathcal{X}, \mathcal{P}) .$$

The three components have the following meaning.

- **The Ignorance Space** $\mathcal{D}$: The ignorance space $\mathcal{D}$ is the set that contains all those values of D which cannot be excluded in a given situation. The set $\mathcal{D}$ reflects the existing ignorance and, hence, indicates what still can be learned. Thus, by the ignorance space $\mathcal{D}$, designing purposeful learning processes becomes possible. Note that not to state explicitly the ignorance means to assume complete knowledge and to cease learning. The establishment of a continuous learning process is only possible if the existing ignorance is quantitatively given. The ignorance space also illustrates that learning consists of excluding "what is not" and not in assuming "what is."

- **The Variability Function** $\mathcal{X}$: Any subset of $\mathcal{D}$ represents a certain level of knowledge or, equivalently, of ignorance. Moreover, it determines the set of the future outcomes of the random variable X. The set of these values represents the range of variability of X. The variability function $\mathcal{X}$ is therefore defined for subsets of the ignorance space $\mathcal{D}$. The images of $\mathcal{X}$ are the corresponding ranges of variability of X. Thus, for $\mathcal{D}_0 \subset \mathcal{D}$ the set $\mathcal{X}(\mathcal{D}_0)$ consists of those values of X which have to be taken into account because they might occur.

- **The Random Structure Function** $\mathcal{P}$: The value d of D does not only determine the range of variability $\mathcal{X}(\{d\})$ of the random variable, but also the corresponding probability distribution, i.e., the random structure on $\mathcal{X}(\{d\})$. The random structure function $\mathcal{P}$ is therefore defined for subsets of the ignorance space $\mathcal{D}$ with images being the corresponding class of probability distributions of X. For $d_0 \in \mathcal{D}_0 \subset \mathcal{D}$, the probability distribution $\mathcal{P}(\{d_0\})$ is denoted by $P_{X|d_0}$.

4.2 Estimating Probabilities

Consider a given population and a specified contagious disease and let the fraction infected p in the population be of interest that shall be determined by a measurement procedure. The necessary measurement process consists of a random sample representing n persons that must be selected randomly from the population. The indicator variable X_i for an ith person to be selected adopts the value 1 in the case of an infection otherwise the value 0. The indicator variables of all persons to be selected represent the sample $(X_1, \ldots, X_n)$.

The sample function is given by the sum $X = \sum_{i=1}^{n} X_i$ with the deterministic variable D that determines the probability distribution of X given by the unknown fraction infected p and the known sample size n.

4.2.1 Bernoulli Space

In this case the Bernoulli space $\mathbb{B}_{X,D}$ is extremely simple:

- **Ignorance Space** $\mathcal{D}$: If nothing is known about the fraction infected p except that it is not 0 (there is no infected person) and not 1 (all persons are infected), then the ignorance space is given as

$$\mathcal{D} = \{(n, p) \mid 0 < p < 1\}.$$

- **Variability Function** $\mathcal{X}$:

$$\mathcal{X}(\{(n, p)\}) = \{0, 1, \ldots, n\}, \text{ for } (n, p) \in \mathcal{D}.$$

- **Random Structure Function** $\mathcal{P}$:

$$\mathcal{P}(\{(n, p)\}) = P_{X|(n,p)}, \text{ for } (n, p) \in \mathcal{D}$$

with

$$P_{X|(n,p)}(\{x\}) = \binom{n}{x} p^x (1 - p)^{n-x}, \text{ for } x \in \mathcal{X}(\{(n, p)\}).$$

The above specified Bernoulli space describes the measurement process for the unknown value p and can be used for deriving a measurement procedure.

4.2.2   The Stochastic Measurement Procedure

A stochastic measurement procedure $C_D^{(\beta)}$ is a function that assigns to each observed result $\{x\}$ of the sample function a measurement result $C_D^{(\beta)}(\{x\})$ that is a subset of the ignorance space $\mathcal{D}$. The stochastic measurement procedure is defined by its measurement results as follows:

$$C_D^{(\beta)}(\{x\}) = \left\{ (n, p) \mid x \in A_X^{(\beta)}((n, p)) \right\},$$

where β denotes a lower bound for the probability of obtaining a correct result and $A_X^{(\beta)}((n, p))$ is a prediction with respect to the random variable $X \mid (n, p)$ that will occur with probability of at least β. The graphical representation of the measurement procedure $C_D^{(\beta)}$ for $\beta = 0.95$ and $n = 136$ is given in Fig. 1. This yields a correct result with probability of at least 95% (a result is correct if it contains the actual but unknown value of the fraction infected). For the observation $x = 62$, the measurement result is indicated explicitly, representing the following measurement result:

$$C^{(0.95)}(\{62\}) = \{p \mid 0.37 \leq p \leq 0.54\}.$$

The interpretation of the result is straightforward: the unknown value of the fraction infected is between 37 and 54%. If the accuracy of the result is not good enough, a larger sample size n must be used. The measurement result $C^{(0.95)}(\{x\})$ improves the Bernoulli space by reducing the corresponding ignorance space. The improved

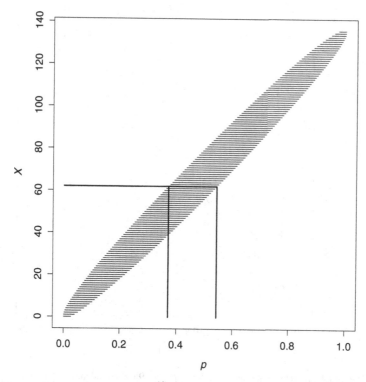

Fig. 1 The graphical representation of $C_D^{(\beta)}$ of Sect. 4.2.2 with reliability level $\beta = 0.95$ and sample size $n = 136$

Bernoulli space can then be used for making predictions that are as reliable as required and as precise as possible.

The contrast with standard epidemiological methods is made clear by considering the problem of comparing two probabilities p_1 and p_2, each associated with a Bernoulli space of the type just defined. If we let x_1 and x_2 be two corresponding, and independent, observations, we may construct a new product Bernoulli space with measurement procedure

$$C_D^{(\beta)}(\{x_1, x_2\}) = C_D^{(\beta')}(\{x_1\}) \times C_D^{(\beta')}(\{x_2\}),$$

the appropriate adjustment of the reliability level being $\beta' = \beta^{1/2}$.

Next, suppose we construct a level β confidence interval for the odds ratio. Although this is nominally an inference involving one parameter, such a confidence interval also implies a confidence set for the parameter vector (p_1, p_2), given by the area bounded by the contours implied by the interval's endpoints. If in reality $(p_1, p_2) = (0.1, 0.25)$, we might see observations $(x_1, x_2) = (18, 34)$ when $n = 136$. A level $\beta = 95\%$ confidence interval for the odds ratio based on Fisher's

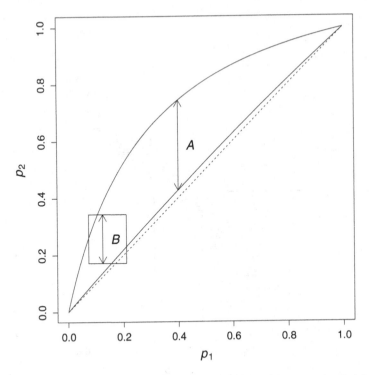

Fig. 2 The graphical representation of a level $\beta = 95\%$ confidence set implied by odds ratio confidence interval $(0.23, 0.89)$ (area indicated by arrow A), and measurement result $C_D^{(\beta)}(\{18, 34\})$ with reliability level $\beta = 0.95$ (area indicated by arrow B). The identity is indicated by the dashed line

exact test is given by $(0.23, 0.89)$. The measurement result $C_D^{(\beta)}(\{18, 34\})$, and the confidence set based on the odds ratio confidence interval are shown in Fig. 2.

The comparison is instructive. Formally, the null hypothesis $H : p_1 = p_2$ is (correctly) rejected by the odds ratio inference, but not by measurement $C_D^{(\beta)}(\{18, 34\})$, in the sense that it is intersected by the identity. In fact, it is not hard to verify that the odds ratio test is, at least for this example, more powerful (that is, possesses the smaller Type II error). However, the problem with this inference is revealed in Fig. 2. The confidence set includes an unhelpfully large variety of models, including some for which both p_1 and p_2 are arbitrarily close to 1, or to 0. This has practical implications. For example, if we report an odds ratio significantly different from 1, we are usually doing so with the intention of impacting public health or clinical practice in some way. This, in turn, carries the implication that p_1 or p_2 is large enough to be of interest in the first place. However, our confidence set does not support this conclusion, in exactly the same sense that $C_D^{(\beta)}(\{18, 34\})$ does not eliminate the possibility that $p_1 = p_2$. So what can we conclude?

At this point, common sense can be expected to take over, so let us follow the implications. The data is clearly incompatible with the more problematic models in the confidence set. This is quite true, but we need to acknowledge what we are doing. We are making the confidence set strictly smaller (or, equivalently, making a second inference statement). Certainly, the theory of hypothesis testing supports such a procedure, but is unequivocal about the cost, which is an increase in the observed significance level. That is, we are trimming the confidence set to more resemble $C_D^{(\beta)}(\{18, 34\})$ while adjusting the observed significance level. This last adjustment may very well force the confidence set to cross the identity. The final result is a confidence set which does not differ from $C_D^{(\beta)}(\{18, 34\})$ in any important way.

Elart von Collani died on February 25, 2017. We thank Claudia von Collani for permission to include his paper in this book.

References

1. Bernoulli, J. (1713). *The art of conjecturing, together with letter to a friend on sets in court tennis*. Baltimore: Johns Hopkins University Press. Translated and edited E.D. Sylla, 2006.
2. Bernoulli, J. (1713). *Ostwald's Klassiker der exakten Wissenschaften Nr. 108: Wahrscheinlichkeitsrechnung. 3/4 Teil mit dem Anhange: Brief an einen Freund "uber das Ballspiel. (Jeu de Paume)*. W. Engelmann, Leipzig. Translated and edited R. Haussner, 1899.
3. Meng, X.-L. (2009). Desired and feared - What do we do now and over the next 50 years? *The American Statistician, 63*, 202–210.
4. von Collani, E. (2004). Empirical stochastics. In E. von Collani (Ed.), *Defining the science of stochastics. Sigma series in stochastics* (vol. 1, pp. 175–214). Lemgo: Heldermann Verlag.
5. von Collani, E. (2004). Theoretical stochastics. In E. von Collani (Ed.) *Defining the science of stochastics. Sigma series in stochastics* (vol. 1, pp. 147–174). Lemgo: Heldermann Verlag.

Branching Processes: A Personal Historical Perspective

Peter Jagers

Abstract This article is a slightly edited and updated version of an evening talk during the random trees week at the Mathematisches Forschungsinstitut Oberwolfach, January 2009. It gives a—personally biased—sketch of the development of branching processes, from the mid nineteenth century to 2010, emphasizing relations to bioscience and demography, and to society and culture in general.

Keywords Branching processes · Mathematical history

1 Introduction and Summary

Few mathematical or even scientific fields cherish their past like branching processes. Ted Harris's classical treatise from 1963 [18] opens by a terse but appetizing two-page flashback. Three years later, David Kendall's elegant overview was published, and like Charles Mode in his monograph [38], I could borrow from that for the historical sketch in my 1975 book [22], but also add some observations of my own.

At that time we all knew that the French, notably de Candolle and Bienaymé, had considered the nobility and family extinction problem before Galton publicized it. In my introduction, I speculated about connections between Bienaymé and the demographer Benoiston de Châteauneuf, who had been studying old French noble families. The plausibility of such contact was subsequently corroborated by Chris Heyde and Eugene Seneta in their "I.J. Bienaymé: Statistical Theory Anticipated" [19]. They were actually relatives.

Heyde and Seneta also showed that Bienaymé was not only first at formulating the mathematical problem, but indeed knew its solution already in 1845 [2]. The

P. Jagers (✉)
Department of Mathematical Sciences, Chalmers University of Technology, Gothenburg, Sweden

University of Gothenburg, Gothenburg, Sweden
e-mail: jagers@chalmers.se

© Springer Nature Switzerland AG 2020
A. Almudevar et al. (eds.), *Statistical Modeling for Biological Systems*,
https://doi.org/10.1007/978-3-030-34675-1_18

original publication has not been found, but as pointed out by Bru et al. [3] (see also the monograph by Iosifescu et al. [20]) there is a proof in a treatise by A.A. Cournot [10], published only 2 years after Bienaymé's communication. Though this is not explicitly stated, it seems plausible that Cournot reproduced Bienaymé's argument. (The book by Iosifescu and co-authors also presents an intriguing discussion by Bienaymé arguing that the limited size of mankind (at that time!) should show that human mean reproduction must have varied above and below one in historical time.)

There are good reasons for branching processes to keep its heritage alive. Not only is the background in the frequent disappearance of family names, even in growing populations, picturesque and easily understood, it is also something that could not have been explained by prevailing—and long dominating—deterministic population theory. Indeed, it provides convincing arguments for a stochastic population theory, and not only for "small" populations. In spite of its alluring disguise, the family extinction problem concerns an important and basic feature of population development, the frequent extinction of family lines and, as a consequence, ubiquitous shared ancestry.

It also tells a story of interplay between mathematics, natural science, culture, and society. Indeed, listen to Galton's classical formulation in Educational Times 1873, initiating modern theory:

> "PROBLEM 4001: A large nation, of whom we will only concern ourselves with adult males, N in number, and who each bear separate surnames colonise a district. Their law of population is such that, in each generation, a_0 percent of the adult males have no male children who reach adult life; a_1 have one such male child; a_2 have two; and so on up to a_5 who have five.
>
> Find (1) what proportion of their surnames will have become extinct after r generations; and (2) how many instances there will be of the surname being held by m persons."

Rarely does a mathematical problem convey so much of the flavor of its time, colonialism and male supremacy, and maybe an underlying concern for a diminished fertility of noble families, opening the doors for the genetically dubious lower classes.

It reveals a mathematical theory initiated not by mathematicians but by a broad savant, Francis Galton, a polyhistor well versed in mathematics but primarily if anything, a biologist. We see an example falsifying both extremist views on science, that of a pure science, and in particular mathematics, devoid of political meaning and implications; and that degrading science and scientific development into a purely social phenomenon. Indeed, in branching processes, they all meet: pure mathematical development, biology, physics, and demography, and the concoction is spiced to perfection by the social and cultural context in which it is formed.

As is well known, Watson determined the extinction probability as a fixed point of the reproduction generation function f. He observed that 1 is always such a fixed point, $f(1) = 1$, and from this he and Galton [43] concluded, seemingly without hesitation: "All the surnames, therefore, tend to extinction in an indefinite time, and this result might have been anticipated generally, for a surname lost can never be recovered, and there is an additional chance of loss in every successive generation. This result must not be confounded with that of the extinction of the male population; for in every binomial case where q is greater than 2 we have $t_1 + 2t_2 + \cdots + qt_q > 1$, and, therefore an indefinite increase of male population."

It is strange that so intelligent a pair as Galton and Watson (the latter was a clergyman who had been second Wrangler at Cambridge, and carried on mathematics and physics as a Rector and even was awarded an honorary D. Sc. by his alma mater) could have presented, and even believed in such a verbiage. It is even stranger that it took more than 50 years to rectify it, in particular since Bienaymé had already published a correct statement of the extinction theorem. I always thought the reason simply was that people of the time just did not notice, or bother about, such a mathematical trifle. But according to Heyde and Seneta, "its implications were strongly doubted" already at the time of publication.

And indeed, I checked an almost contemporary and nonmathematical criticism quoted by them, by a Swedish historian or political scientist, Pontus Fahlbeck. He was a commoner who married a Finnish baroness and became the author of a monumental two-volume treatise on the Swedish aristocracy [16]. There he gives a correct, verbal description of the relationship between growth of the whole and frequent extinction of separate family lines. He writes—somewhat condescendingly or intimately, it may seem: "Galton, who with characteristic curiosity considered the question, has tried to investigate to what extent families … must die out, with the help of a competent person." Fahlbeck then recounts examples considered by Galton, showing that "the tendency is the extinction of all." (The account is not completely lucid.) This is followed by a sequel of questions, and a reassuring answer:

"If this course of events be based on a mathematical law, then it should be as necessary, or not? And what then about our general conclusions, that no necessity forces extinction? Is there not in this a contradiction, which if both arguments are right, as they undoubtedly are, leads to what philosophers call an antinomy? However, mathematical calculations, as applied to human matters, may seem unrelenting but are actually quite innocuous. The necessity lies buried in them like an electrical current in a closed circuit, it cannot get out and has no power over reality." ([16, pp. 133–135], my translation).

As is well known, it was another polyhistor, J.B.S. Haldane, chemist, physiologist, geneticist, statistician, and prolific political writer in the New Statesman as well as the Daily Worker (he was a notable member of the intellectual British left of the 1930s and 1940s, beautifully described by Doris Lessing, among others) who got things basically right, although the final formulation was printed slightly later [42]. If the mean number of children is less than, or equal to, one, then Galton and Watson were right, but if it exceeds one, then there is another and smaller positive fixed point, which yields the correct extinction probability $q < 1$.

In many realistic situations, however, this extinction probability though less than one, will be large together with the mean reproduction $m = f'(1)$. Indeed, values of 0.75 and 2, respectively, e.g., are obtained for realistic reproductive patterns among men, or for that sake women, in historical times, and similarly among animals in wild life.

Lecturing in Peking in October 2008, I met with a cute illustration of this, which may well have occurred to some of you. In the China Daily I read that Kung Techen, who was the 77th great... grandson of Confucius (Kung Futse) had died on Taiwan at the age of 89. Yes, same surname inherited from father to son for more than 75 generations. Since Confucius (500 B.C.), China's population has undergone a tremendous growth, but as we all know, there are few Chinese family names. Indeed, Wikipedia tells us that three surnames (partly different in different parts of the country) are carried by some 30% of the population. In Korea the situation is even more extreme; half the population has one of the names Kim, Lee, or Park.

Thus, branching processes were born out of a social and demographic context. Its first fundamental result, the extinction theorem, has relevance far beyond that, in explaining homogeneity in large populations, as well as part of the more than frequent extinction in the course of natural evolution. Indeed, in 1991 the paleontologist David Raup claimed that 99.99% of all species, ever existing on our earth are extinct now.

When branching processes reappeared in scientific literature, between the World Wars, the impetus came from genetics (Fisher and Haldane) and biology more generally. Haldane deduced his approximation for the survival probability, still very important for the consideration of fresh, slightly fitter mutants in a resident population. In Russia, Kolmogorov coined the term branching process itself.

After World War II, the nuclear age had arrived. In Stalin's Moscow, Kolmogorov and his disciples, people like the Yaglom twin brothers and B. A. Sevastyanov, tried to pursue their research as a purely mathematical undertaking. But of no or little avail. Sevastyanov's thesis was classified, while being written, and since he himself was not deemed reliable he was not allowed to keep it. Instead it was locked up in a safe in the library. In the mornings it was handed it out to its author, who worked on it until five, when he had to return it.

Kolmogorov and others, including I believe Sacharov, protested, and finally the ban was lifted [44]. Things had become easier than in the 1930s.

In the USA, meanwhile, Ted Harris was employed by the Rand Corporation, an integral part of the military-industrial complex, and his work on electron–photon cascades and Galton–Watson processes with continuous type space (energy) was clearly inspired by nuclear physics. But both he and Sevastyanov saw themselves as mathematicians, though working on a pattern relevant for natural science. Sevastyanov even takes a rather purist stance; I have heard him saying that mathematics is nothing but mathematics, a somewhat unexpected opinion from a mathematician who is neither an algebraist nor a topologist, not even a pure analyst, and who actually after his thesis worked several years in the secret military part of the Steklov Institute, the so called "Box." Maybe it comes naturally to someone who has devoted his life to mathematics in the overly politicized Soviet Union.

With such leaders, it is not surprising that the 1950s and 1960s were an era of mathematization. Time structure was added to the simply reproductive branching process in what Bellman and Harris called *age-dependent* processes, depicting populations where individuals could have variable life spans, but split into a random number of children at death, independently of age. Sevastyanov introduced truly age-dependent branching processes, where reproduction probabilities could depend upon the mother's age at splitting.

The processes thus arising were not Markovian in real time, but could be analytically treated using renewal properties, and the then remarkable renewal theory, which had recently been established by Feller and others. Another development retained the Markov property, but viewed population evolution as occurring in real time, thus establishing connection to the elementary birth-and-death processes that were flourishing in semi-applied literature.

These approaches however remained in a sort of physical world, far from animal or even plant population dynamics, in the sense that they all considered childbearing through splitting only, like fission, cell division, or molecular replication. Or, for the classical Galton–Watson process, there was the alternative interpretation of disregarding time, and just count generations, as though they did not overlap in real time. The only exceptions were the models from the birth-and-death sphere, where exponentially distributed life spans allowed alternative interpretations. That also led to the first model of populations where individuals could give birth during their lives, Kendall's generalized birth-and-death process [31].

The first monographs, Harris's from 1963 [18] and Sevastyanov's from 1971 [41], as well as Athreya and Ney's from 1972 [1], however stayed firmly in the tradition of physical splitting. Branching processes remained separated from the deterministic differential equations and matrix formulations, and of population dynamics and mathematical demography. It was the rise of point process theory that rendered the formulation of general branching processes natural, so as to depict populations where individuals can give birth repeatedly, in streams of events formed by a point process, and possibly even of various types. In 1968 time was ripe, and Crump's and Mode's article and mine introducing such general branching processes appeared simultaneously in the winter 1968–1969 [11, 12, 21]. Mine was also part of my Ph.D. thesis, defended in October 1968, fortunately. In those times in Sweden, formal novelty was insisted upon for theses. In spite of the enormous friendliness of my polite Japanese opponent, Kiyosi Ito himself, I might have encountered difficulties, had the local mathematics (analysis) professors known that some Americans had done the same, skeptical towards probability theory, as they were. The status of probability within mathematics has certainly changed since then!

The advent of general branching processes meant that branching processes now embraced virtually all mathematical population theory. The dominating mathematical population framework since more than a century was the *stable population theory*, dating back to Quetelet and Lotka [34, 35]. Its real father or forerunner was, however, Euler who deduced its main findings, the exponential increase of population size and how the ensuing stable age distribution is determined by survival and reproduction rates, already around 1750 [15]. As I pointed out in my 1975 book [22], the apologetically minded Euler even used rapid population growth as an argument against "those incredulous men" who would not believe that the sons of Adam could have populated the earth during the 5000 years since Adam and Eve were ousted from Eden. Anyhow, his contributions were long forgotten in the demographic and mathematical biology communities.

Stable population theory is deterministic but based upon a probabilistic formulation of individual life events. All its findings could now be strictly proven in terms of general branching processes, and basic concepts like average age at childbirth given an interpretation. In the quite different area of cell and tumor biology, the concept of balanced exponential growth turned out to coincide with demographic stable population growth, and readily availed itself to analysis in terms of age-dependent binary splitting processes. A substantial part of my 1975 book was devoted to this area, further developed in the rich production by Andrei Yakovlev and Nick Yanev (cf. [44]).

The general stabilization of population composition could however, it turned out, be brought one step further: stable population theory and cell kinetics had only considered the distribution of age and phases (like mitosis) in old populations. Those are what could be called individual properties: it is your age and nobody else's. In a population there are however also important relational properties.

My research into such matters started in a quaint manner. In my youth, Gothenburg had a well-known doctor in charge of the city alcoholic polyclinic. Now that he had retired in the late 1970s he took up a research idea that he had toyed with for some time. He had made the observation that an astonishing proportion of his patients were first-born.

In the literature, he found that not only Gothenburg alcoholics, but also poets, statesmen, and people suffering from various mental disorders had been found often to be first-born. Galton had even claimed that the first-borns were the motor of history, since they were more often "men of note." The retired doctor realized that the apparent over-representation of first-borns could be an artifact, and performed a primitive but adequate simulation experiment. He drew the family trees of an invented but realistic population on a long white paper tablecloth. Towards the end of the paper roll, he then sampled individuals at random, or at least haphazardly. Many were first-borns. Now he wanted to discuss with me.

I found the probability of being first-born in an old single-type supercritical general branching process. It is $E[e^{-rT}]$ where r is the Malthusian parameter and T the mother's age at her first bearing. Since the Malthusian parameter equals ln 2 divided by the doubling time and the latter is usually larger than age at first bearing, the probability of being first-born tends to be larger than 0.5, even in populations with large broods, litters, or families [23]. (In an old critical population the probability of being first-born is larger than one over the expected non-null sibship size, due to Jensen's inequality.) The question showed the appropriateness of general branching processes; in the classical splitting framework it could not even have been posed, because all children are born together.

The important point is, however, that being first-born is not a property of your own life and birth-time. It concerns your relationship to your sibship. Thus, this simple observation led on to an investigation of how the whole pedigree, family structure, and type distribution in multi-type populations stabilize during exponential growth. A strict framework for general branching processes in abstract type spaces was formulated, related to branching tree ideas due to Neveu and Chauvin [5, 40]. In these, type distribution and ancestry, and hence mutational history could be traced backward in a Markov renewal structure. Our group published a whole sequel of papers on these topics during the 1980s and 1990s [23, 24, 28, 39], and indeed a final (?) attempt to popularize the admittedly heavy theory by restriction to discrete time more recently [29]. Stable pedigrees and backward times was virgin land, the only exceptions being the investigations by Bühler into the family structure of Bellman–Harris processes [4] and later by Joffe and Waugh into kin numbers in Galton–Watson processes [30].

In the meantime, deterministic population dynamics had advanced through work by eminent mathematicians like Odo Diekmann [13, 14] and Mats Gyllenberg, inspired by the biologist Hans Metz [37]. They had realized that the differential equations formulations they had been brought up with were becoming a straitjacket, and turned to semigroups of positive operators, yielding a theory corresponding to the Markov renewal theory of expectations of multi-type general branching. However, they took a further step, considering the feedback loop individual $\to$ population $\to$ environment $\to$ individual. Through this theory, *structured population dynamics*, they were able to analyze the fascinating new ideas that Metz and his followers had advanced to explain evolution, under the name of *adaptive dynamics*.

This was a new challenge to branching processes, and is being met in a series of path-breaking papers by Sylvie Méléard, Nicholas Champagnat, Amaury Lambert, and their co-workers in terms of a birth-and-death formulation [6–9] but also allowing age-structure [36]. On our side, we have tried to formulate models investigating populations in habitats with a carrying capacity, i.e., a threshold above which reproduction is sub-critical, thus approaching the consistency of adaptive dynamics, and in particular the problem of *sympatric speciation*, that is how successive small mutations can lead to new species, and their coexistence. This is done in a toy model of binary Galton–Watson style processes [33] and for more general branching populations [26].

The general problem of interaction in population dynamics is elusive. On the one hand, the very concept of a population builds upon individuals in some sense being the agents, those changing the population by their actions. The branching process idea is to make this vague idea of "individual initiative" precise by sharpening it into the requirement of stochastic independence between individuals. This is proper as an idealization, but obviously takes us far from reality. In some cases that can be remedied, as in the models considered by Méléard, Champagnat, and Lambert, in our recent papers, or in the population size dependence studied by [32], Klebaner, and others, which allows an understanding of the linear growth occurring in the famous polymerase chain reactions, PCR [17]. But a real liberation from independence, e.g., replacing it by conditional (given population history) exchangeability in some form, yet remains out of reach.

This overview has centered on my own interests, branching processes as a form of theoretical biology. In particular it has focused on the supercritical case, which was the main interest of my own expansive youth. Several of my recent papers, befittingly, deal with the path and time to extinction [27] and the general issue of persistence versus extinction [25]. However, most of the revival branching processes and related areas experienced in the 1990s, and which continues to this day has a different character. Mainly it is purely mathematical; partly it is inspired by computer algorithms. The whole area of super-processes and measure-valued

Markov branching processes would belong to the former realm, whereas the study of random trees though certainly pure mathematics also has drawn upon both phylogenetics and computer science. But these are areas where others have more insight than I.

References

1. Athreya, K. B., & Ney, P. E. (1972). *Branching processes*. Berlin: Springer.
2. Bienaymé, I. J. (1845). De la loi de multiplication et de la durée des familles. *Société philomathique de Paris Extraits, 5,* 37–39.
3. Bru, B., Jongmans, F., & Seneta, E. (1992). I.J. Bienaymé: Family information and proof of the criticality theorem. *International Statistical Review, 60,* 177–183.
4. Bühler, W. J. (1971). Generations and degree of relationship in supercritical Markov branching processes. *Zeitschrift für Wahrscheinlichkeitstheorie und Verwandte Gebiete, 18,* 141–152.
5. Bühler, W. J. (1972). The distribution of generations and other aspects of the family structure of branching processes. In *Proceedings of the Sixth Berkeley Symposium on Mathematical Statistics and Probability, Volume 3: Probability Theory* (pp. 463–480). Berkeley: University of California Press.
6. Champagnat, N. (2006). A microscopic interpretation for adaptive dynamics trait substitution sequence models. *Stochastic Processes and their Applications, 116,* 1127–1160.
7. Champagnat, N., Ferrière, R., & Méléard, S. (2006). Unifying evolutionary dynamics: From individual stochastic processes to macroscopic models. *Theoretical Population Biology, 69,* 297–321.
8. Champagnat, N., Ferrière, R., & Méléard, S. (2008). Individual-based probabilistic models of adaptive evolution and various scaling approximations. In *Seminar on stochastic analysis, random fields and applications V. Progress in probability* (Vol. 59, pp. 75–113). Basel: Springer.
9. Champagnat, N., & Lambert, A. (2007). Evolution of discrete populations and the canonical diffusion of adaptive dynamics. *The Annals of Applied Probability, 17,* 102–155.
10. Cournot, A. A. (1847). *De l'origine et des limites de la correspondence entre l'algèbre at la géométrie*. Paris: Hachette.
11. Crump, K. S., & Mode, C. J. (1968). A general age-dependent branching process I. *Journal of Mathematical Analysis and Applications, 24,* 494–508.
12. Crump, K. S., & Mode, C. J. (1969). A general age-dependent branching process II. *Journal of Mathematical Analysis and Applications, 25,* 8–17.
13. Dieckmann, U., & Doebeli, M. (1999). On the origin of species by sympatric speciation. *Nature, 400,* 354–357.
14. Dieckmann, U., & Law, R. (1996). The dynamical theory of coevolution: A derivation from stochastic ecological processes. *Journal of Mathematical Biology, 34,* 579–612.
15. Euler, L. (1767). Recherches génerales sur la mortalité et la multiplication du genre humain. *Memoires de l'academie des sciences de Berlin, 16,* 144–164.
16. Fahlbeck, P. E. (1898). *Sveriges adel: statistisk undersökning öfver de å Riddarhuset introducerade ätterna (The Swedish nobility, a statistical investigation of the families of the house of nobility)* (Vols. 1–2). Lund: C. W. K. Gleerup.
17. Haccou, P., Jagers, P., & Vatutin, V. A. (2005). *Branching processes: Variation, growth, and extinction of populations*. Cambridge: Cambridge University Press.
18. Harris, T. E. (1963). *The Theory of Branching Processes*. Berlin: Springer. Reprinted by Courier Dover Publications, 1989.
19. Heyde, C. C., & Seneta, E. (1977). *I.J. Bienaymé: Statistical theory anticipated*. New York: Springer.

20. Iosifescu, M., Limnios, N., & Oprisan, G. (2007). *Modèles stochastiques*. Paris: Hermes Lavoisier.
21. Jagers, P. (1969). A general stochastic model for population development. *Scandinavian Actuarial Journal, 1969*, 84–103.
22. Jagers, P. (1975). *Branching processes with biological applications*. London: Wiley.
23. Jagers, P. (1982). How probable is it to be firstborn? and other branching process applications to kinship problems. *Mathematical Biosciences, 59*, 1–15.
24. Jagers, P. (1989). General branching processes as Markov fields. *Stochastic Processes and their Applications, 32*, 183–212.
25. Jagers, P. (2011). Extinction, persistence, and evolution. In F. A. Chalub & J. F. Rodrigues (Eds.) *The mathematics of Darwin's legacy* (pp. 91–104). Basel: Springer.
26. Jagers, P., & Klebaner, F. C. (2011). Population-size-dependent, age-structured branching processes linger around. *Journal of Applied Probability, 48*, 249–260.
27. Jagers, P., Klebaner, F. C., & Sagitov, S. (2007). On the path to extinction. *Proceedings of the National Academy of Sciences of the United States of America, 104*, 6107–6111.
28. Jagers, P., & Nerman, O. (1996). The asymptotic composition of supercritical, multi-type branching populations. In *Séminaire de Probabilités XXX* (pp. 40–54). Berlin: Springer.
29. Jagers, P., & Sagitov, S. (2008). General branching processes in discrete time as random trees. *Bernoulli, 14*, 949–962.
30. Joffe, A., & Waugh, W. A. O. (1982). Exact distributions of kin numbers in a Galton-Watson process. *Journal of Applied Probability, 19*, 767–775.
31. Kendall, D. G. (1948). On the generalized "birth-and-death" process. *Annals of Mathematical Statistics, 19*, 1–15.
32. Kersting, G. (1992). Asymptotic Gamma distributions for stochastic difference equations. *Stochastic Processes and their Applications, 40*, 15–28.
33. Klebaner, F. C., Sagitov, S., Vatutin, V. A., Haccou, P., & Jagers, P. (2011). Stochasticity in the adaptive dynamics of evolution: The bare bones. *Journal of Biological Dynamics, 5*, 147–162.
34. Lotka, A. J. (1934). *Théorie analytique des associations biologiques* (Vol. 1). Paris: Hermann.
35. Lotka, A. J. (1939). *Théorie analytique des associations biologiques* (Vol. 2). Paris: Hermann.
36. Méléard, S., & Tran, C. V. (2009). Trait substitution sequence process and canonical equation for age-structured populations. *Journal of Mathematical Biology, 58*, 881–921.
37. Metz, J. A., Geritz, S. A., Meszéna, G., Jacobs, F. J., & Van Heerwaarden, J. S. (1996). Adaptive dynamics, a geometrical study of the consequences of nearly faithful reproduction. In S. J. van Strien & S. M. Verduyn Lunel (Eds.), *Stochastic and spatial structures of dynamical systems* (Vol. 45, pp. 183–231). Amsterdam: North-Holland.
38. Mode, C. J. (1971). *Multitype branching processes: Theory and applications*. New York: Elsevier.
39. Nerman, O., & Jagers, P. (1984). The stable doubly infinite pedigree process of supercritical branching populations. *Zeitschrift für Wahrscheinlichkeitstheorie und Verwandte Gebiete, 65*, 445–460.
40. Neveu, J. (1986). Arbres et processus de Galton-Watson. *Annales de l'Institut Henri Poincaré Probabilités et Statistiques, 2*, 199–207.
41. Sevastyanov, B. A. (1971). *Vetvyashchiesya Protsessy (Branching processes)*. Moscow: Nauka.
42. Steffensen, J. F. (1930). Om Sandssynligheden for at Afkommet uddør. *Matematisk Tidsskrift B*, 19–23.
43. Watson, H. W., & Galton, F. (1875). On the probability of the extinction of families. *Journal of the Anthropological Institute of Great Britain and Ireland, 4*, 138–144.
44. Yakovlev, A. Y., & Yanev, N. M. (1989). *Transient processes in cell proliferation kinetics. Lecture notes in biomathematics* (Vol. 82). Berlin: Springer.

Principles of Mathematical Modeling in Biomedical Sciences: An Unwritten Gospel of Andrei Yakovlev

Leonid Hanin

Abstract This article describes Dr. Andrei Yakovlev's unique philosophy of mathematical modeling in biomedical sciences. Although he never formulated it in a systematic way, it has always been central to his work and manifested amply in the course of the author's 22-year research collaboration with this visionary scholar. We address methodological tensions between mathematics and biomedical sciences, epistemological status of mathematical models, and various methodological questions of a more practical nature arising in mathematical modeling and statistical data analysis including model selection, model identifiability, and concordance between the model and the observables.

The vast legacy of Andrei Yakovlev does not only consist of ideas, methods, and results scattered over his more than 200 papers and books and generously disseminated in his lectures. These were precious fruits on the tree of his life; the tree itself, however, was fed by the root system of his intellect, personality, ethics, and philosophy of science. The roots were mostly hidden, and only his family, close friends, and co-workers could at times see their manifestation. I was privileged to be his friend and collaborator for 22 years, and I feel a pressing duty to write about one of the focal points of the inner laboratory of his mind—his philosophy of mathematical modeling.

Andrei belonged to the generation that became active in the 1960s and 1970s and has a paradoxical mindset that is idealistic and technocratic at the same time. Embodying this mindset, he believed that a day will come when biologists and medical doctors will work side by side with mathematicians and statisticians and be guided by mathematical models. Only time will tell whether this almost

---

Dedicated to the memory of Andrei Yakovlev (1944–2008).

L. Hanin (✉)
Department of Mathematics and Statistics, Idaho State University, Pocatello, ID, USA
e-mail: hanin@isu.edu

© Springer Nature Switzerland AG 2020
A. Almudevar et al. (eds.), *Statistical Modeling for Biological Systems*,
https://doi.org/10.1007/978-3-030-34675-1_19

biblical vision is prophetic or utopian. Whatever the truth may be, this dream made methodology of mathematical modeling in biomedical sciences a constant presence on Andrei's intellectual radar screen. Andrei never articulated his thoughts about mathematical modeling in any systematic way, perhaps leaving this endeavor for a future time. His sparks of insight into this subject are dispersed over many of his papers, books, and grant proposals. His modeling ideas originated from a variety of sources: his research experience, trials and errors, the work of others, and his ongoing, sometimes heated and emotional, debates with his colleagues, co-workers, and disciples. Perhaps even more importantly, they were an extension of his personality, the historical and cultural circumstances of his life, and his worldview.

Over more than two decades Andrei and I discussed the methodology of mathematical modeling and associated statistical issues on several dozen occasions. Quite often, I debated on the side of pure mathematics, arguing that abstract mathematical ideas may by themselves, due to their depth and generality, capture some important aspects of biological reality. He was amazingly receptive to such arguments, which is almost unbelievable for someone who started his professional life as an M.D., then went on to obtain a Ph.D. in biology, and only later became a Sc.D. in mathematics. However, his main concern was for uncovering the laws of nature operative in the biomedical milieu, and I must confess that in our joint work it was usually I who had to walk 80% of the way toward common methodological ground.

What follows are my thoughts on mathematical modeling in biology and medicine inspired by our discussions. Some of them are interpretations and extrapolations of Andrei's ideas that came to light in our collaborations and debates. Of necessity, they are filtered through the prism of my own vision, and any deviation from what he could have intended is my sole responsibility. He himself would probably have formulated them quite differently, and undoubtedly in a more coherent way.

1 Mathematics and Biology: A Tough Marriage

Let me begin with a famous quote from Eugene Wigner, a 1963 Nobel laureate in physics. In his 1960 article entitled *The Unreasonable Effectiveness of Mathematics* [8], he wrote:

> The miracle of the appropriateness of the language of mathematics for the formulation of the laws of physics is a wonderful gift, which we neither understand nor deserve.

This prompts a natural question: does Wigner's observation about the language of mathematics hold true for biology, a field that now occupies an even more prominent position within the natural sciences than physics occupied some 100, or even 50 years ago? From my experience working with biologists and attending numerous conferences where mathematicians and biomedical scientists have met face to face, the answer is decidedly negative. In spite of considerable mutual attraction—and even greater drive on the part of biology and medicine to become more quantitative—there is a certain degree of separation and tension between the two sciences, cultures, and intellectual worlds of mathematics and biology. Summarizing his extensive biological experience and contrasting the logic of mathematics with that of biology, I.M. Gelfand once remarked in his characteristically blunt manner that "in biology, quite a different logic is at work; it does not require mathematics far beyond multiplication table." It should not come, then, as a big surprise that no prominent biologists are to be found among the most high powered mathematicians nor has the reverse ever occurred. By contrast, there is no dearth of great mathematicians (such as Euler, Gauss, Lagrange, Poincaré, and von Neumann) who made fundamental contributions to physics, and likewise, many renowned physicists were first-rate mathematicians (think of Archimedes, Newton, Laplace, Fourier and, closer to our time, Wigner, to name just a few). Interestingly enough, the contribution of physicists to biology was also critical: the advent of modern molecular biology and genetics would have been substantially delayed if it were not for Schrödinger, Delbrück, and Gamov. Finally, as far as the impact of biologists on quantitative sciences other than mathematics is concerned, the most notable example is probably R.A. Fisher, a geneticist and evolutionary biologist who became one of the founders of modern mathematical statistics.

The most typical misgivings that biologists and medical doctors have about mathematical models is that they are not comprehensive and detailed enough to encompass all aspects of biomedical processes that these models seek to describe. However, as Andrei used to emphasize, a mathematical model of a specific biological process is not intended to incorporate all contemporary biological findings, concepts, and hypotheses. To be useful, it must capture the most salient features of the process being studied while leaving inessential details aside. In this way, a mathematical model helps to purify, generate, and validate only the central and most basic biological concepts and hypotheses.

A general principle driving empirical sciences (and minds) is, "the more details the better." However, in many cases, knowledge of detailed mechanisms is quite unnecessary. For example, consider modeling the response of a homogeneous cell population to fractionated radiation, assuming that—as it happens in a typical clinical setting—successive doses are separated by at least 1 day. In principle, the event of cell survival depends on a whole host of factors, the most prominent probably being the action of DNA repair mechanisms. These are enzymatic biochemical processes that allow cells to repair (and sometimes misrepair) various types of radiation-induced lesions to one or several DNA bases, as well as DNA double-strand breaks. Damage repair processes, however, operate on the time scale of minutes to hours, so that by the time of the next exposure all transient processes of inactivation and recovery of damaged cells will have been completed. As a result, no matter how complex the underlying biological processes may be, cell survival can be characterized with reasonable accuracy by a single parameter: cell survival probability as a function of the radiation dose. Empirical scientists might dismiss such arguments as "mathematical tricks." And yet, such "tricks," based on fundamental biological insights, lie at the heart of successful mathematical modeling.

The greatest paradox of modern biology is that, although it claims to be reducible to chemistry and physics, it has never formulated its laws in such generality and with such rigor and quantitative precision as to even approach those of chemistry, and especially physics. One can even argue that the laws of biology simply do not exist, if by "law" we mean something simple, general, precise, and immutable. Every known rule in biology has exceptions and intricacies. Consider, for example, the two most general principles of genetics and molecular biology: (1) that genes encode proteins, and (2) that triplets of successive RNA bases (codons) code for amino acids. Alas, even they are not universal, and are more complex than they may seem. In fact, the same gene may encode many different proteins as a result of at least two mechanisms—alternative splicing and the presence of multiple promoters. The correspondence between codons and amino acids is also quite tricky: first, the three end-of-transcription codons (UAA, UAG, UGA) do not code for amino acids; second, the remaining 61 codons code for 20 amino acids, so that more than one codon (actually, up to six codons) may encode the same amino acid; and finally, some species utilize slightly modified genetic codes. As a result, the spirit of basic biological rules (1) and (2) is quite different from that of the laws of physics and chemistry as we know them; unlike the latter laws of nature, the former rules resemble somewhat arbitrary (or so they seem) instructions for a discrete automaton.

Still, are there universal laws of life? Is there fundamental simplicity lurking behind the enormous diversity and complexity of biological phenomena? I believe the answer to both questions is affirmative. But to uncover these simple and irreducible first principles of living organisms, biology has to part ways with reductionism, accept the existence of non-physical forms of life-specific energy, and discover the laws by which they work. (Having written this, I can almost see Andrei frowning at me for such a suggestion.) Only then will mathematics be able to demonstrate the full power of its language, ideas, and methods in the biomedical arena. But until then, mathematicians should be very careful in selecting biological objects to model, so as to combine biomedical relevance with mathematical tractability.

2 Mathematical Models as Axiomatic Systems

For a natural scientist, all theoretical constructs—concepts, principles, rules, methods, hypotheses, interpretations—are liable to change as new experimental findings emerge that do not fit the old theories. These changes of theoretical framework constitute the evolution (and sometimes revolutions) of science. This only makes it more tempting to apply this paradigm of change to mathematical models. Over the last 25 years, I have served as a reviewer to more than 30 biomathematical, biostatistical, and biomedical journals. During that time, I encountered many instances where natural scientists and half-baked mathematicians had dealt frivolously with mathematical models by changing, suppressing, or adding model assumptions in the process of model building or deriving model-based results. It is precisely such manipulations that could invalidate the inferences drawn from mathematical models. These manipulations misconstrue the nature and function of mathematical modeling and account, in part, for the bad press that mathematics sometimes receives within biomedical sciences and beyond. Once model assumptions are formulated and agreed upon, the model becomes an axiomatic system from which implications are derived with the same mathematical rigor one would expect in pure mathematics. Of course, the model assumptions may change as new data, theories, and hypotheses come to light. This change, in turn, will engender new models and new axiomatic systems.

Unlike natural sciences, mathematics—when done correctly—makes a transition from assumptions (axioms) to conclusions (theorems) with impeccable accuracy, and only this makes mathematical models a solid rock in the sea of natural sciences. The main methodological difference between them is that in the natural sciences, at least in principle, every conclusion, theoretical result, or conjecture is open to experimental verification, while in mathematics such an empirical criterion of truth simply does not exist. The only tool available to a mathematician is rigorous non-contradictory arguments based on formal logic or its intuitive counterparts.

According to Aristotle, nature abhors a vacuum; so does science. The gap between mathematics and natural sciences is gradually being filled through the use of numerical analysis, algorithms and computational tools, theoretical computer science, and statistics. Some of their methods and results are quite rigorous and belong within the realm of mathematics. Others are based on intuition and computational experience rather than proofs, and thus are more heuristic; essentially, they are a part of the natural sciences, except for their experiments, that are done in silico. Over time, these heuristic methods will be formalized, and rigorous results will emerge about their domain of applicability, rates of convergence, computational complexity, optimality, statistical power, and properties of their solutions.

Scientists willing to engage in mathematical modeling should be proficient with logic, axiomatic systems, and mathematics in general; also, they must have a rigorous training in probability and statistics. Practically, this amounts to having at least a Master's degree in mathematics or its autodidactic equivalent. They should learn and know mathematics because, unlike empirical sciences, it teaches us to look for the ultimate reasons behind all things in the Universe. Let me illustrate this point with an example drawn from my experience as a reviewer. Many mathematical models and statistical methods employ Markov chains. The defining Markov property states that the probability of any future event given the present and the past is equal to the probability of the future event given the present alone. Those who have not had adequate training in probability and stochastic processes believe that this loose statement is the correct definition of the Markov property. Wrong! A critical missing assumption here is that the "present" is a *particular* state of the Markov chain, rather than an arbitrary event from the σ-algebra of the present. Here is a simple example showing that the "naïve" definition is false. Consider a Markov chain $\{X_n : n \geq 0\}$ on a two-element state space $\{a, b\}$ with the identity transition probability matrix and the uniform initial distribution. Then $Pr(X_2 = a \mid X_1 = a$ or $b, \ X_0 = a) = 1$ while $Pr(X_2 = a \mid X_1 = a$ or $b) = 0.5$.

3 A Mathematical Modeler's Dilemma: Deterministic or Stochastic?

As a medical doctor and cell biologist by primary training, Andrei dealt with various unobservable critical microevents, inter alia: spontaneous mutations or those caused by radiation or carcinogens, formation of misrepaired or irreparable DNA lesions and chromosomal aberrations, malignant transformation of cells, events of metastasis shedding by the primary tumor, and inception of metastases at various secondary sites. After a period of time during which the effects of these events remain latent—hours, days, months, years, and sometimes decades—these events bring about observable outcomes such as cell death, clinically manifest disease, detectable primary cancer, its recurrence, or metastasis. Clearly, the only way to establish quantitative relationships between unobservable microevents and their remote outcomes is through mathematical modeling. Parameters of such models include the timing of critical microevents and their rates of occurrence, as well as rates and other characteristics of various intermediate processes. When contemplating such a model, one critical question is whether the model should be deterministic or stochastic. The enormous diversity, variability, and flexibility of biological processes and numerous sources of uncertainty and noise led Andrei consistently to favor stochastic models. Always keeping an eye on elucidating biological mechanisms, he had mechanistic models as his first choice. However, in many cases, biological mechanisms are too complex to be accounted for explicitly, or may be largely unknown, gene expression being a typical example. Here, Andrei would call upon his other lifetime passion: statistical models and methods.

Undoubtedly, there are many areas where deterministic models, usually based on ordinary or partial differential equations, provide an accurate description of biomedical processes: biochemical reactions, diffusion, transport phenomena, motility, photo- and chemotaxis, transmission of neural signals, blood circulation, imaging, etc. Conversely, there are several important areas where stochastic models are indispensable:

1. Biological processes influenced by inherently stochastic quantum effects. As an example, this is the case in microdosimetry, which accounts for the stochasticity of specific energy deposition to sensitive sites within a cell and produces accurate models of irradiated cell survival [3].
2. Occurrence of all kinds of low probability—high impact microevents.

3. Kinetics of randomly changing discrete quantities, including dynamics of small populations.
4. Processes whose mechanisms have yet to be discovered.

In general, stochastic models are more flexible (and partly because of this, more parsimonious) than more mechanistically driven and rigid deterministic models that describe the same phenomena. Also, stochastic models lend themselves to a larger variety of statistical methods of fitting, parameter estimation, and hypothesis testing than the corresponding deterministic models, whose fitting to data is essentially limited to regression (typically, in the least squares setting). On balance, however, I think that the dichotomy between deterministic and stochastic modeling is not as clear cut as it is thought to be. One reason is that deterministic models are quite often augmented with stochastic components. For example, biochemical reactions of the complement system may be combined with information about the distribution of the random number of antigens on the membrane of pathogenic cells; photo- and chemotaxis models contain Brownian motion terms; and in general, many deterministic models have stochastic terms accounting for various sources of random fluctuation, noise, and measurement errors. Another reason is that many deterministic models arise as approximations to certain stochastic models. As a typical example, a deterministic model may describe the time course of the expected value of some quantity whose evolution is governed by a stochastic process that grounds a more general stochastic model. The latter model, then, reveals not only the evolution of the mean, but also of the variance, correlation structure, and other characteristics of the process in question. In such cases, however, the deterministic approximation may prove to be more computationally tractable.

4 A Devil in the Corner: Model Non-identifiability

In November 1995, Andrei called my office in Detroit to discuss the two-stage model of carcinogenesis. This elegant mechanistic model developed by Moolgavkar, Venzon, and Knudson (the MVK model) [6, 7] posits the following two-stage process of spontaneous or induced carcinogenesis: first, susceptible stem cells acquire one pre-malignant mutation according to a Poisson process with rate ν, forming a population of *initiated cells* that undergoes a birth and death Markov process of clonal expansion with rates of division α and death (or differentiation) β; second, initiated cells acquire through asymmetric division with rate μ the second mutation that converts them into fully malignant cells that grow into detectable tumors over some deterministic or random time. Since this time is typically small relative to the time to appearance of the first malignant cell, the latter can be viewed as the observed time to tumor. The distribution of this time was computed in closed form as a function of parameters ν, α, β, μ [5, 9]. This computation opened up the prospect of estimating these unobservable yet biologically important parameters from time to tumor data. No attempts, however, have succeeded in doing so, and

Andrei wondered why. His intuition was that the problem lay not in the set-up of the experiments or the statistical data analysis but in the structure of the model itself. On the same day, I called him back with the results: even if the distribution of the time to the first malignant cell is fully known, the four model parameters are not jointly identifiable from this distribution; in fact, only three combinations of these four parameters can be computed. These results were published in [2] (and in a more heuristic way, independently discovered in [4]). That was the first of my many encounters with the phenomenon of model non-identifiability.

Consider a deterministic or stochastic model that depends on a parameter set θ and produces a certain output (which can be a number, a function, or a probability distribution). If any two different parameter sets lead to distinct outputs, the model is called *identifiable*. In other words, a model is non-identifiable if there exist two distinct parameter sets that generate the same output (typically, in this case there is an infinite set of vectors of model parameters giving rise to the same output). Thus, model identifiability is an intrinsic, structural property of the model and is only indirectly related to estimation of model parameters. The relation is that, for a non-identifiable model, no method whatsoever would allow us to compute or estimate model parameters from the observed output while model identifiability may lead, at least in principle, to the possibility of parameter estimation. However, the quality of such an estimation depends on the extent and precision of our knowledge of the output (which is usually partial and approximate) and the nature of the method employed.

Let me illustrate the notion of model non-identifiability with a simple example. Imagine a mass m on an ideal spring with spring constant k. In the absence of air resistance, the oscillation of the mass is described by the function

$$X(t) = x_0 \cos \omega t + \frac{v_0}{\omega} \sin \omega t, \quad \omega = \sqrt{\frac{k}{m}},$$

where $X(t)$ is the position of the mass at time t, relative to the position of static equilibrium, $x_0 = X(0)$ is the initial position of the mass, and $v_0 = X'(0)$ is its initial velocity. The model of oscillations depends on parameters m, k, x_0, v_0. However, it is clear that no observation of the oscillation process would allow us to determine parameters m and k separately; what can only be determined is their combination $\omega^2 = k/m$. Thus, the model is non-identifiable, and the simplest set of its identifiable parameters is $k/m, x_0, v_0$.

As shown in [1], every non-identifiable model may be re-parameterized to become identifiable, and its original set of n parameters can be expressed as functions of the new set of m identifiable parameters, where $m \leq n$. In some particular cases, as in the above example of simple harmonic oscillation, some of the model parameters are identifiable from the output, while for other parameters, only their simple combinations (such as ratios, sums, or products) are identifiable. A more common situation is that only complex combinations of many or all model parameters allow identification, and hence estimation, from the observed output. Such overparameterized models have little utility for furthering our understanding of biomedical processes they intend to describe, and should therefore be avoided. It is also worth noting that non-identifiability does not preclude assessing a model's adequacy by computing its goodness of fit to a given sample of output values. But in such a case, computing the best fitting set of original parameters is impossible.

In summary, identifiability sets up a natural limit to the complexity of a model used to describe a given set of output variables. If the number of model parameters exceeds some critical value beyond which the model becomes non-identifiable, its utility drops off. This serves as a good illustration of the principle of parsimony that Andrei used as a guide in his scientific endeavors.

5 Mathematical Models and Biological Reality

As much as Andrei loved mathematical modeling, the idea of modeling for its own sake was totally foreign, if not repellent, to him. His chief concern was whether a model had a chance of generating new biomedical knowledge. If for any reason this was unlikely, he would not attempt modeling in the first place. He never thought highly of models of a specific biological phenomenon that arose from "general considerations," calling them "mathematical weeds." He strongly believed and passionately advocated that every modeling effort should start with an in-depth exploration of the biological or medical problem at hand by reading original literature, including experimental works, and consulting with biological and medical experts.

A principal touchstone for the feasibility of mathematical modeling is availability of experimental or observational data related to the effects that the model is meant to describe. Moreover, the quantity and structure of biological data determine the model's outputs, mathematical nature, and complexity. As Andrei put it once (and over time, I came to embrace this notion as well), "A mathematical model should be tailored to observations available in a given biomedical setting." In essence, a mathematical model should provide a formula, or algorithm, for (the probability of) a generic observation. If observations are sparse or subject to large errors, a detailed multi-parametric model would only be inadequate for one of two reasons: either it would be non-identifiable or the estimates of model parameters would not be reliable enough due to large variances. Thus, the data that a model is designed to describe serves as an integral part of the model. This observation suggests a means of overcoming model non-identifiability: to extend the set of model outputs while keeping the parameter set unchanged, so that the model becomes identifiable. This tack requires, of course, an appropriate extension of the set of observables.

It is also important to ensure that parameters incorporated into the model under consideration are estimable by means of modern statistical techniques. Whether or not this requirement can be met depends not only on the model structure but also on the availability of experimental data adequate for estimation purposes. In our joint research with Andrei, on many more occasions than we would be merely amused by, this seemingly trivial requirement proved to be a major impediment to our modeling efforts. Some experimental data sets failed to form a random sample from a probability distribution because of either stochastic dependence or heterogeneity of underlying distributions. In many data sets, the replication number of the experiments or observations was not large enough to produce any statistically significant estimates of model parameters. Other data sets rendered even the most simplistic models for the corresponding theoretical output non-identifiable. Finally, in many instances, data sets were so complex that they resisted description by any known mathematically or computationally tractable model. It was sad and vexing to witness how laborious, costly experiments resulted in no new biological knowledge because mathematical and statistical aspects of the problem were ignored.

My foregoing remarks point to the following conclusion: for biomedical experiments to be useful, those who plan them should consult mathematicians and statisticians prior to making any decisions about the design of the experiments. Andrei's vision of mathematicians and biomedical scientists working in concert is not that utopian, after all. Quite the contrary, it is an imperative for the increasingly quantitative world of biomedicine.

In mathematical modeling of biomedical processes, the main challenge is to strike a delicate balance between biological realism of the proposed model and its mathematical (or at least computational) feasibility and identifiability. To gauge if a given model exhibits the right balance, two benchmarks should be invoked. One is the model's adequacy and predictive power—tangible characteristics measured by the goodness of fit to the data. The other is intangible and elusive, yet more important: a model must meet the criteria of simplicity, symmetry, parsimony, generality, and beauty. Creating such a model is an art open only to a few genuine masters. One of them was Andrei Yakovlev.

Acknowledgements I would like to thank my wife Marina and my sons Mark and Boris for numerous suggestions on the substance and style of this paper. Their fond recollections of Andrei Yakovlev served as emotional markers that helped me recreate his thoughts about mathematical modeling. I am also grateful to Jack Hall for his insightful comments.

References

1. Hanin, L. G. (2002). Identification problem for stochastic models with application to carcinogenesis, cancer detection and radiation biology. *Discrete Dynamics in Nature and Society, 7*, 177–189.
2. Hanin, L. G., & Yakovlev, A. Y. (1996). A nonidentifiability aspect of the two-stage model of carcinogenesis. *Risk Analysis, 16*, 711–715.
3. Hanin, L. G., & Zaider, M. (2010). Cell-survival probability at large doses: An alternative to the linear-quadratic model. *Physics in Medicine & Biology, 55*, 4687–4702.
4. Heidenreich, W. F. (1996). On the parameters of the clonal expansion model. *Radiation and Environmental Biophysics, 35*, 127–129.
5. Kopp-Schneider, A., Portier, C. J., & Sherman, C. D. (1994). The exact formula for tumor incidence in the two-stage model. *Risk Analysis, 14*, 1079–1080.

6. Moolgavkar, S. H., & Knudson, A. G. (1981). Mutation and cancer: a model for human carcinogenesis. *Journal of the National Cancer Institute, 66,* 1037–1052.
7. Moolgavkar, S. H., & Venzon, D. J. (1979). Two-event models for carcinogenesis: incidence curves for childhood and adult tumors. *Mathematical Biosciences, 47,* 55–77.
8. Wigner, E. P. (1960). The unreasonable effectiveness of mathematics in the natural sciences. *Communications on Pure and Applied Mathematics, 13,* 1–14.
9. Zheng, Q. (1994). On the exact hazard and survival functions of the MVK stochastic carcinogenesis model. *Risk Analysis, 14,* 1081–1084.

Appendix: Publications of Andrei Yakovlev

Books

1. Yakovlev, A.Y. and Zorin, A.V. *Computer Simulation in Cell Radiobiology*, Series "Lecture Notes in Biomathematics", vol. 74, Berlin-Heidelberg-New York, Springer-Verlag, **1988**.
2. Yakovlev, A.Y. and Yanev, N.M. *Transient Processes in Cell Proliferation Kinetics*, Series "Lecture Notes in Biomathematics", vol. 82, Berlin-Heidelberg-New York, Springer-Verlag, **1989**.
3. Hanin, L.G., Pavlova, L.V. and Yakovlev, A.Y. *Biomathematical Problems in optimization of Cancer Radiotherapy*, CRC Press, Florida, USA, **1994**.
4. Yakovlev, A.Y. and Tsodikov, A.D. *Stochastic Models of Tumor Latency and Their Biostatistical Applications*, World Scientific Publ., Singapore, **1996**.

Peer-Reviewed Papers Published in English

1. Yakovlev, A.Yu., Zorin, A.V. and Isanin, N.A. The kinetic analysis of induced cell proliferation, *J. Theoretical Biology*, **1977**, vol. 64, pp. 1–25.
2. Yakovlev, A.Yu., Malinin, A.M., Terskikh, V.V. and Makarova, G.F. Kinetics of induced cell proliferation at steady-state conditions of cell culture, *Cytobiologie*, **1977**, vol. 14, pp. 279–283.
3. Yakovlev, A.Y., Hanson, K.P., Lepekhin, A.F., Zorin, A.V. and Zhivotovsky, B.D. Probabilistic characteristics for the life-span of RNA molecules. Kinetic analysis of RNA specific radioactivity, *Studia Biophysica*, **1982**, vol. 88, pp. 195–212.
4. Yanev, N.M. and Yakovlev, A.Yu. On the distribution of marks over a proliferating cell population obeying the Bellman-Harris branching process, *Mathematical Biosciences*, **1985**, vol. 75, pp. 159–173.

© Springer Nature Switzerland AG 2020
A. Almudevar et al. (eds.), *Statistical Modeling for Biological Systems*,
https://doi.org/10.1007/978-3-030-34675-1

5. Dimitrov, B.N., Rachev, S.T. and Yakovlev, A.Yu. Maximum likelihood estimation of the mortality rate function, *Biometrical J.*, **1985**, vol. 27, pp. 317–326.

6. Kadyrova, N.O. and Yakovlev, A.Yu. On the parametric approach to survival data analysis. I. Radiobiological applications, *Biometrical J.*, **1985**, vol. 27, pp. 441–451.

7. Zorin, A.V. and Yakovlev, A.Yu. The properties of cell kinetics indicators. A computer simulation study, *Biometrical J.*, **1986**, vol. 28, pp. 347–362.

8. Tanushev, M., Myasnikova, E.M. and Yakovlev, A.Yu. A simple survival model for combined action of two damaging agents, *Biometrical J.*, **1987**, vol. 29, pp. 93–102.

9. Zorin, A.V. and Yakovlev, A. Simulation of in vitro kinetics of normal and irradiated cells. I. Description and substantiation of a simulation model, *Syst. Anal. Model. Simul.*, **1987**, vol. 4, pp. 133–145.

10. Zorin, A.V., Zherbin, E.A. and Yakovlev, A.Yu. Simulation of in vitro kinetics of normal and irradiated cells. II. Analysis of experimental evidence, *Syst. Anal. Model. Simul.*, **1987**, vol. 4, pp. 147–163.

11. Rachev, S.T., Yakovlev, A.Yu., Kadyrova, N.O. and Myasnikova, E.M. On the statistical inference from survival experiments with two types of failure, *Biometrical J.*, **1988**, vol. 30, pp. 835–842.

12. Yakovlev, A.Yu. and Rachev, S.T. Bounds for crude survival probabilities within competing risks framework and their statistical application, *Statistics and Probability Letters*, **1988**, vol. 6, pp. 389–394.

13. Rachev, S.T. and Yakovlev, A.Yu. Theoretical bounds for the tumor treatment efficacy, *Syst. Anal. Model. Simul.*, **1988**, vol. 5, pp. 37–42.

14. Gusev, Yu.V. and Yakovlev, A.Yu. Simulation modeling of controlled cellular systems, *Automation and Remote Control*, **1988**, vol. 49, Part 2, pp. 914–921.

15. Rachev, S.T. and Yakovlev, A.Yu. Bounds for the probabilistic characteristics of latent failure time within the competing risks framework, *SERDICA Bulgaricae mathematicae publicationes*, **1988**, vol.14, pp. 325–332.

16. Rachev, S.T., Yakovlev, A.Yu. and Kadyrova, N.O. A concept of δ – stochastic ordering and its application to the competing risks model, *SERDICA Bulgaricae mathematicae publicationes*, **1989**, vol. 15, pp. 28–31.

17. Rachev, S.T., Kadyrova, N.O. and Yakovlev, A.Y. Maximum likelihood estimation of the bimodal failure rate for censored and tied observations, *Statistics*, **1989**, vol. 20, pp. 135–140.

18. Kadyrova, N.O., Rachev, S.T. and Yakovlev, A.Y. Isotonic maximum likelihood estimation of the bimodal failure rate – a computer-based study, *Statistics*, **1989**, vol. 20, pp. 271–278.

19. Gusev, Yu.V., Yakovlev, A.Yu. and Yanev, N.M. Computer simulation of cell renewal in small intestine of intact and irradiated animals, *Syst. Anal. Model. Simul.*, **1989**, vol. 6, pp. 293–307.

20. Pavlova, L.V., Hanin, L.G. and Yakovlev, A.Yu. Accurate upper bounds for the functional of effectiveness of tumor radiation therapy, *Automat. Remote Control*, **1990**, vol. 51, Part 2, pp. 973–982.

21. Goot, R.E., Rachev, S.T., Yakovlev, A.Yu., Kadyrova, N.O. and Zharinov, G.M. Some statistical tests associated with a concept of δ – stochastic ordering of random variables, *SERDICA Bulgaricae mathematicae publicationes*, **1990**, vol. 16, pp. 240–245.

22. Tsodikov, A.D. and Yakovlev, A.Y. On the optimal policies in cancer screening, *Mathematical Biosciences*, **1991**, vol. 107, pp. 21–45.

23. Pavlova, L.V., Khanin, L.G. and Yakovlev, A.Yu. Optimal fractionation of radiation dose by the criterion of differential survival probabilities of normal and tumor cells, *Automat. Remote Control*, **1991**, vol. 52, Part 2, pp. 257–265.

24. Tsodikov, A.D. and Yakovlev, A.Y. Optimal preventive examination strategy with competing risks, *Automat. Remote Control*, 1991, vol. 52, Part 2, pp. 385–391.

25. Yakovlev A., Tsodikov A.D. and Bass L. A stochastic model of hormesis, *Mathematical Biosciences*, **1993**, vol. 116, pp. 197–219.

26. Klebanov L.B., Rachev S.T. and Yakovlev A.Yu. A stochastic model of radiation carcinogenesis: Latent time distributions and their properties, *Mathematical Biosciences*, **1993**, vol. 113, pp. 51–75.

27. Klebanov, L.B., Rachev, S.T. and Yakovlev, A.Y. On the parametric estimation of survival functions, *Statistics and Decisions*, **1993**, suppl. 3, pp. 83–102.

28. Hanin, L.G., Rachev, S.T. and Yakovlev, A.Y. On the optimal control of cancer radiotherapy for non-homogeneous cell populations, *Advances in Applied Probability*, **1993**, vol. 25, pp. 1–23.

29. Rachev, S.T. and Yakovlev, A.Yu. A random minima scheme and carcinogenic risk estimation, *The Mathematical Scientist*, **1993**, vol. 18, pp. 20–36 (invited paper).

30. Yakovlev, A.Yu. Comments on the distribution of clonogens in irradiated tumors, Letter to the Editor, *Radiation Research*, **1993**, vol. 134, pp. 117–120.

31. Yakovlev, A.Y., Goot, R.E. and Osipova, T.T. The choice of cancer treatment based on covariate information, *Statistics in Medicine*, **1994**, vol. 13, pp. 1575–1581.

32. Yakovlev, A.Yu. Comments on survival models allowing for cure rates, *Statistics in Medicine*, **1994**, vol. 13, pp. 983–986.

33. Asselain, B., Fourquet, A., Hoang, T., Myasnikova, E. and Yakovlev, A.Y. Testing independence of competing risks: An application to the analysis of breast cancer recurrence, *Biometrical J.*, **1994**, vol. 36, pp. 465–473.

34. Yakovlev, A.Y., Tsodikov, A.D. and Anisimov, V.N. A new model of aging: Specific versions and their application, *Biometrical J.*, **1995**, vol. 137, pp. 435–448.

35. Tyurin, Yu.N., Yakovlev, A.Y., Shi, J. and Bass, L. Testing a model of aging in animal experiments, *Biometrics*, **1995**, vol. 51, pp. 363–372.

36. Tsodikov, A.D., Asselain, B., Fourquet, A., Hoang T. and Yakovlev, A.Y. Discrete strategies of cancer post-treatment surveillance. Estimation and optimization problems, *Biometrics*, **1995**, vol. 51, pp. 437–447.

37. Rachev, S.T., Wu, C. and Yakovlev, A.Y. A bivariate limiting distribution of tumor latency time, *Mathematical Biosciences*, **1995**, vol. 127, pp. 127–147.

38. Yakovlev, A.Y. Threshold models of tumor recurrence, *Mathematical and Computer Modelling*, **1996**, vol. 23, pp. 153–164 (invited paper).

39. Ivankov, A.A., Tsodikov, A.D. and Yakovlev, A.Y. Computer simulation of tumor recurrence, *Journal of Biological Systems*, **1996**, vol. 4, pp. 291–302.

40. Hoang, T., Tsodikov, A.D., Yakovlev, A.Yu. and Asselain, B. A parametric analysis of tumor recurrence data, *Journal of Biological Systems*, **1996**, vol. 4, pp. 391–403.

41. Yakovlev, A.Y. and Polig, E. A diversity of responses displayed by a stochastic model of radiation carcinogenesis allowing for cell death, *Mathematical Biosciences*, **1996**, vol. 132, pp. 1–33.

42. Myasnikova, E.M., Rachev, S.T. and Yakovlev, A.Y. Queueing models of potentially lethal damage repair in irradiated cells, *Mathematical Biosciences*, **1996**, vol. 135, pp. 85–109.

43. Hanin, L.G., Klebanov, L.B. and Yakovlev, A.Y. Randomized multihit models and their identification, *Journal of Applied Probability*, **1996**, vol. 33, pp. 458–471.

44. Hanin, L.G. and Yakovlev, A.Y. A nonidentifiability aspect of the two-stage model of carcinogenesis, *Risk Analysis*, **1996**, vol. 16, pp. 711–715.

45. Asselain, B., Fourquet, A., Hoang, T., Tsodikov, A.D. and Yakovlev, A.Y. A parametric regression model of tumor recurrence: An application to the analysis of clinical data on breast cancer, *Statistics and Probability Letters*, **1996**, vol. 29, pp. 271–278.

46. Yakovlev, A.Y., Hanin, L.G., Rachev, S.T. and Tsodikov, A.D. A distribution of tumor size at detection and its limiting form, *Proc. Natl. Acad. Sci. USA*, **1996**, vol. 93, pp. 6671–6675.

47. Boucher, K. and Yakovlev, A.Y. Estimating the probability of initiated cell death prior to tumor induction, *Proc. Natl. Acad. Sci. USA*, **1997**, vol. 94, pp. 12776–12779.

48. Hanin, L.G., Rachev, S.T., Tsodikov, A.D. and Yakovlev, A.Y. A stochastic model of carcinogenesis and tumor size at detection, *Advances in Applied Probability*, **1997**, vol. 29, pp. 607–628.

49. Tsodikov, A.D., Asselain, B. and Yakovlev, A.Y. A distribution of tumor size at detection: An application to breast cancer data, *Biometrics*, **1997**, vol. 53, 1495–1502.

50. Kruglikov, I.L., Pilipenko, N.I., Tsodikov, A.D. and Yakovlev, A.Y. Assessing risk with doubly censored data: An application to the analysis of radiation-induced thyropathy, *Statistics and Probability Letters*, **1997**, vol. 32, pp. 223–230.

51. Yakovlev, A.Y., Müller, W. A., Pavlova, L.V. and Polig, E. Do cells repair precancerous lesions induced by radiation? *Mathematical Biosciences*, **1997**, vol. 142, pp. 107–117.

52. Yakovlev, A.Y. and Pavlova, L.V. Mechanistic modeling of multiple tumori-genesis: An application to data on lung tumors in mice exposed to urethane, *Annals of the NY Academy of Sciences*, **1997**, vol. 837, pp. 462–469.

53. Tsodikov, A.D., Loeffler, M. and Yakovlev, A.Y. Assessing the risk of secondary leukemia in patients treated for Hodgkin's disease. A report from the International Database on Hodgkin's Disease, *Journal of Biological Systems*, **1997**, vol. 5, pp. 433–444.

54. Boucher, K., Pavlova, L.V. and Yakovlev, A.Yu. A model of multiple tumorigenesis allowing for cell death: Quantitative insight into biological effects of urethane, *Mathematical Biosciences*, **1998**, vol. 150, pp. 63–82.

55. Yakovlev, A.Y., Boucher, K., M. Mayer-Proschel and Noble, M. Quantitative insight into proliferation and differentiation of oligo dendrocyte type 2 astrocyte progenitor cells in vitro, *Proc. Natl. Acad. Sci. USA*, **1998**, vol. 95, pp. 14164–14167.

56. Tsodikov, A.D., Loeffler, M. and Yakovlev, A.Y. A parametric regression model with time dependent covariates: An application to the analysis of secondary leukemia. A report from the International Database on Hodgkin's disease, *Statistics in Medicine*, **1998**, vol. 17, pp. 27–40.

57. Yakovlev, A.Yu., Mayer-Proschel, M. and Noble, M. A stochastic model of brain cell differentiation in tissue culture, *J. Mathematical Biology*, **1998**, vol. 37, pp. 49–60.

58. Yakovlev, A.Yu., Tsodikov, A.D., Boucher, K. and Kerber, R. The shape of the hazard function in breast carcinoma: curability of the disease revisited, *Cancer*, **1999**, vol. 85, pp. 1789–1798.

59. Yakovlev, A.Yu., Boucher, K. and DiSario, J. Modeling insight into spontaneous regression of tumors, *Mathematical Biosciences*, **1999**, vol. 155, pp. 45–60.

60. von Collani, E., Tsodikov, A., Yakovlev, A., Mayer-Proschel, M. and Noble, M. A random walk model of oligodendrocyte generation in vitro and associated estimation problems, *Mathematical Biosciences*, **1999**, vol. 159, pp. 189–204.

61. Boucher, K., Yakovlev, A.Y., Mayer-Proschel, M. and Noble, M. A stochastic model of temporally regulated generation of oligodendrocytes in vitro, *Mathematical Biosciences*, **1999**, vol. 159, pp. 47–78.

62. Myasnikova, E.M., Asselain, B. and Yakovlev, A.Yu. Spline-based estimation of cure rates: An application to the analysis of breast cancer data, *Mathematical and Computer Modelling*, **2000**, vol. 32, pp. 217–228 (invited paper).

63. Yakovlev, A., von Collani, E., Mayer-Proschel, M. and Noble, M. Stochastic formulations of a clock model for temporally regulated generation of oligodendrocytes in vitro, *Mathematical and Computer Modelling*, **2000**, vol. 32, pp. 125–137 (invited paper).

64. Zorin, A.V., Yakovlev, A.Y., Mayer-Proschel, M. and Noble, M. Estimation problems associated with stochastic modeling of proliferation and differentiation of O-2A progenitor cells in vitro, *Mathematical Biosciences*, **2000**, vol. 167, pp. 109–121.

65. Bartoszyński, R., Edler, L., Hanin, L., Kopp-Schneider, A., Pavlova,L., Tsodikov, A., Zorin, A. and Yakovlev, A. Modeling cancer detection: Tumor size as a source of information on unobservable stages of carcinogenesis, *Mathematical Biosciences*, **2001**, vol. 171, pp. 113–142.

66. Gregori, G., Hanin, L., Luebeck, G., Moolgavkar, S. and Yakovlev, A. Testing goodness of fit with stochastic models of carcinogenesis, *Mathematical Biosciences*, **2001**, vol. 175, pp. 13–29.

67. Hanin, L.G., Zaider, M. and Yakovlev, A.Y. Distribution of the number of clonogens surviving fractionated radiotherapy: A longstanding problem revisited, *International Journal of Radiation Biology*, **2001**, vol. 77, pp. 205–213.

68. Tsodikov, A., Dicello, J., Zaider, M., Zorin, A. and Yakovlev, A. Analysis of a hormesis effect in the leukemia caused mortality among atomic bomb survivors, *Human and Ecological Risk Assessment*, **2001**, vol. 7, pp. 829–847 (invited paper).

69. Zaider, M., Zelefsky, M.J., Hanin, L.G., Tsodikov, A.D., Yakovlev, A.Y. and Leibel, S.A. A survival model for fractionated radiotherapy with an application to prostate cancer, *Physics in Medicine and Biology*, **2001**, vol. 46, pp. 2745–2758.

70. Tsodikov, A.D., Bruenger, F., Lloyd, R., Polig, E., Miller, S. and Yakovlev, A.Y. Modeling and analysis of the latent period of osteosarcomas induced by incorporated 239Pu: The role of immune responses, *Mathematical and Computer Modelling*, **2001**, vol. 33, pp. 1377–1386.

71. Hanin, L.G., Tsodikov, A.D. and Yakovlev, A.Y. Optimal schedules of cancer surveillance and tumor size at detection, *Mathematical and Computer Modelling*, **2001**, vol. 33, pp. 1419–1430.

72. Trelford, J., Tsodikov, A.D. and Yakovlev, A.Y. Modeling posttreatment development of cervical carcinoma: Exophytic or endophytic – does it matter? *Mathematical and Computer Modelling*, **2001**, vol. 33, pp. 1439–1443.

73. Szabo, A. and Yakovlev, A.Y. Preferred sequences of genetic events in carcinogenesis: Quantitative aspects of the problem, *Journal of Biological Systems*, **2001**, vol. 9, pp. 105–121.

74. Zorin, A.V., Tsodikov, A.D., Zharinov, G.M. and Yakovlev, A.Y. The shape of the hazard function in cancer of the cervix uteri, *Journal of Biological Systems*, **2001**, vol. 9, pp. 221–233.

75. Boucher, K., Zorin, A.V., Yakovlev, A.Y., Mayer-Proschel, M. and Noble, M. An alternative stochastic model of generation of oligodendrocytes in cell culture, *J. Mathematical Biology*, **2001**, vol. 43, pp. 22–36.

76. Szabo, A., Boucher, K., Carroll, W.L., Klebanov, L.B., Tsodikov, A.D. and Yakovlev, A.Y. Variable selection and pattern recognition with gene expression data generated by the microarray technology, *Mathematical Biosciences*, **2002**, vol. 176, pp. 71–98.

77. Szabo, A., Boucher, K., Jones, D., Klebanov, L.B., Tsodikov, A.D. and Yakovlev, A.Y. Multivariate exploratory tools for microarray data analysis, *Biostatistics*, **2003**, vol. 4, pp. 555–567.

78. Tsodikov, A.D., Ibrahim, J.G. and Yakovlev, A.Y. Estimating cure rates from survival data: An alternative to two-component mixture models, *Journal of the American Statistical Association*, **2003**, vol. 98, pp. 1063–1078.

79. Hanin, L.G. and Yakovlev, A.Y. Multivariate distributions of clinical covariates at the time of cancer detection, *Statistical Methods in Medical Research*, **2004**, vol. 13, pp. 457–489.

80. Xiao, Y., Frisina, R., Gordon, A., Klebanov, L. and Yakovlev, A. Multivariate search for differentially expressed gene combinations, *BMC Bioinformatics*, **2004**, vol. 5, Article 164.

81. Boucher, K., Asselain, B., Tsodikov, A.D. and Yakovlev, A.Y. Semiparametric Versus Parametric Regression Analysis Based On The Bounded Cumulative Hazard Model: An Application To Breast Cancer Recurrence, In: *Semiparametric Models and Applications to Reliability, Survival Analysis and Quality of Life*, **2004**, M. Nikulin, N. Balakrishnan, M. Mesbah and N. Limnious, eds., Birhäuser, pp. 399–418.

82. Berry, D. A., Cronin, K. A., Plevritis, S. K., Fryback, D. G., Clarke, L., Zelen, M., Mandelblatt, J., Yakovlev, A. Y., Habbema, J. D. F. and Feuer, E. J. Effect of screening and adjuvant therapy on mortality from breast cancer, *New England Journal of Medicine*, **2005**, vol. 353, pp. 1784–1792.

83. Hyrien, O., Mayer-Pröschel, M., Noble, M. and Yakovlev, A.Y. Estimating the life-span of oligodendrocytes from clonal data on their development in cell culture, *Mathematical Biosciences*, **2005**, vol. 193, pp. 255–274.

84. Hyrien, O., Mayer-Pröschel, M., Noble, M. and Yakovlev, A.Y. A stochastic model to analyze clonal data on multi-type cell populations, *Biometrics*, **2005**, vol. 61, pp. 199–207.

85. Qiu, X., Klebanov, L. and Yakovlev, A.Y. Correlation between gene expression levels and limitations of the empirical Bayes methodology for finding differentially expressed genes, *Statistical Applications in Genetics and Molecular Biology*, **2005**, vol. 4, Article 34.

86. Qiu, X., Brooks, A.I., Klebanov, L. and Yakovlev, A. The effects of normalization on the correlation structure of microarray data, *BMC Bioinformatics*, **2005**, vol. 6, Article 120.

87. Yanev, N., Jordan, C.T., Catlin, S. and Yakovlev, A. Two-type Markov branching processes with immigration as a model of leukemia cell kinetics, *Proceedings of the Bulgarian Academy of Sciences*, **2005**, vol. 58, pp. 1025–1032.

88. Zorin, A.V., Edler, L., Hanin, L.G. and Yakovlev, A.Y. Estimating the natural history of breast cancer from bivariate data on age and tumor size at diagnosis, In: *Quantitative Methods for Cancer and Human Health Risk Assessment*, **2005**, L. Edler and C.P. Kitsos, eds, Wiley, NY, pp. 317–327.

89. Almudevar, A., Klebanov, L.B., Qiu, X., Salzman, P. and Yakovlev, A.Y. Utility of correlation measures in analysis of gene expression, *NeuroRx*, **2006**, vol. 3, pp. 384–395 (invited paper).

90. Hanin, L., Hyrien, O., Bedford, J. and Yakovlev, A. A comprehensive model of irradiated cell populations in culture, *Journal of Theoretical Biology*, **2006**, vol. 239, pp. 401–416.

91. Hanin, L., Miller, A., Zorin, A. and Yakovlev, A. The University of Rochester model of breast cancer detection and survival, *Journal of the National Cancer Institute*, **2006**, Monograph 36, pp. 86–95.

92. Hyrien, O., Ambeskovic, I., Margot Mayer-Proschel, M., Noble, M. and Yakovlev, A. Stochastic modeling of oligondendrocyte generation in cell culture: Model validation with time-lapse data, *BMC Theoretical Biology and Medical Modelling*, **2006**, vol. 3, Article 21.

93. Hyrien, O., Mayer-Proschel, M., Noble, M. and Yakovlev, A. The statistical analysis of longitudinal clonal data on oligodendrocyte generation, *WSEAS Transactions on Biology and Biomedicine*, **2006**, vol. 3, pp. 238–243.

94. Klebanov, L., Gordon, A., Xiao, Y., Land, H. and Yakovlev, A. A permutation test motivated by microarray data analysis, *Computational Statistics and Data Analysis*, **2006**, vol 150, pp. 3619–3628.

95. Klebanov, L., Jordan, C. and Yakovlev, A. A new type of stochastic dependence revealed in gene expression data, *Statistical Applications in Genetics and Molecular Biology*, **2006**, vol. 5, Article 7.

96. Klebanov, L. and Yakovlev, A. Treating expression levels of different genes as a sample in microarray data analysis: Is it Worth a Risk? *Statistical Applications in Genetics and Molecular Biology*, **2006**, vol. 5, Article 9.

97. Qiu, X., Xiao, Y., Gordon, A. and Yakovlev, A. Assessing stability of gene selection in microarray data analysis, *BMC Bioinformatics*, **2006**, vol. 7, Article 50.

98. Qiu, X. and Yakovlev, A. Some comments on instability of false discovery rate estimation, *Journal of Bioinformatics and Computational Biology*, **2006**, vol. 4, pp. 1057–1068.

99. Xiao, Y., Gordon, A. and Yakovlev, A. The L_1-version of the Cramer-von-Mises test for two-sample comparisons in microarray data analysis, *EURASIP Journal of Bioinformatics and Computational Biology*, **2006**, Article ID 85769, pp. 1–9.

100. Yakovlev, A. Y. and Yanev, N. M. Branching stochastic processes with immigration in analysis of renewing cell populations, *Mathematical Biosciences*, **2006**, vol. 203, pp. 37–63.

101. Yakovlev, A. and Yanev, N. Label distributions in branching stochastic processes, *Proceedings of the Bulgarian Academy of Sciences*, **2006**, vol. 59, pp. 1123–1130.

102. Klebanov, L. and Yakovlev, A. A new approach to testing for sufficient follow-up in cure-rate analysis, *Journal of Statistical Planning and Inference*, **2007**, vol. 137, pp. 3557–3569 (invited paper).

103. Gordon, A., Glazko, G., Qiu, X. and Yakovlev, A. Control of the mean number of false discoveries, Bonferroni, and stability of multiple testing, *The Annals of Applied Statistics*, **2007**, vol. 1, pp. 179–190.

104. Klebanov, L., Qiu, X., Welle, S. and Yakovlev, A. Statistical methods and microarray data, *Nature Biotechnology*, **2007**, vol. 25, pp. 25–26.

105. Klebanov, L., Glazko, G., Salzman, P., Yakovlev, A. and Xiao, H. A multivariate extension of the gene set enrichment analysis, *Journal of Bioinformatics and Computational Biology*, **2007**, vol. 5, pp. 1139–1153.

106. Klebanov, L. and Yakovlev, A. Is there an alternative to increasing the sample size in microarray studies? *Bioinformation*, **2007**, vol. 1, pp. 429–431.

107. Klebanov, L. and Yakovlev, A. How high is the level of technical noise in microarray data? *Biology Direct*, **2007**, vol. 2, Article 9.

108. Klebanov, L. and Yakovlev, A. Diverse correlation structures in microarray gene expression data and their utility in improving statistical inference, *Annals of Applied Statistics*, **2007**, vol. 1, pp. 538–559.

109. Chen, L., Klebanov, L. and Yakovlev, A.Y. Normality of gene expression revisited, *Journal of Biological Systems*, **2007**, vol. 15, 39–48.

110. Qiu, X. and Yakovlev, A. Comments on probabilistic models behind the concept of false discovery rate, *Journal of Bioinformatics and Computational Biology*, **2007**, vol. 5, pp. 963–975.

111. Yakovlev, A. and Yanev, N. Branching populations of cells bearing a continuous label, *PLISKA Studia Mathematica Bulgarica*, **2007**, vol. 18, pp. 387–400.

112. Hanin, L.G. and Yakovlev, A.Y. Identifiability of the joint distribution of age and tumor size at detection in the presence of screening, *Mathematical Biosciences*, **2007**, vol. 208, pp. 644–657.

113. Yakovlev, A. Y. and Yanev, N. M. Age and residual lifetime distributions for branching processes, *Statistics and Probability Letters*, **2007**, vol. 77, pp. 503–513.

114. Xiao, Y., Gordon, A. and Yakovlev, A. A C++ program for the Cramér-von Mises two-sample test, *Journal of Statistical Software*, **2007**, vol. 17, pp. 1–15.

115. Klebanov, L., Chen, L. and Yakovlev, A. Revisiting adverse effects of cross-hybridization in Affymetrix gene expression data: do they matter for correlation analysis? *Biology Direct*, **2007**, vol. 2, Article 28.

116. Klebanov, L., Qiu, X. and Yakovlev, A. Testing differential expression in non-overlapping gene pairs: A new perspective for the empirical Bayes method, *Journal of Bioinformatics and Computational Biology*, **2008**, vol. 6, pp. 301–316.

117. Yakovlev, A. Y. , Stoimenova, V. K. and Yanev, N. M. Branching models of progenitor cell populations and estimation of the offspring distributions, *Journal of the American Statistical Association*, **2008**, vol. 103, pp. 1357–1366.

118. McMurray, H.R., Sampson, E.R., Compitello, G., Kinsey, C., Newman, L., Smith, B., Chen, S.R., Klebanov, L., Salzman, P., Yakovlev, A. and Land, H. Synergistic response to oncogenic mutations defines gene class critical to cancer phenotype. *Nature*, **2008**, vol. 453, pp. 1112–1116.

119. Klebanov, L., Qiu, X. and Yakovlev, A. Testing differential expression in nonoverlapping gene pairs: a new perspective for the empirical Bayes method. *Journal of Bioinformatics and Computational Biology*, **2008**, vol. 6, pp. 301–316.

120. Li, X., Nott, S.L., Huang, Y., Hilf, R., Bambara, R.A., Qiu, X., Yakovlev, A., Welle, S. and Muyan, M. Gene expression profiling reveals that the regulation of estrogen-responsive element-independent genes by 17 beta-estradiol-estrogen receptor beta is uncoupled from the induction of phenotypic changes in cell models. *Journal of Molecular Endocrinology*, **2008**, vol. 40, pp. 211–229.

121. Hu, R., Qiu, X., Glazko, G., Klebanov, L. and Yakovlev, A. Detecting intergene correlation changes in microarray analysis: a new approach to gene selection. *BMC Bioinformatics*, **2009**, vol. 10, Article 20.

122. Yakovlev, A. and Yanev, N. Relative frequencies in multitype branching stochastic processes, *Annals of Applied Probability*, **2009**, vol. 19, pp. 1–14.

123. Hanin, L., Awadalla, S.S., Cox, P., Glazko, G. and Yakovlev, A. Chromosome-specific spatial periodicities in gene expression revealed by spectral analysis. *Journal of Theoretical Biology*, **2009**, vol. 256, pp. 333–342.

124. Gordon, A., Chen, L., Glazko, G. and Yakovlev, A. Balancing type one and two errors in multiple testing for differential expression of genes. *Computational Statistics and Data Analysis*, **2009**, vol. 53, pp. 1622–1629.

125. Yakovlev, A. and Yanev, N. Limiting distributions for multitype branching processes. *Stochastic Analysis and Applications*, **2010**, vol. 28, pp. 1040–1060.

126. Chen, L., Klebanov, L., Almudevar, A., Proschel, C. and Yakovlev, A. A study of the correlation structure of microarray gene expression data based on mechanistic modelling of cell population kinetics. *Modeling and Inference in Biomedical Science: In Memory of Andrei Yakovlev*, A. Almudevar, W. J. Hall and D. Oakes, ed., This Volume

Peer-Reviewed Papers Translated into English

127. Zorin, A.V., Isanin, N.A. and Yakovlev, A.Y. A study of the kinetics of cell transition on DNA synthesis in the regenerating liver, *Proc. Acad. Sci. USSR*, **1973**, vol. 212, pp. 226–228 (English transl.).

128. Gusev, Yu.V., Zorin, A.V. and Yakovlev, A.Y. Simulation of cell systems in stable and developing tissues, *Ontogenesis*, **1986**, vol. 17, pp. 565–577 (English transl.).

129. Gusev, Yu.V. and Yakovlev, A.Y. Computer simulation of cell renewal in small intestine epithelium of intact and irradiated animals, *Proc. Acad. Sci. USSR*, **1987**, vol. 294, pp. 712–715 (English transl.).

130. Rachev, S.T., Klebanov, L.B. and Yakovlev, A.Y. An estimate of the rate of convergence to the limit distribution for the minima scheme for random number of identically distributed random variables. In: *Stability Problems for Stochastic Models*, Moscow, VNIISI, **1988**, pp. 120–124 (English transl. in J. Soviet Mathem.).

Peer-Reviewed Papers Translated into English

131. A steady state of cell populations. Necessary and sufficient conditions of a steady state, in a narrow sense, and of an exponential state, *Cytology*, **1971**, vol. 13, pp. 844–849.

132. Some possibilities to explore the processes underlying tissue homeostasis, *Cytology*, **1971**, vol.13, pp. 1417–1425.

133. Some problems of the kinetics of normal cell populations. Steady – state and exponential growth (with V. A. Guschhin), *Biophysics and Theoretical Biology*, Proceeding of the University of Tbilisi, **1971**, vol. 139, pp. 79–96.

134. Modeling the humoral regulation of cell proliferation in adult animal tissues, *Cytology*, **1972**, vol. 14, pp. 685–698.

135. On the modelling of the radiation-induced mitotic block, *Cytology*, **1973**, vol. 15, pp. 616–619.

136. The twenty four hours' rhythm of proliferation and parameters of the mitotic cycle phases, *Cytology*, **1973**, vol. 15, pp. 473–475.

137. Modeling some features of the cellular systems' dynamic variability (with A.V. Zorin), *Cytology*, **1974**, vol. 16, pp. 941–949.

138. On the sorption properties of cells in the regenerating liver (with N.A. Isanin, G. B. Kuznetsova), *Cytology*, **1974**, vol. 16, pp. 1420–1423.

139. The mistery of chalones (popular paper), *Priroda (Nature)*, **1974**, pp. 78–85.

140. The kinetic analysis of the cells starting DNA replication in the systems stimulated to proliferate. I. Properties of a mathematical model of cell kinetics (with A.V. Zorin, N.A. Isanin), *Cytology*, **1975**, vol. 17, pp. 667–673.

141. The kinetic analysis of the cells starting DNA replication in the systems stimulated to proliferate. II. The theoretical basis of the method (with A.V. Zorin, N.A. Isanin), *Cytology*, **1975**, vol. 17, pp. 776–782.

142. The kinetic analysis of the cells starting DNA replication in the systems stimulated to proliferate. III. The regenerating liver (with A.V. Zorin, N.A. Isanin), *Cytology*, **1975**, vol. 17, pp. 783–790.

143. Some peculiarities of the process of autophagocytic vacuole formation in hepatocytes of the regenerating rat liver (with N.N. Petrovichev), *Cytology*, **1975**, vol. 17, pp. 1087–1089.

144. Labeled mitoses curve in different states of cell proliferation kinetics. I. General principles of mathematical modelling (with A.M. Malinin), *Cytology*, **1976**, vol. 18, pp. 1270–1277.

145. Labeled mitoses curve in different states of cell proliferation kinetics. II. Mathematical modeling on the basis of transient phenomena in cell kinetics (with A.M. Malinin), *Cytology*, **1976**, vol. 18, pp. 1330–1338.

146. Labeled mitoses curve in different states of cell proliferation kinetics. III. Computer simulation (with A.M. Malinin), *Cytology*, **1976**, vol. 18, pp. 1464–1469.

147. The pleotypic control of molecular events in the cell cycle (with N.A. Isanin), *Cytology*, **1977**, vol. 19, pp. 463–473.

148. The involvement of the cell lysosomal apparatus in the regulation of cell proliferation (with N.A. Isanin), *Cytology*, **1977**, vol. 19, pp. 575–584.

149. Labeled mitoses curve in different states of cell proliferation kinetics. IV. Some additional remarks on the method of "flux distributions", *Cytology*, **1978**, vol. 20, pp. 589–592.

150. Labeled mitoses curve in different states of cell proliferation kinetics. V. The influence of diurnal rhythm of cell proliferation on the shape of the labeled mitoses curve (with A.F. Lepekhin, A.M. Malinin), *Cytology*, **1978**, vol. 20, pp. 630–635.

151. The dynamic reserving of hepatocytes as a mechanism of maintenance of the regenerating liver specialized functions, *Cytology*, **1979**, vol. 21, pp. 1243–1252.

152. The dynamics of induced cell proliferation within the model of branching stochastic process. I. Numbers of cells in successive generations (with N. M. Yanev), *Cytology*, **1980**, vol. 22, pp. 945–953.

153. A new approach to the evaluation of probabilistic characteristics of RNA molecules life-span in eukaryotic cells (with K.P. Hanson, A.F. Lepekhin, B. D. Zhivotovsky), *Vestnik of the Academy of Medical Sciences of the USSR*, **1980**, pp. 89–91.

154. A simple model of reliability to provide the liver specialized functions in the course of its regeneration after partial hepatectomy (with T. V. Antipova), *Cytology*, **1981**, vol. 23, pp. 473–476.

155. On the simulation of reliability of cell renewal (with A.V. Zorin), *Cytology*, **1982**, vol. 24, pp. 110–114.

156. Dynamics of induced cell proliferation within the model of branching stochastic process. II. Some characteristics of the cell cycle temporal organization (with N. M. Yanev), *Cytology*, **1983**, vol. 25, pp. 818–826.

157. Simulation modeling of tumor cell population kinetics after irradiation (with A.V. Zorin, V.A. Gushchin, F.A. Stephanenko, O.N. Cherepanova), *Experimental Oncology*, **1983**, vol. 5, pp. 27–30.

158. Kinetics of proliferative processes induced by phytohemagglutinin in irradiated lymphocytes, *Radiobiology*, **1983**, vol. 23, pp. 449–453.

159. Computer simulation of tumour initiation in continuously renewing tissues (with Yu.V. Gusev, E.A. Zherbin), *Problems in Oncology*, **1984**, vol. 30, pp. 74–79.

160. Basic problems in statistical analysis of survival data for oncological patients (with N.O. Kadyrova, T.V. Antipova, E.A. Zherbin), *Problems in Oncology*, **1984**, vol. 30, pp. 24–32.

161. Nonparametric survival data analysis for oncological patients (with N.O. Kadyrova), *Problems in Oncology*, **1984**, vol. 30, pp. 35–48.

162. Parametric analysis of the survival rate of animals after a double exposure to ionizing radiation (with N.O. Kadyrova, S.F. Vershinina, L.V. Pavlova), *Radiobiology*, **1985**, vol. 25, pp. 698–700.

163. On the interpretation of the recovery of cells from potentially lethal radiation damage in stationary cell culture (with V.A. Guschin, A.V. Zorin, F.A. Stephanenko), *Studia Biophysica*, **1985**, vol. 107, pp. 195–203.

164. A semistochastic model of tumour growth and a new parametric family of distributions for the life-time of tumour horsts (with T.V. Antipova, L.V. Pavlova), *Studia Biophysica*, **1986**, vol. 116, pp. 51–58.

165. A study of the factors influencing the clonogenic capacity of irradiated cells by the methods of computer simulation (with S.A. Danielyan, A.V. Zorin), *Radiobiology*, **1986**, vol. 26, pp. 516–520.

166. Simulation of the postirradiation recovery of the "crypt-villus" system (with Yu.V. Gusev), *Radiobiology*, **1987**, vol. 27, pp. 748–753.

167. A new family of distributions for parametric analysis of data on the life-time of tumour hosts (with T.V. Kamynina, V.N. Anisimov, N.V. Zhukovskaya), *Problems in Oncology*, **1987**, vol. 33, pp. 23–28.

168. Parametric analysis of survival of cancer patients: the results of radiotherapy of patients with cancer of the cervix uteri (with T.V. Kamynina, G.M. Zharinov), *Problems in Oncology*, **1987**, vol. 33, pp. 51–55.

169. Comparison of survival curves based on the model of proportional risks (with E.M. Myasnikova, G.M. Zharinov), *Problems in Oncology*, **1988**, vol. 34, pp. 18–23.

170. Estimation of parameters of the experimental tumor growth kinetics (with T.V. Kamynina, V.B. Klimovich), *Experimental Oncology*, **1988**, vol. 10, pp. 66–69.

171. Analysis of the survival rate after a combined radiation effect. Sinergism or antagonism of the effects of two factors (with E.M. Myasnikova, S.T. Rachev, E.I. Komarov, I.V. Remizova, R.S. Budagov, L.N. Chureyeva), *Radiobiology*, **1988**, vol. 28, pp. 478–483.

172. A method for estimation of the reproductive cell death probability (with P.V. Balykin, R.E. Goot, A.V. Zorin, M.S. Tanushev, N.M. Yanev), *Studia Biophysica*, **1988**, vol. 123, pp. 117–124.

173. Theoretical bounds for the efficacy of radiation therapy (with S.T. Rachev), *Medical Radiology*, **1988**, pp. 17–21.

174. Nonparametric interval estimates of the reliability of systems within the competing risks framework (with S.T. Rachev, Yu.N. Andreev), *Communication Technique*, **1989**, pp. 58–68.

175. The efficiency of early detection of oncological diseases and temporal characteristics of tumor development (with N.P. Napalkov, A.D. Tsodikov), *Problems in Oncology*, **1990**, vol. 36, pp. 1293–1300.

176. optimization of time instants for monitoring of the population under risk of oncological diseases (with N.P. Napalkov, L.V. Petukhov, A.D. Tsodikov), *Problems in Oncology*, **1991**, vol. 37, pp. 144–151.

177. An approach to the problem of individual choice of cancer treatment (with R.E. Goot, G.M. Zharinov, T.T. Osipova), *Problems in Oncology*, **1992**, pp. 667–671.

178. Parametric estimation of the duration of tumor latency (with A.D. Tsodikov), *Problems in Applied Mathematics*, **1992**, St. Petersburg Technical University, pp. 133–140.
179. Identification of a multi-hit model for nonhomogeneous cell populations (with L.V. Pavlova, L.G. Hanin), *Radiobiology*, **1992**, vol. 32, pp. 785–787.
180. The whys and wherefores of a parametric estimator of the clonogen progression time with simulation (with A.A. Ivankov and A.D. Tsodikov), *Problems in Oncology*, **1993**, vol. 39, pp. 220–225.
181. The proliferative potential of a treated tumor. Validation of parametric methods by computer simulations (with A.A. Ivankov and A.D. Tsodikov), *Problems in Oncology*, **1996**, vol. 42, pp. 34–39.
182. The parametric statistical analysis of data on breast cancer recurrence after the radical treatment (with A. Ivankov, A. Tsodikov, T. Yakimova), *Problems in Oncology*, **1996**, vol. 42, pp. 85–87.
183. Parametric analysis of survival data on patients treated for cervical cancer by fractionated radiotherapy. A new method and its applications (with Zorin, A.D., Tsodikov, A.D., Hanin, L.G., Zharinov, G.M., Zaikin, G.V.), *Problems in Oncology*, **2001**, vol. 47, pp. 307–311.

Book Chapters

184. On the parametric analysis of survival data in radiobiology and oncology (with N.O. Kadyrova, T.V. Antipova), in "Reliability of Biological Systems", Naukova Dumka, Kiev, **1985**, pp. 33–40 (in Russian).
185. A study of stochastic stability of controlled cell systems by means of computer simulation (with Yu. V. Gusev), *Theory of Complex Systems and Methods for Their Modelling*, Institute for Systems Studies, **1985**, pp. 154–162 (in Russian).
186. On the parametric analysis of survival data in radio-oncological research (with T.V. Antipova), In: *Application of Mathematical and Physico-Technical Methods in Roentgenology*, **1985**, pp. 11–16, Obninsk (in Russian).
187. Theoretical bounds for the radiosensibilization of tumors (with S.T. Rachev), In *optimization of Spatial and Temporal Parameters of Radiation Therapy*, **1986**, pp. 22–31, Obninsk (in Russian).
188. Tests for comparing the efficiency of different methods of radiation therapy (with R.E. Goot, S.T. Rachev, N.O. Kadyrova), *ibid*, **1986**, pp. 19–22.
189. Rachev, S.T., Goot, R.E., Kadyrova, N.O. and Yakovlev, A.Yu. Some statistical tests for comparing survival curves, Proceedings of the International Seminar *Statistical Data Analysis*, Varna, **1987**, pp. 193–198.
190. Tanushev, M.S., Yanev, N.M. and Yakovlev, A.Yu. The Bellman-Harris branching processes and distributions of marks in proliferating cell populations, Proc. *Ist World Congress of the Bernuelli Society*, VNU-Press, Holland, **1988**, vol. 2, pp. 725–728.

191. Rachev, S.T. and Yakovlev, A.Yu. Some problems of the competing risks theory (invited lecture), Lecture Notes, V International Summer School on *Probability Theory and Mathematical Statistics*, Varna, 1985, Publishing House of the Bulgarian Academy of Sciences. Sofia, **1988**, pp. 171–187.

192. Hanin, L.G., Rachev, S.T., Goot, R.E. and Yakovlev, A.Yu. Precise upper bounds for the functionals describing the tumor treatment efficiency, *Lecture Notes in Mathematics* (Springer-Verlag), **1989**, vol. 1421, pp. 50–67.

193. Tsodikov, A.D. and Yakovlev, A.Y. A minimax strategy of screening within the competing risks framework, In *Proc. of IV National Summer School on Application of Mathematics in Engineering*, pp. 78–84, **1990**, Varna, Bulgaria.

194. Tsodikov, A., Yakovlev, A. and Petukhov, L. Some approaches to screening optimization, In: Proc. Seminaire *Statistique Des Processus En Milieu Medical*, **1990–91**, University Paris-V, pp. 1–48.

195. Hanin, L.G., Klebanov, L.B., Pavlova, L.V. and Yakovlev, A.Yu. A randomized multihit model of irradiated cell survival and its identification, In: Proc. *Seminaire Statistique Des Processus En Milieu Medical*, University Paris-V, **1992**, pp. 71–96.

196. Ivankov, A., Hoang, T., Loeffler, M., Asselain, B., Tsodikov, A. and Yakovlev, A. Distribution of clonogens progression time – A computer simulation study, In: Proc. Seminaire *Statistique des Processus en Milieu Medical*, University Paris-V, **1992**, pp. 287–294.

197. Hoang, T., Tsodikov, A.D. and Yakovlev, A. Modeling and estimation of a long risk of breast cancer recurrence, In: Proc. Seminaire *Statistique Des Processus En Milieu Medical*, Universite Rene Descartes (Paris V), B. Bru, C. Huber, B. Prum (eds), **1992**, ibid, pp. 247–260 (in French).

198. Rivkind, V.J., Seregin, G.A. and Yakovlev, A.Y. The measure theory and its applications. *Text-book*, 98 p., **1992**, St. Petersburg Technical University (in Russian).

199. Klebanov, L.B. and Yakovlev, A.Yu. A stochastic model of radiation carcinogenesis, *Lecture Notes in Mathematics* (Springer-Verlag), **1993**, vol. 1546, pp. 89–99.

200. Yakovlev, A.Yu., Asselain, B., Bardou, V.J., Fourguet, A., Hoang, T., Rochefordiere, A. and Tsodikov, A.D. A simple stochastic model of tumor recurrence and its applications to data on the premenopausal breast cancer, in *Biometrie et Analyse de Donnees Spatio-Temporelles*, French Biometrical Society, **1993**, No 12, pp. 66–82.

201. Ivankov, A.A., Asselain, B., Fourquet, A., Hoang, T., Yakimova, T.P., Tsodikov, A.D. and Yakovlev, A.Yu. Estimating the growth potential of a treated tumor from time to recurrence observations, In: Proc. Seminaire *Statistique Des Processus En Milieu Medical*, University Paris-V, **1993**, pp. 65–94.

202. Kadyrova, N.O., Hoang, T. and Yakovlev, A.Yu. A computer program for the isotonic maximum likelihood estimation of a bimodal hazard function from censored observations, In: Proc. Seminaire *Statistique Des Processus En Milieu Medical*, University Paris-V, **1993**, pp. 277–302.

203. Yakovlev, A.Yu. Statistical methods for the carcinogenic risk assessment, In: *KfK-Seminarreihe Aktuelle Forschungsgebiete in der Mathematik*, R. Seifert and T. Westermann, eds, Karlsruhe, Germany, **1994**, pp. 63–91.
204. Hoang, T., Tsodikov, A.D., Yakovlev, A. Y. and Asselain, B. Modelling breast cancer recurrence, In *Mathematical Population Dynamics: Analysis of Heterogeneity*, Vol. 2: Carcinogenesis and Cell & Tumor Growth, O. Arino, D. Axelrod, M. Kimmel, Eds., Wuerz Publications, Winnipeg, Manitoba, Canada, **1995**, pp. 283–296.
205. Yakovlev, A., Müller, W.A., Pavlova, L.V. and Polig, E. A stochastic model of carcinogenesis allowing for cell death. Statistical and biological considerations in: *Advances in Mathematical Population Dynamics - Molecules, Cells and Man*, Arino, O., Axelrod, D., Kimmel, M. eds, World Scientific, Singapore, **1997**, pp. 271–282.

Selected Inventions

1. Szabo, A., Tsodikov, A., Yakovlev, A., Boucher, K., Jones, D. A., and Carroll, W. Methods for Identifying Differentially Expressed Genes by Multivariate Analysis of Microarray Data. Pub. No.: US 2004/0265830 Al, Dec. 30, 2004.
2. Szabo, A., Boucher, K., Jones, D., Klebanov, L., Tsodikov, A. and Yakovlev, A. Methods for Identifying Large Subsets of Differentially Expressed Genes Based on Multivariate Microarray Data Analysis. Pub. No.: US 2006/0088831 Al, Apr. 27, 2006.

Index

A

Age and residual lifetime distributions
  age-dependent process, 12
  Bellman–Harris branching process, 11
  branching processes, 4
  cell populations, 9
  definition, 10–11
  integral equations, 10
  MBPHPI, 12–13
  standard deviations, 54
Age-dependent processes
  asymptotic behavior, 29–37
  Bellman–Harris branching process, 22
  biological
    applications, 22
    motivation, 23–24
  branching (*see* Branching processes)
  equations, 24–29
  homogeneous Poisson immigration, 39–41
  model, 24–29
  multitype population, 17–18
  notation, 24–29
  simulation study, 42–44
  statistical inference, 21, 37–39
Ambiguity
  advisory board, 301
  epidemiological measurements, 303–304
  statistics, 302–303
Analysis of variance (ANOVA) model
  inference, 283–284
  longitudinal data setting, 284–285
  mean and variance, 282–283
  missing data, 284–285

  simulation, 285–286
  statistical methods, 282
  variance homogeneity, 282
Archimedean copulas, 197, 202, 259, 261–264,
      272, 274, 276–277
Artificial mixtures, 191–193
Asymptotic normality, 4, 181, 216, 220

B

Bellman–Harris branching process, 4–6, 11,
      22, 24–26, 53
Biomedical sciences, 321–330
Branching processes
  age-dependent, 9–13, 17–18, 314
  biological applications, 4
  cell
    cycle temporal organization, 5–6
    proliferation kinetics, 3
  estimation theory, 13–16
  extinction probability, 312, 313
  label distributions, 7–9
  limit theorems, 13–16
  mathematical population theory, 315
  mitotic division, 4
  nobility and family extinction, 311, 312
  nonmathematical criticism, 313
  path-breaking papers, 317
  plant population dynamics, 315
  pulse-labeled discrete marks, 6–7
  residual lifetime distributions, 9–13
  social and demographic context, 314
  theory, 5

Breast cancer
  data, 162
  one- and two-sample problems, 185
  therapy, 158

C

Cancer data analysis
  ACS, 118
  colon cancer (*see* Colon cancer)
  construction of likelihoods, 122–126
  CPS I, 118
  epidemiology multistage models, 119
  mathematical issues, 120–122
  MSCE model, 119
  number and size distribution, 127
  rate ratios, 118
  2SCE, 119
Causal inference, 138, 154
Cell proliferation
  biological parameters, 126
  cycle temporal organization, 5–6
  kinetics (*see* Branching processes)
  label distribution, 9
Cellular kinetics
  gene expression levels, 55–56
  simulation study, 56–58
Center for the Health Assessment of
    Mothers and Children of Salinas
    (CHAMACOS), 139, 143–144,
    147–149, 154
Clonal expansion model, 119, 123, 129, 131
Colon cancer
  folate
    fortification, 128–131
    supplementation, 130
  MSCE model, 127
  epidemiologic studies, 131
  screening, 128
Confidence bands
  for e, 179–182
  illustration, 182–184
  and inference, 184–185
  mean residual life, 171
Consistency, 39, 126, 142, 170, 317
Copula models, 200–203, 262, 264, 274, 276
Correlation analysis, 60
Correlation between true and false discoveries,
    68–72
  biological conditions, 66
  discussion, 75–77
  FDR, 64, 65
  microarray analysis, 73–74
  multiple testing design, 73–74

  number of errors, 64
  parametric model with two t-tests, 66–67
  PCER/PFER, 64, 65
  permutation-based extension, 65
  simulation results, 74–75
  testing statistical significance, 63
  two t-statistics, 67–68
Correlation structure
  cell
    mixtures, 48–53
    population kinetics (*see* Cellular
      kinetics)
  characteristics, 327
  microRNAs, 48
Count response, 285
Cure rate, 158, 161, 165

D

Diffusion process, 246, 247, 249, 250, 252
Discrete and continuous labels
  artificial mixture approach, 197
  Bellman–Harris branching processes, 4
  biological experiments, 22
  branching populations, 6–7
  covariate vector, 199

E

Environmental risk factors
  algorithm (*see* Machine-learning)
  cancer risk, 131
  CHAMACOS data set, 139
  population effects, 138
Epidemiology
  ambiguity, 300–304
  disease risks, 138
  environmental, 148
  human ignorance, 298
  illness/disease, 298–299
  multistage models, 119, 131
  predictions, 297
  proportional hazards model, 119
  quantification, 299–300
  uncertainty and, 297–300
Euler–Maruyama scheme, 247–249

F

False discovery rate (FDR), 50, 51, 64–66, 91,
    108, 113
Familywise error rate (FWER)
  arbitrary dependence structure, 65
  FDR, 64

generalization, 88
hypothesis, 82
monotone procedure, 89
MTPs, 64
null hypothesis, 149
Flexible design
decision-making, 212
GSD and AD, 213
numerical results, 233–241
PFS, 212
survival endpoints, 214
two-stage procedure, 212–213
Flow cytometry, 9
Folate supplementation, 129–131
Frailties
Clayton's model—gamma distributed,
265–266
model-generating function, 204
univariate model, 260

G
Gene expression data
correlation (*see* Correlation structure)
gene set analysis, 113–114
Generalized family-wise error rate, 86–87
bounds, 92–94
false rejections, 82
Gene set analysis (GSA)
algorithm, 294
gene pathway correlation, 293
Maxmean statistic, 292
univariate *p*-value, 292–293
Gene set enrichment analysis (GSEA), 113,
290, 291

H
Heterogeneity
cell type pattern, 56
correlations, 48
pairwise sample, 48
variance inflation, 52
decomposing sources, 60
subjects, 58–59
Holm procedure
extensions, 89
optimality, 88–89

I
Immigration
age-dependent, 39–41

branching (*see* Branching processes)
cell evolution, 12
non-homogeneous Poisson, 4, 12, 13
parameter, 13
Improved local linearization (ILL) method,
250–254
Inference
ad hoc approaches, 154
ANOVA, 283–284
asymptotic results, 16
BGW processes, 16
candidate explanatory variables, 139
cell kinetics parameters, 24
confidence bands, 184–185
multitype branching processes, 4
permutation test, 146–147
statistical, 37–39
variance homogeneity, 282

K
Kendall distribution, 270–271, 274–275
Kernel density estimator, 246, 249, 250

L
Label distributions, 4, 7–9
Large number of ancestors
estimation theory, 13–16
false discoveries, 64
gene expression, 48
limit theorems, 13–16
LLN, 283
Latent time distribution model
breast cancer data, 162
Cox model, 159, 163–165
dataset, 162–163
discussion, 165–166
parametric survival models, 158–159
PHH, 158
survival data analysis, 158
time-dependent covariates, 160
Yakovlev model, 160–162, 165
Leukemia
blast cells, 23
computation time, 113
dataset, 74
granulocytic, 171
Life expectancy, 170
Limiting distributions, 16, 170
Limiting Gaussian process, 170, 173, 248
Local linearization method, 246

M
Machine-learning
  algorithm, 138–139
  background, 139–141
  CHAMACOS data, 139, 143–144
  data
    analysis, 138
    structure, 141–143
  discussion, 153–154
  DR-IPW estimation, 145–146
  epidemiology, 138
  joint inference, 148–149
  parameters, 141–143
  results, 149–153
  single exposure inference, 146–147
Maximum likelihood estimation (MLE)
  artificial mixtures, 191–193
  gamma and positive stable shared frailty
    models, 203–205
  numerical experiments, 206–207
  simulations, 206–207
Mean residual life
  alternative conditions, 176–177
  bivariate, 187
  censored data, 185
  chronic granulocytic leukemia, 171
  confidence bands (*see* Confidence bands)
  convergence, 171–176
  examples, 178–179
  inference, 184–185
  life processes, 169
  median and quantile residual life functions,
    185
  monotone and ordered, 186
  semiparametric models, 186
  variance, 170
Mechanistic modeling
  cell
    division stage, 54
    type mixtures, 55
  gamma parameters, 54
  gene expression levels, 55–56
  simulation study, 56–58
  statistical effect, 53
Microarray analysis
  calcium signaling pathway, 290
  gene expression (*see* Gene expression data)
  genomic data, 64
  glial cell populations, 50
  GSA (*see* Gene set analysis (GSA))
  testing, 73–74, 289
Missing data
  Laplace transform, 193
  longitudinal data, 284–285

  mixture model, 208
  original target model, 192
Monotone procedure
  "all-or-nothing" theorem, 94–95
  Bonferroni, 90
  FWER, 92–94
  optimality, Holm Procedure, 88–89
  quasi-thresholds, 90–91
  sharp inequalities, 91–92
Multinomial regression, 197–199
Multiple comparisons
  correlation (*see* Correlation between true
    and false discoveries)
  gene expression (*see* Gene expression data)
Multiple hypothesis testing, 106–109
  applications, 82
  theory and methods, 82
Multiple testing procedures (MTPs)
  application, 66
  comparison of procedures, 88
  DNA microarray experiments, 81–82
  FWER/FDR, 64, 86–87
  hypotheses, 64, 82
  microarray analysis, 73–74
  monotonicity, 84
  PFER, 87–88
  procedures
    step-down, 84–86
    step-up, 84–86
  TSUD, 86
  uninformed, 83–84
Multistage carcinogenesis
  cancer epidemiology, 119
  clinical trials, 118
  colon cancer (*see* Colon cancer)
  likelihoods, 122–126
  mathematical issues, 120–121
  number and size distribution, 127
  rate ratios, 118
  2SCE model, 119
  statistical approaches, 119
Multitype branching processes, 4, 16

N
Next generation sequencing tests, 293–294
Nonlinear transformation models (NTM),
  199–200, 202

O
Oligodendrocytes, 18, 23
Overdispersion, 285, 286

P
Pathway analysis, 290, 295
Per-family error rate (PFER)
  false
    positives, 64
    rejections, 87
  PCER, 65
  probability distribution, 95
Personal perspective, 311–318, 322
Plateau models, 158, 166
Positive dependence, 65, 74, 77, 197, 202
Principles of mathematical modeling
  biological reality, 329–330
  and biology, 322–324
  modeler's dilemma, 326–327
  models, 324–326
  non-identifiability, 327–329
  paradoxical mindset, 321
Proportional hazards (PH) model
  baseline conditions, 260
  cohort and case-control studies, 118
  Schoenfeld's graph, 164
  score statistics, 215–216
  statistical approaches, 119
  survival distribution, 159, 215–216
Pseudo-likelihood, 38, 44, 250

Q
Quasi-EM (QEM), 193–197, 202–204,
    207–209
Quasi-likelihood, 22, 39, 44

R
Ranking, 64, 108, 138
Renewal equation, 11, 25, 32, 34–36
Retinopathy, 205–206
RNA-seq
  microarray data, 290, 295
  next generation sequencing technologies,
    290, 293–294

S
Sample size recalculation
  average testing power, 77
  boxplots, 51
  density plots, 70
  determination, 221
  graphical representation, 307
  ncc-RCC trial, 218
  sample properties, 39
  simulation, 52, 236, 240
  SPRT, 104

Screening
  algorithms, 138
  colon cancer, 128
  and prevention strategies, 119
  two-screen protocol, 130
Sequential hypothesis testing, 102
Sequential probability ratio test (SPRT),
    102–103
  confidence sets, 114
  fixed level tests, 100–101
  gene set analysis, 113–114
  hybrid test, 102
  Monte Carlo hypothesis tests, 99–100
  multiple hypothesis tests, 106–109
  optimal design
    constrained optimization, 110
    numerical example, 112, 113
    solution method, 111–112
  significance level, 100
  single hypothesis test, 104–106
Simulated maximum likelihood (SML)
    methods, 246, 248–250
Simulation studies
  application, 230–231
  design, 222–223
  evaluation, 229–230
  results, 223–229
Statistical inference, 16, 18, 24, 37–39
Stem cells, 53–55, 119, 123, 127, 327
Step-down procedure, 84–86, 88–91, 108,
    109
Step-up procedure, 84–86
  monotonicity, 82
  MTP, 108
  TSUD, 86
Stochastic differential equations (SDEs)
  estimation methods
    Euler method, 247–248
    SML, 248–250
  HIV dynamic model, 255–256
  parameter estimation methods, 246
  practical applications, 245
Stochastic models
  Bernoulli space, 304–305
  branching processes, 6
  cell proliferation kinetics (*see* Cell
    proliferation)
  differential equation, 246–247
  non-informative censoring, 215
  probabilities
    Bernoulli space, 305–306
    stochastic measurement procedure,
      306–309
  transcription regulation, 60

Survival
   adaptive design approaches
      CRP principle, 219–220
      P-value combination rules, 218–219
      sample size determination, 221
   analysis, 158, 261
   Cox proportional hazards model, 215–216
   statistical decision, 216–217

T
Time-dependent covariate
   model
      Cox, 159, 160, 164
      MSCE, 126
      parameters, 121
      Poisson process, 26
      time-varying intensities, 132
Truncation
   Clayton's model, 265–266
   exterior power families, 269
   higher dimensions, 276–277
   interior power families, 269

   invariance, 271–273
   inverse Gaussian model, 266
   Kendall distribution, 274–275
   logarithmic series distribution, 267
   non-negative survival time, 259–262
   Poisson model, 267
   positive stable distributions, 267–269
   semi-invariance, 259
   uniqueness, 262–265
Tumor recurrence data, 161, 326
Two-stage clonal expansion (2SCE), 119–122,
   124–127

U
Unwritten Gospel, 321–330

Y
Yakovlev, Andreï
   mathematical modeling, 321–330
   parametric models, 166
   tumor growth model, 160

Printed in the United States
by Baker & Taylor Publisher Services